T5-CVD-910

R00176 34771

CHICAGO PUBLIC LIBRARY
HAROLD WASHINGTON LIBRARY CENTER
R0017634771

REF
QD
502
.B47
Cop.1

Bernasconi, Claude F.

Relaxation kinetics

DATE DUE

REF
QD
502
.B47
Cop.1

FORM 125 M

Business/Science/Technology
Division

The Chicago Public Library

DEC 17 1977

Received

Relaxation Kinetics

Relaxation Kinetics

Claude F. Bernasconi

Division of Natural Sciences
University of California
Santa Cruz, California

1976

ACADEMIC PRESS New York San Francisco London
A Subsidiary of Harcourt Brace Jovanovich, Publishers

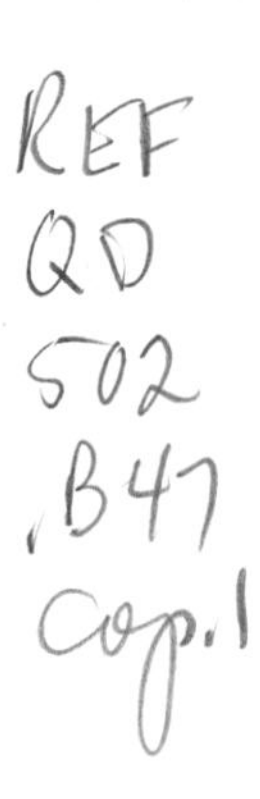
REF
QD
502
.B47
Cop.1

COPYRIGHT © 1976, BY ACADEMIC PRESS, INC.
ALL RIGHTS RESERVED.
NO PART OF THIS PUBLICATION MAY BE REPRODUCED OR TRANSMITTED IN ANY FORM OR BY ANY MEANS, ELECTRONIC OR MECHANICAL, INCLUDING PHOTOCOPY, RECORDING, OR ANY INFORMATION STORAGE AND RETRIEVAL SYSTEM, WITHOUT PERMISSION IN WRITING FROM THE PUBLISHER.

ACADEMIC PRESS, INC.
111 Fifth Avenue, New York, New York 10003

United Kingdom Edition published by
ACADEMIC PRESS, INC. (LONDON) LTD.
24/28 Oval Road, London NW1

Library of Congress Cataloging in Publication Data

Bernasconi, Claude F
Relaxation kinetics.

Includes bibliographical references and indexes.
1. Chemical reaction, Rate of. I. Title.
QD502.B47 541'.394 75-30463
ISBN 0-12-092950-3

PRINTED IN THE UNITED STATES OF AMERICA

BUSINESS/SCIENCE/TECHNOLOGY DIVISION
THE CHICAGO PUBLIC LIBRARY
DEC 17 1977

Contents

Preface ix

Acknowledgments xi

PART I THEORY OF CHEMICAL RELAXATION 1

Chapter 1 Basic Principles

1.1 Linearization of Rate Equations 3
1.2 Relaxation Time 6
1.3 Transient and Stationary Relaxation Methods 9
Problems 9

Chapter 2 Relaxation Times in Single-Step Systems

2.1 The A + B $\rightleftharpoons$ C System 11
2.2 Other Single-Step Systems 13
Problems 17
References 19

Chapter 3 Relaxation Times in Two-Step Systems

3.1 General Considerations Regarding Multistep Systems 20
3.2 The A + B $\rightleftharpoons$ C $\rightleftharpoons$ D System 21
3.3 Some Other Two-Step Systems 31
Problems 36
References 38

Chapter 4 Relaxation Times in Common Multistep Systems

4.1 General System. Castellan's Treatment 40
4.2 The A + B $\rightleftharpoons$ C $\rightleftharpoons$ D $\rightleftharpoons$ E (+F) System 42
4.3 The E + S $\rightleftharpoons$ ES $\rightleftharpoons$ EP $\rightleftharpoons$ P + E System (Enzyme Reactions) 49
4.4 Cyclic Reaction Schemes 51
4.5 Miscellaneous Multistep Schemes. Two Calculated Examples from Organic Chemistry 62
Problems 72
References 74

Chapter 5 What Is a Small Perturbation?

5.1 Conditions for Linearization in the A + B $\rightleftharpoons$ C System 76
5.2 Other Systems 80
Problems 84
References 84

Chapter 6 Relaxation Amplitudes in Single-Step Systems

6.1 The Extent of Equilibrium Displacement 85
6.2 Determination of ΔH from Relaxation Amplitudes 91
6.3 Determination of K from Relaxation Amplitudes 93
Problems 97
References 97

Chapter 7 Relaxation Amplitudes in Multistep Systems

7.1 Two-Step Systems with Rapid Equilibration of One Step 99
7.2 Normal Modes of Reactions 104
7.3 Applications of Normal Mode Analysis 112
Problems 128
References 129

Chapter 8 Complete Solution of the Relaxation Equation

8.1 Derivation of the Complete Relaxation Equation 130
8.2 Transient and Forced Solutions as Special Cases of the Complete Solution 133
8.3 Complete Solution for Some Common Step Functions (Transient Relaxation Methods) 133
8.4 Forced Solution for Oscillating Forcing Function (Stationary Relaxation Methods) 137
Problems 140
Reference 140

Chapter 9 Evaluation of Relaxation Times from Experimental Relaxation Curves

9.1 One Relaxation Time 141
9.2 Two or More Relaxation Times 142
9.3 Mean Relaxation Times 148
Problems 156
References 157

Chapter 10 Chemical Relaxation in Complex Systems

10.1 Binding of Small Molecules to Multiunit Enzymes 158
10.2 Cooperative Conformational Transitions of Linear Biopolymers (Helix–Coil Transition of Polypeptides) 164
10.3 Cooperative Binding of Small Molecules to Linear Biopolymers 170
10.4 Molecular Aggregation Phenomena. Micelle Formation 172
References 176

PART II EXPERIMENTAL TECHNIQUES AND APPLICATIONS 178

Chapter 11 The Temperature-Jump Method

11.1 The Temperature Pulse 181
11.2 Detection of Concentration Changes 189
11.3 The Combination Stopped-Flow–Temperature-Jump Apparatus 196
11.4 Temperature-Jump Equipment and Its Operation; Commercial Products 198

11.5 Applications of the Temperature-Jump Method 202
Problems 216
References 217

Chapter 12 Pressure-Jump Methods

12.1 Principles and Apparatus 222
12.2 Shock Wave Apparatus 227
12.3 Applications 228
Problems 230
References 230

Chapter 13 The Electric Field-Jump Method

13.1 Principles 232
13.2 Experimental Techniques 234
13.3 Applications 236
Problem 238
References 238

Chapter 14 The Concentration-Jump Method

14.1 Principles 240
14.2 The Stopped-Flow Technique 240
14.3 Applications 241
References 243

Chapter 15 Ultrasonic Techniques

15.1 Principles 244
15.2 General Experimental Considerations and Treatment of Data 251
15.3 Experimental Methods 253
15.4 Applications of the Ultrasonic Methods 256
Problems 265
References 267

Chapter 16 Stationary Electric Field Methods

16.1 Principles and Experimental Techniques 271
16.2 Applications 276
References 278

Index 279

Preface

The terms "relaxation kinetics" or "chemical relaxation" are usually associated with the study of very rapid reactions. This is because most experimental techniques based on the principle of chemical relaxation were developed specifically to measure the rate of very fast reactions. In its broadest sense, the term "chemical relaxation" is used to describe the self-adjustment of a perturbed molecular system to its thermal equilibrium. The time ("relaxation time") required in this self-adjustment is related to the specific rates of the chemical reactions involved. Hence, one can obtain a great deal of kinetic information by measuring relaxation times. All one has to do is to perturb the system, let it self-readjust to its (new) equilibrium state, and find some suitable device for monitoring this process as a function of time.

The great success of relaxation kinetics in measuring the rates of very fast reactions is due to the availability of a number of techniques able to perturb a chemical equilibrium extremely rapidly—e.g., in 1 μsec or even much less—much more rapidly than one could possibly mix two reagents even in a fast mixing device such as a stopped-flow apparatus. It should be pointed out, however, that the principles of chemical relaxation can be applied to slow reactions as well.

One special feature of relaxation kinetics is that one deals with rate processes of systems that are close to equilibrium. This has the great advantage of reducing all rate equations, regardless of order, to linear relationships and thus allows the treatment of complex systems in a relatively straightforward fashion; evidently this is just as big an advantage in the study of slow as it is for fast reactions.

The principles of chemical relaxation and the special techniques for the study of fast reactions have been applied mainly to problems in inorganic chemistry such as metal complex formations, to proton transfer reactions, and to a variety of biochemical and biophysical processes involving proteins and nucleic acids. Organic chemists have thus far been rather reluctant in using these techniques, although it is clear that many mechanistic problems will be solved only by measuring, with relaxation methods, the rates by which reactive intermediates are formed and disappear. Part of the problem may be that no comprehensive treatment of the subject has been available that talks the language of the organic or physical organic chemist. This book, which is an outgrowth from a course offered at the University of California at Santa Cruz, is an attempt to remedy the situation.

Apart from being more up to date, this book contains relatively little that has not already been said in the "bible" of the relaxation kineticist, viz. Eigen and DeMaeyer's 1963 article in Volume VIII of "Technique of Organic Chemistry," to which I will refer frequently. However, the present book will be more detailed and explicit in the treatment of the theory of chemical relaxation, because I felt this would be beneficial to the novice with little background in mathematics. At the same time, there is an emphasis on material of direct practical applicability, and thus the book should also become a valuable addition to the library of every researcher active in the field.

On the other hand, a number of topics treated in the "bible" or elsewhere will get little attention, partly in order to keep the size of the book at a manageable level, partly because they were felt to be of lesser importance. In view of an abundance of reviews on specific experimental methods, I shall mainly discuss their most important aspects and frequently refer to more detailed descriptions in the literature.

A word about the organization of this book. With regard to the theoretical part a mastery of Chapters 1, 2, 3, 9 and a qualitative understanding of Chapters 5 and 6 are sufficient theoretical background for the treatment of relatively simple reaction systems if only the use of transient relaxation methods (temperature-jump, pressure-jump, electric field jump, concentration-jump) is contemplated. Chapter 8 (particularly Section 8.4) needs to be consulted for the theory of stationary relaxation techniques (ultrasonic and stationary electric field methods). The other theoretical chapters deal with more complex systems and other more sophisticated aspects of chemical relaxation (relaxation amplitudes). Note also that at the end of most chapters a number of exercise problems are given.

Acknowledgments

My foremost thanks go to Professor Manfred Eigen whom I had the privilege to work with at the Max Planck Institute for Biophysical Chemistry in Göttingen and who awakened my interest in the subject of this book. I am also grateful for his comments about the manuscript. My special thanks also go to Professors G. Schwarz and J. F. Bunnett, as well as to my students D. Bear, J. Gandler, A. Lauritzen, M. Muller, and H.-C. Wang for reading and commenting on various parts of the manuscript. Furthermore, I am indebted to Professors J. Crossley, L. De Maeyer, E. M. Eyring, P. Hemmes, H. Hoffman, J. E. Rassing, V. C. Reinsborough, Z. A. Schelly, L. J. Slutsky, J. Stuehr, K. Tamm, D. Thusius, E. Wyn-Jones, T. Yasunaga, and R. Zana, to Drs. W. Knoche, C. R. Rabl, and B. H. Robinson for preprints, original drawings, or photographs of some illustrations, and to Professor E. L. Menger for help in proofreading. Last but not least, I wish to thank my wife, Regula, who bore with me during times when this book seemed to get more than its fair share of my attention.

Part I | Theory of Chemical Relaxation

Chapter 1 | Basic Principles

1.1 Linearization of Rate Equations

Consider a chemical equilibrium such as the common system

$$A + B \underset{k_{-1}}{\overset{k_1}{\rightleftharpoons}} C \tag{1.1}$$

Let us recall that a chemical equilibrium is a dynamic affair, i.e., A reacts continuously with B to form C but at the same time an equal number of molecules of C decompose to regenerate A and B. The equilibrium state of this and any other chemical system not only depends on the chemical identity of the species involved but is determined by a set of external parameters such as the temperature, pressure, solvent, stoichiometric concentrations, etc. When this equilibrium state is perturbed, for instance by the addition of more reactants, dilution, a change of pH, or a change in temperature, pressure, etc., the system adjusts itself to the changed set of external parameters. The adjustment process usually manifests itself by a measurable change in the concentrations of some or all species partaking in the equilibrium. In the case of system 1.1 such an adjustment might lead to some excess production of C at the expense of A and B (the equilibrium is said to be shifted to the right) if for instance some reactant A or B were added from the outside to the system or if the equilibrium constant $K_1 = k_1/k_{-1}$ were increased as a consequence of a rise in temperature. A shift toward increasing the concentration of A and B at the expense of C (shift to the left) would occur if raising the temperature would decrease K_1 or if some C could be added from the outside.

It is evident that the adjustment of the system to the new equilibrium conditions is a consequence of the dynamic nature of chemical equilibria. The rate of this adjustment or the rate of "chemical relaxation" is therefore determined by the rate of the very reactions which make up the equilibrium. Thus, by measuring the rate of chemical relaxation or the so-called relaxation time one can obtain the necessary information for an evaluation of k_1 and k_{-1}.

We shall illustrate how the relationships between the relaxation time and the rate coefficients for reaction 1.1 are derived for a temperature-jump experiment.

Our starting point is the general rate equation

$$-dc_A/dt = -dc_B/dt = dc_C/dt = k_1 c_A \cdot c_B - k_{-1} c_C \tag{1.2}$$

which describes the dynamic behavior of the system under any set of conditions. At equilibrium Eq. 1.2 becomes

$$-dc_A/dt = k_1 \bar{c}_A \cdot \bar{c}_B - k_{-1} \bar{c}_C = 0 \tag{1.3}$$

The bar over the concentration symbols denotes equilibrium concentrations, related by the mass law expression

$$\bar{c}_C/\bar{c}_A \cdot \bar{c}_B = k_1/k_{-1} = K_1 \tag{1.4}$$

Since in a relaxation experiment the system relaxes from an initial equilibrium state to a final equilibrium state we introduce the following symbols:

$\bar{c}_A{}^i, \bar{c}_B{}^i, \bar{c}_C{}^i$	Equilibrium concentration at initial temperature T_i
$k_1{}^i, k^i_{-1}; K_1{}^i = k_1{}^i/k^i_{-1}$	Rate and equilibrium constants at T_i
$\bar{c}_A{}^f, \bar{c}_B{}^f, \bar{c}_C{}^f$	Equilibrium concentrations at final temperature, $T_f = T_i + \Delta T$
$k_1{}^f, k^f_{-1}; K_1{}^f = k_1{}^f/k^f_{-1}$	Rate and equilibrium constants at $T_f = T_i + \Delta T$

In terms of these symbols Eq. 1.3 can be written as

$$-dc_A/dt = k_1{}^i \bar{c}_A{}^i \cdot \bar{c}_B{}^i - k^i_{-1} \bar{c}_C{}^i = 0 \tag{1.5}$$

or

$$-dc_A/dt = k_1{}^f \bar{c}_A{}^f \cdot \bar{c}_B{}^f - k^f_{-1} \bar{c}_C{}^f = 0 \tag{1.6}$$

Equation 1.5 describes the state before the temperature jump, Eq. 1.6 the state reached after chemical relaxation is complete.

Let us now discuss the events during chemical relaxation. We assume that the temperature jump is instantaneous on the time scale of the ensuing relaxation process (for situations where this assumption is not justified see Chapter 8). If we now consider the state of the system immediately following the temperature jump, i.e., before relaxation sets in, we realize that the concentrations are still those referring to the initial equilibrium state (before the temperature jump). On the other hand, we are at T_f and so the rate constants are $k_1{}^f$ and k^f_{-1}. Hence the rate equation becomes

$$-dc_A/dt = k_1{}^f \bar{c}_A{}^i \cdot \bar{c}_B{}^i - k^f_{-1} \bar{c}_C{}^i \tag{1.7}$$

Note that for $K_1{}^f \neq K_1{}^i$, which is usually the case, we have $dc_A/dt \neq 0$ in Eq. 1.7.

As time progresses relaxation sets in and the concentrations of A, B, and C approach their final equilibrium values $\bar{c}_A{}^f$, $\bar{c}_B{}^f$, and $\bar{c}_C{}^f$. Thus, Eq. 1.7 is a special case of the more general equation

$$-dc_A/dt = k_1{}^f c_A c_B - k^f_{-1} c_C \tag{1.8}$$

which, for example, may refer to a state where relaxation has already made some progress but is not complete yet.

For a formal description of what occurs during chemical relaxation it is useful to express the (time-dependent) concentrations c_A, c_B, and c_C in terms of their new equilibrium values and their (time-dependent) deviations from these equilibrium values

$$c_A = \bar{c}_A^f + \Delta c_A \tag{1.9a}$$

$$c_B = \bar{c}_B^f + \Delta c_B \tag{1.9b}$$

$$c_C = \bar{c}_C^f + \Delta c_C \tag{1.9c}$$

According to the principle of mass conservation (or mass balance), we have

$$c_A + c_C = \bar{c}_A^f + \bar{c}_C^f \tag{1.10a}$$

$$c_B + c_C = \bar{c}_B^f + \bar{c}_C^f \tag{1.10b}$$

Substituting Eqs. 1.9 for c_A, c_B, and c_C, respectively, leads to

$$\Delta c_A + \Delta c_C = 0 \tag{1.11a}$$

$$\Delta c_B + \Delta c_C = 0 \tag{1.11b}$$

or

$$\Delta c_A = \Delta c_B = -\Delta c_C = x \tag{1.12}$$

Hence Eqs. 1.9 become

$$c_A = \bar{c}_A^f + x \tag{1.13a}$$

$$c_B = \bar{c}_B^f + x \tag{1.13b}$$

$$c_C = \bar{c}_C^f - x \tag{1.13c}$$

Since the relation

$$\begin{aligned} dc_A/dt &= dc_B/dt = -dc_C/dt = d(\bar{c}_A^f + x)/dt \\ &= d\bar{c}_A/dt + dx/dt = dx/dt \end{aligned} \tag{1.14}$$

holds (note $d\bar{c}_A/dt = 0$), Eq. 1.8 can be written as

$$\begin{aligned} dx/dt &= -k_1^f(\bar{c}_A^f + x)(\bar{c}_B^f + x) + k_{-1}^f(\bar{c}_C - x) \\ &= -k_1^f\bar{c}_A^f\bar{c}_B^f + k_{-1}^f\bar{c}_C^f - k_1^f(x)^2 - [k_1^f(\bar{c}_A^f + \bar{c}_B^f) + k_{-1}^f]x \end{aligned} \tag{1.15}$$

The first two terms vanish because of Eq. 1.6 and one obtains

$$dx/dt = -[k_1^f(\bar{c}_A^f + \bar{c}_B^f) + k_{-1}^f]x - k_1^f(x)^2 \tag{1.16}$$

which can be further simplified if only small equilibrium perturbations are considered. In most treatments of the subject a perturbation is said to be small when $|x| = |\Delta c_j| \ll \bar{c}_j^f$ immediately following the perturbation for all species partaking in the equilibrium. Although in Chapter 5 the question of small perturbations is dealt with more extensively, let us adopt this simple definition for the time being. In that case the square term $k_1^f(x)^2$ is negligibly small compared to the other terms and Eq. 1.16 becomes

$$dx/dt = -[k_1(\bar{c}_A + \bar{c}_B) + k_{-1}]x \tag{1.17}$$

It is this deletion of the square term which is called the linearization of the rate equation. Note that for simplicity we have omitted the superscript f in Eq. 1.17. Henceforth this superscript will often not be included explicitly; it is to be understood that unless otherwise stated equilibrium concentrations, equilibrium constants, and rate constants will always refer to the final state.

Equation 1.17 can be written as

$$dx/dt = -(1/\tau)x \tag{1.18}$$

with

$$1/\tau = k_1(\bar{c}_A + \bar{c}_B) + k_{-1} \tag{1.19}$$

Several features deserve to be noted. Equation 1.17 or 1.18 is independent of the sign of x; i.e., it does not matter whether the original equilibrium was shifted to the right or to the left, the rate of approach to the new equilibrium is the same. This also means that we could have chosen $x = \Delta c_C$ instead of $x = \Delta c_A$. Hence instead of Eq. 1.18 we can also write, even more generally,

$$d\,\Delta c_j/dt = -(1/\tau)\,\Delta c_j \tag{1.20}$$

where j refers to any of the reacting species A, B, or C. This is a characteristic feature of chemical relaxation as long as the perturbation is small. Note, however, that for *large* perturbations dx/dt *does* depend on the sign of x as is obvious from Eq. 1.16 since a change in the sign of x only changes the sign of the dx/dt and of the $[k_1(\bar{c}_A + \bar{c}_B) + k_{-1}]x$ terms, but has no effect on the $k_1(x)^2$ term (for more details see Chapter 5).

1.2 Relaxation Time

Using the same principles illustrated above for the A + B $\rightleftharpoons$ C system, one can linearize the rate equation of any one-step equilibrium reaction and write it in the form of Eq. 1.18 or 1.20 (the expression for $1/\tau$ depends on the particular reaction scheme; see Chapter 2). Equations 1.18 or 1.20 are the familiar equations for first-order kinetic processes; they state that the rate with which a small deviation from the equilibrium vanishes is proportional to the extent of this deviation. τ is called the *relaxation time* of the system. Integration of Eq. 1.18

$$\int_{x_0}^{x} x^{-1}\,dx = -\tau^{-1}\int_0^t dt$$

leads to

$$\ln(x/x_0) = -t/\tau \tag{1.21}$$

or

$$x = x_0 \exp(-t/\tau) \tag{1.22}$$

with

$$x_0 = \bar{c}_A^{\,i} - \bar{c}_A^{\,f} \tag{1.23}$$

Similarly, from Eq. 1.20 we obtain

$$\Delta c_j = \Delta c_j^0 \exp(-t/\tau) \tag{1.24}$$

or

$$c_j = \bar{c}_j^{\mathrm{f}} + \Delta c_j^0 \exp(-t/\tau) \tag{1.25}$$

with

$$\Delta c_j^0 = \bar{c}_j^{\mathrm{i}} - \bar{c}_j^{\mathrm{f}} \tag{1.26}$$

Figure 1.1 shows how the various quantities defined above are related. The relaxation time corresponds to the time needed for Δc_j to decrease by a factor e

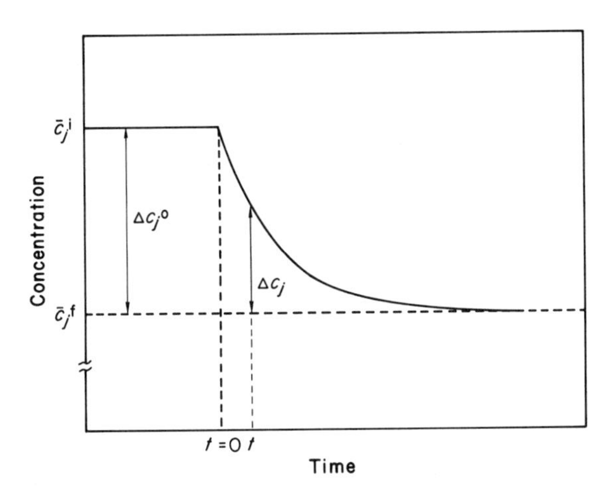

FIGURE 1.1. Chemical relaxation according to Eq. 1.25.

(base of natural logarithms, $e = 2.718\ldots$). This is best appreciated by setting $t = \tau$ in Eq. 1.24, which leads to

$$\Delta c_j = \Delta c_j^0/e = \Delta c_j^0/2.718 = 0.368\,\Delta c_j^0$$

At this point it should be briefly mentioned that a number of authors prefer to express x or Δc_j with reference to the equilibrium condition prevailing *before* the perturbation. In this case we have

$$c_{\mathrm{A}} = \bar{c}_{\mathrm{A}}^{\mathrm{i}} + \Delta c_{\mathrm{A}} = \bar{c}_{\mathrm{A}}^{\mathrm{i}} + x = \bar{c}_{\mathrm{A}}^{\mathrm{f}} + x - x_\infty \tag{1.27a}$$

$$c_{\mathrm{B}} = \bar{c}_{\mathrm{B}}^{\mathrm{i}} + \Delta c_{\mathrm{B}} = \bar{c}_{\mathrm{B}}^{\mathrm{i}} + x = \bar{c}_{\mathrm{B}}^{\mathrm{f}} + x - x_\infty \tag{1.27b}$$

$$c_{\mathrm{C}} = \bar{c}_{\mathrm{C}}^{\mathrm{i}} + \Delta c_{\mathrm{C}} = \bar{c}_{\mathrm{C}}^{\mathrm{i}} - x = \bar{c}_{\mathrm{C}}^{\mathrm{f}} - x + x_\infty \tag{1.27c}$$

with

$$x = c_{\mathrm{A}} - \bar{c}_{\mathrm{A}}^{\mathrm{i}} \tag{1.28}$$

$$x_\infty = \bar{c}_{\mathrm{A}}^{\mathrm{f}} - \bar{c}_{\mathrm{A}}^{\mathrm{i}} \tag{1.29}$$

Linearization of the rate equation leads to

$$dx/dt = -(1/\tau)(x - x_\infty) = (1/\tau)(x_\infty - x) \tag{1.30}$$

while integration

$$\int_0^x \frac{dx}{x_\infty - x} = \frac{1}{\tau}\int_0^t dt$$

leads to

$$\ln \frac{x_\infty - x}{x_\infty} = -\frac{t}{\tau} \tag{1.31}$$

or

$$x = x_\infty[1 - \exp(-t/\tau)] \tag{1.32}$$

In Chapters 2–7, we always express x or Δc_j with reference to the *final* equilibrium (Eq. 1.9). In Chapter 8 we encounter situations where the definition $\Delta c_j = c_j - \bar{c}_j^{\,i}$ may be more convenient.

Turning our attention now to the physical meaning of the relaxation time we see that its reciprocal value has the same meaning as any other rate constant of a first-order process, be it an irreversible first-order reaction, a radioactive decay, or related physical phenomena. The most common method of evaluating τ is based on Eq. 1.21 or, using logarithms to the base 10 and rearranging,

$$\log x = \log x_0 - t/2.303\tau \tag{1.33}$$

According to this equation a plot of log x versus time is linear with a slope of $-1/2.303\tau$.

As is well known and pointed out in almost every kinetics textbook, there are two features about Eq. 1.33 which make the evaluation of τ particularly easy.

1. The true zero point ($t = 0$) of the relaxation curve need not be known, it can be chosen arbitrarily (see Problems).

2. In most experiments one does not monitor actual concentrations but a physical property proportional to the concentrations, such as absorbance or conductance. It is a unique feature of first-order reactions that the proportionality constants between the physical property and the concentration (say the molar extinction coefficients) of the respective species do not need to be known. This is true whether only one or all species present in the system contribute to this physical property. Let us illustrate this with reaction 1.1 and assume we measure the absorbance. If all three species absorb, we have for the optical density

$$\text{OD} = l(\epsilon_A c_A + \epsilon_B c_B + \epsilon_C c_C) \tag{1.34}$$

and

$$\Delta\text{OD} = l(\epsilon_A \Delta c_A + \epsilon_B \Delta c_B + \epsilon_C \Delta c_C) \tag{1.35}$$

where l is the path length and the ϵ_j are the molar extinction coefficients. Recalling Eq. 1.12 and expressing x by Eq. 1.22 leads to

$$\Delta\text{OD} = l(\epsilon_A + \epsilon_B - \epsilon_C)x_0 \exp(-t/\tau) \tag{1.36}$$

As long as $\epsilon_A + \epsilon_B - \epsilon_C \neq 0$ there will in fact be a change in OD associated with chemical relaxation. Plotting log(ΔOD) versus time according to

$$\log(\Delta\text{OD}) = \log l + \log(\epsilon_A + \epsilon_B - \epsilon_C) + \log x_0 - t/2.303\tau \qquad (1.37)$$

again affords a straight line of slope $-1/2.303\tau$. The same holds true if a quantity proportional to ΔOD, e.g., the millivolt output of an oscilloscope, is plotted versus time (see Section 9.1).

To conclude let us now turn to the relation between the relaxation time and kinetic parameters of the reaction. Equation 1.19 relates $1/\tau$ to the rate coefficients of the reaction, which are the quantities of main interest to the kineticist. Note that the forward and reverse reactions contribute additively to τ^{-1}, another characteristic feature of chemical relaxation. Thus, it is the faster of the two processes which contributes most to τ^{-1}.

1.3 Transient and Stationary Relaxation Methods

Relaxation kineticists have used two different types of experimental approaches in studying chemical relaxation. The first involves a single perturbation of a chemical system at equilibrium, brought about by a sudden change of an external parameter such as temperature, pressure, or concentration, as described in the preceding section. Here the relaxation time is measured by monitoring the rate of change in the concentration of any one or several of the species immediately following the perturbation. Such methods are known as transient methods, jump methods, or step-function techniques, and Eq. 1.22 (or 1.24 or 1.32) is called the transient solution of the relaxation equation. The derivations that were illustrated for the temperature-jump method are of course exactly analogous to those for all transient techniques.

The second class of techniques is known as stationary methods. Here the chemical system is subjected to an oscillating forcing function or oscillating perturbation such as a sound wave which produces temperature and pressure fluctuations in the solution, or an oscillating electrical field. Interaction between the forcing function and the chemical system can lead to an oscillation in the position of the chemical equilibrium, with a phase lag that depends on the relation between the chemical relaxation time and the frequency of the forcing function. At the same time energy is absorbed by the system. The relaxation time can be determined by an analysis of the absorbed energy or the phase lag as a function of the frequency.

The mathematical treatment of the chemical relaxation phenomenon for stationary methods is more complicated and involves the forced solution of the complete relaxation equation; it is presented in Chapter 8.

Problems

1. Show that the true $t = 0$ point need not be known when evaluating the relaxation time from an exponential curve.

2. According to the principles discussed in this chapter, the rate of the equilibration of reaction

$$A \underset{k_{-1}}{\overset{k_1}{\rightleftharpoons}} B$$

can be studied by monitoring the relaxation which follows a perturbation of the equilibrium. Show that one could also measure the rate of equilibration by starting the reaction with pure A, i.e., at $t = 0$, $c_A = [A]_0$ and $c_B = 0$.

3. In order to be able to linearize the rate equation for the reaction $A + B \rightleftharpoons C$, one must restrict oneself to small perturbations. Show that for the reaction

$$A + B \underset{k_{-1}}{\overset{k_1}{\rightleftharpoons}} C + D$$

one obtains a linearized rate equation even for large perturbations provided that $k_1 = k_{-1}$.

4. Derive Eq. 1.30.

5. Express c_j as a function of t, τ, $\bar{c}_j^{\mathrm{f}}$ and Δc_j^{∞}.

Chapter 2 | *Relaxation Times in Single-Step Systems*

2.1 The A + B ⇌ C System

The reciprocal relaxation time for the title system has been derived in Section 1.1 (Eq. 1.19) and is given by

$$1/\tau = k_1(\bar{c}_A + \bar{c}_B) + k_{-1} \tag{2.1}$$

As is evident from Eq. 2.1 k_1 and k_{-1} can in principle be evaluated from the concentration dependence of τ, τ^{-1} being a linear function of $(\bar{c}_A + \bar{c}_B)$ (Fig. 2.1).

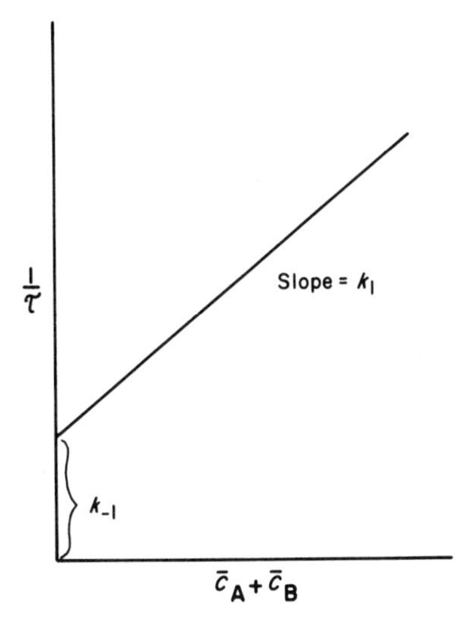

FIGURE 2.1. Determination of rate constants from the concentration dependence of $1/\tau$ for the A + B ⇌ C system.

In order to obtain highly accurate values for both k_{-1} (intercept) and k_1 (slope) it is desirable to cover an extended concentration range, since best accuracy for k_{-1} is generally obtained from a series of data where $k_{-1} > (\gg) k_1(\bar{c}_A + \bar{c}_B)$, whereas for good precision in k_1 one wants data extending up to concentrations where $k_1(\bar{c}_A + \bar{c}_B) > (\gg) k_{-1}$.

In most practical situations it is the stoichiometric rather than the equilibrium concentrations of A and B which are known, unless the equilibrium constant is known independently or $\bar{c}_A$ and $\bar{c}_B$ can be measured directly. Thus Eq. 2.1 appears to be of rather limited practical value. However, there are several ways out of this difficulty.

1. When the situation permits the best thing to do is to keep the concentration of one of the reactants quasi constant, i.e., independent of the position of the equilibrium, for instance by having it present in a large excess over the other (pseudo-first-order conditions). Thus, for $[A]_0 \gg [B]_0$, where $[A]_0 = \bar{c}_A + \bar{c}_C$ and $[B]_0 = \bar{c}_B + \bar{c}_C$, $\bar{c}_A \simeq [A]_0 \gg \bar{c}_B$ and Eq. 2.1 becomes

$$1/\tau = k_1[A]_0 + k_{-1} \tag{2.2}$$

k_1 and k_{-1} are determined from a plot of τ^{-1} versus $[A]_0$. Keeping the concentration of A quasi constant by buffering—if, for example, $A = H^+$ or HO^-—has the same effect as long as the buffer base–buffer acid equilibrium is very rapidly established on the time scale of the process under investigation.

Another important consequence of keeping the concentration of A quasi constant is that one is no longer restricted to small equilibrium perturbations. This is because there is no square term in the rate equation as can be shown as follows. Let us define a pseudo-first-order rate coefficient $k_\chi = k_1[A]_0$. Equation 1.15 can then be rewritten as

$$dx/dt = -k_\chi(\bar{c}_B + x) + k_{-1}(\bar{c}_C - x) = -[k_\chi + k_{-1}]x \tag{2.3}$$

which has no square term. Note that Eq. 2.2, which was derived from Eq. 2.1, could have been obtained from Eq. 2.3 as well.

The fact that perturbations of any size are allowed under these conditions can be very useful from an experimental point of view. Larger perturbations lead to larger concentration changes accompanying chemical relaxation and thus usually to a higher precision in the measurements. Provided one does not deal with very rapid reactions one may start a relaxation experiment simply by mixing the reagents A and B, a procedure that is frequently more straightforward than an actual perturbation experiment. For very fast reactions (faster than mixing processes) perturbation techniques as described in Part II must of course be used in any event.

2. Sometimes it is disadvantageous or impractical to choose one of the reactants in a large excess. Assume for example that $k_1 \geq 10^4 k_{-1}$. In order to obtain a reasonably accurate value for k_{-1} it would be desirable to conduct experiments at a concentration of A, the excess component, around 10^{-4} *M* or lower. However, this means, that the concentration of B has to be in the order of 10^{-5} *M* or lower. This is often too little for detecting the concentration changes accompanying chemical relaxation. In other cases the absolute values of k_1 might be very high and thus the term $k_1[A]_0$ so large under all practical pseudo-first-order conditions that τ is immeasurably short for the method at hand. Reducing the concentration of A to the level of B might then suffice to make τ accessible.

In such situations the next best thing to do is to choose $[A]_0 = [B]_0$. It is easily shown (see Problem 2) that in this case Eq. 2.1 can be written as

$$(1/\tau)^2 = 4k_1k_{-1}[A]_0 + (k_{-1})^2 \tag{2.4}$$

Plotting $(1/\tau)^2$ versus $[A]_0$ affords a straight line with a slope of $4k_1k_{-1}$ and an intercept of $(k_{-1})^2$.

If for some reason or another, neither choice of pseudo-first-order conditions or $[A]_0 = [B]_0$ is possible, k_1 and k_{-1} may still be evaluated by an iteration procedure. τ^{-1} is plotted versus $[A]_0 + [B]_0$. At very low concentrations $[A]_0 + [B]_0$ is usually not very different from $\bar{c}_A + \bar{c}_B$ but becomes increasingly so at higher concentrations. ("High" and "low" concentrations are relative terms and depend on the magnitude of the equilibrium constant K_1.) Thus a plot of τ^{-1} versus $[A]_0 + [B]_0$ rises linearly at low concentrations but becomes curved at high concentrations. The slope of the linear portion of this plot gives an approximate value for k_1, whereas the extrapolated intercept is very close to the true k_{-1}. The approximate equilibrium constant K_1 obtained from slope/intercept is then used to calculate approximate equilibrium concentrations of A and B. Now plotting τ^{-1} versus these approximate equilibrium concentrations provides a still better value for k_1 and thus for K_1, which is used to obtain a still better approximation of the true equilibrium concentrations and so on until there is convergence. Two or three such iterations are usually sufficient. For an example see Problem 6.

2.2 Other Single-Step Systems

Applying the same principles used for the derivation of Eq. 1.17 it can easily be shown that for any reaction system of the general form

$$aA + bB + cC + \cdots \underset{k_{-1}}{\overset{k_1}{\rightleftharpoons}} mM + nN + oO + \cdots \tag{2.5}$$

chemical relaxation is always a first-order process and can be described by an equation such as Eq. 1.20, provided the perturbation is small. This is because all higher order terms in the rate equations, regardless of their order, become vanishingly small for small perturbations; i.e., the rate equations can always be linearized. Needless to say, this leads to a very substantial mathematical simplification and is one of the advantages of the relaxation methods. It is comparable to the great advantage of conducting higher order reactions under pseudo-first-order conditions, a device often used by kineticists when measuring rates of irreversible reactions.

It is to be noted that only bimolecular or higher molecular reactions give rise to higher order terms such as $k_1(x)^2$ in Eq. 1.16. Equilibria where the reactions are first order or pseudo first order in both directions need not be linearized; i.e., chemical relaxation is a first-order process regardless of the size of the perturbation.

Some of the most common single-step equilibria are listed in Table 2.1. Note that the relaxation time expressions always contain a term for the forward and one for the reverse reaction in an additive manner. Another notable general feature

Table 2.1. Relaxation Time Expressions for Single-Step Equilibria

	Reaction	Relaxation time
(1)	$A \underset{k_{-1}}{\overset{k_1}{\rightleftharpoons}} B$	$1/\tau = k_1 + k_{-1}$
(2)	$A + B \underset{k_{-1}}{\overset{k_1}{\rightleftharpoons}} C$	$1/\tau = k_1(\bar{c}_A + \bar{c}_B) + k_{-1}$
(3)	$2A \underset{k_{-1}}{\overset{k_1}{\rightleftharpoons}} B$	$1/\tau = 4k_1\bar{c}_A + k_{-1}$
(4)	$nA \underset{k_{-1}}{\overset{k_1}{\rightleftharpoons}} B$	$1/\tau = n^2 k_1(\bar{c}_A)^{n-1} + k_{-1}$
(5)	$A + B \underset{k_{-1}}{\overset{k_1}{\rightleftharpoons}} C + D$	$1/\tau = k_1(\bar{c}_A + \bar{c}_B) + k_{-1}(\bar{c}_C + \bar{c}_D)$
(6)	$A + B \underset{k_{-1}}{\overset{k_1}{\rightleftharpoons}} 2C$	$1/\tau = k_1(\bar{c}_A + \bar{c}_B) + 4k_{-1}\bar{c}_C$
(7)	$A + C \underset{k_{-1}}{\overset{k_1}{\rightleftharpoons}} B + C$ C = catalyst	$1/\tau = (k_1 + k_{-1})\bar{c}_C$
(8)	$A + B + C \underset{k_{-1}}{\overset{k_1}{\rightleftharpoons}} D$	$1/\tau = k_1(\bar{c}_A\bar{c}_B + \bar{c}_A\bar{c}_C + \bar{c}_B\bar{c}_C) + k_{-1}$
(9)	$A + B \rightleftharpoons C$ (several parallel pathways)	$1/\tau = \sum k_i(\bar{c}_A + \bar{c}_B) + \sum k_{-i}$

is that all first-order reactions are represented by their rate coefficients only, whereas the terms associated with a higher order reaction contain some characteristic function of the reactant concentrations as well; the order of their concentration dependence is always $n - 1$, with n being the reaction order. For the most common case, the bimolecular reactions, it is always the sum of the (final) equilibrium concentrations of the reactants multiplied by the rate constant.

Whenever there is a concentration dependence of the relaxation time, it is usually possible to determine k_1 and k_{-1} by measuring τ^{-1} at several concentrations. However, there are often practical problems similar to the ones discussed for the $A + B \rightleftharpoons C$ system and generally a special choice of reaction conditions is called for. A number of these are now discussed along with other interesting features.

Systems $A \rightleftharpoons B$ *and* $A + C \rightleftharpoons B + C$ (C = *Catalyst*). Relaxation experiments can only provide the sum $k_1 + k_{-1}$; independent knowledge of the equilibrium constant is required to obtain k_1 and k_{-1} individually. Perturbations of any size are permitted.

System $2A \rightleftharpoons B$. The factor 4 in the $4k_1\bar{c}_A$ term (Table 2.1) is a consequence of the stoichiometric relationship

$$\Delta c_B = -\tfrac{1}{2}\Delta c_A \tag{2.6}$$

If the relation $\tau^{-1} = 4k_1\bar{c}_A + k_{-1}$ is to be used directly in evaluating k_1 and k_{-1}, the equilibrium concentration of A must be directly measurable or calculable from K_1; in contrast to the $A + B \rightleftharpoons C$ system, one cannot work under pseudo-first-order conditions. However, the relation

$$(1/\tau)^2 = 8k_1k_{-1}[A]_0 + (k_{-1})^2 \tag{2.7}$$

which does not require knowledge of K_1, can be used where $[A]_0$ is the stoichiometric concentration ($[A]_0 = \bar{c}_A + 2\bar{c}_B$). It is derived in a way similar to the derivation of Eq. 2.4.

System $A + B \rightleftharpoons 2C$. This reaction scheme is typical for a large number of disproportionation–comproportionation equilibria arising from electron transfer. A well-known example would be the quinone(A)–hydroquinone(B)–semiquinone (C) redox system. As in the previous system, the factor 4 in the $4k_{-1}\bar{c}_C$ term is a consequence of the stoichiometric relation $\Delta c_C = -2\,\Delta c_A$.

For an evaluation of k_1 and k_{-1} from the concentration dependence of τ^{-1} independent knowledge of the equilibrium constant (or equilibrium concentrations) is usually required, no matter what special conditions are chosen such as keeping A or B quasi constant or having $\bar{c}_A = \bar{c}_B$. However, it is often convenient to have $\bar{c}_A = \bar{c}_B$. This is readily accomplished by making up a solution with $[A]_0 = [B]_0$ or by putting C into the reaction medium where disproportionation of C automatically assures $\bar{c}_A = \bar{c}_B$. In this case one obtains (from the mass law expression)

$$\bar{c}_C = \bar{c}_A K_1^{1/2} \tag{2.8}$$

whereas the mass balance requirements lead to

$$\bar{c}_A + \tfrac{1}{2}\bar{c}_C = [A]_0 \tag{2.9a}$$

or

$$2\bar{c}_A + \bar{c}_C = [C]_0 \tag{2.9b}$$

Thus, one can express the reciprocal relaxation time as

$$\frac{1}{\tau} = \frac{2k_1 + 4k_{-1}K_1^{1/2}}{1 + \frac{1}{2}K_1^{1/2}}[A]_0 \tag{2.10}$$

(in terms of $[A]_0$), or as

$$\frac{1}{\tau} = \frac{2k_1 + 4k_{-1}K_1^{1/2}}{2 + K_1^{1/2}}[C]_0 \tag{2.11}$$

(in terms of $[C]_0$). These equations can be converted into a more useful form, e.g., Eq. 2.11 into

$$\frac{1}{\tau} \cdot \frac{2 + K_1^{1/2}}{2 + 4/K_1^{1/2}} = k_1[C]_0 \tag{2.12}$$

Plotting the left-hand side versus $[C]_0$ provides a straight line with a slope of k_1. Since the line goes through the origin a relatively small number of experiments (theoretically one single experiment) is required to obtain a good value for k_1.

There is a particular feature about the linearization of the rate equation of the system $A + B \rightleftharpoons 2C$ which is noteworthy. Its recognition can sometimes be useful from an experimental point of view. The equation for chemical relaxation is given by

$$d\,\Delta c_A/dt = -[k_1(\bar{c}_A + \bar{c}_B) + 4k_{-1}\bar{c}_C]\,\Delta c_A - [k_1 - 4k_{-1}](\Delta c_A)^2 \tag{2.13}$$

The usual way to assure linearization is to keep the perturbations small so that

$$[k_1 - 4k_{-1}]\,\Delta c_A \ll k_1(\bar{c}_A + \bar{c}_B) + 4k_{-1}\bar{c}_C$$

However, for the special case $k_1 \approx 4k_{-1}$ this inequality holds even for a large Δc_A, and thus leads to linearization regardless of the size of the perturbation. This might appear of rather academic interest since the systems where $k_1 \approx 4k_{-1}$ are not likely to abound. In Chapter 5 it is shown, however, that when this reaction system is coupled to a rapid acid–base equilibrium (which is frequently possible in redox systems), experimental conditions can often be chosen so as to fulfill these special conditions for linearization. The main advantage of working under these conditions is of course the higher accuracy in the measurement of relatively large concentration changes.

Similar considerations apply to other systems where both the forward and reverse reactions are bimolecular, for instance in the system $A + B \rightleftharpoons C + D$. In this case the square term vanishes for $k_1 = k_{-1}$ (see Problem 3, Chapter 1).

System $A + B \rightleftharpoons C + D$. Many substitution reactions and equilibria between general acid–base pairs are of this type. The data are most easily treated when one reactant on each side of the equilibrium can be kept quasi constant. Examples where this is possible might include protonations of carbon bases (B^-) in the presence of a large excess of HA and A^-,

$$B^- + HA \rightleftharpoons BH + A^- \tag{2.14}$$

reversible substitution reactions with a large excess of Y and X over RX and RY,

$$RX + Y \rightleftharpoons RY + X \tag{2.15}$$

addition of amino compounds to a carbonyl group in a pH-buffered system and with the amine in large excess,

$$RNH_2 + {>}C{=}O \rightleftharpoons -\underset{|}{\overset{O^-}{\overset{|}{C}}}-NHR + H^+ \tag{2.16}$$

and many others.

Under such conditions the relaxation time is simply given by

$$1/\tau = k_1[A]_0 + k_{-1}[D]_0 \tag{2.17}$$

where $[A]_0$ and $[D]_0$ are the stoichiometric or buffered equilibrium concentrations of the quasi-constant reactants. k_1 is obtained by plotting τ^{-1} versus $[A]_0$ at constant $[D]_0$, k_{-1} by plotting τ^{-1} versus $[D]_0$ at constant $[A]_0$.

The next best thing is to keep at least one reactant quasi constant, for instance D. This leads to

$$1/\tau = k_1(\bar{c}_A + \bar{c}_B) + k_{-x} \tag{2.18}$$

with $k_{-x} = k_{-1}[D]_0$. Here the same options are available as in the system $A + B \rightleftharpoons C$, i.e., we can use Eq. 2.4 or iteration.

When neither reactant can be kept quasi constant, one should try to have $\bar{c}_A = \bar{c}_B$ and $\bar{c}_C = \bar{c}_D$ which leads to

$$1/\tau = 2k_1[\bar{c}_A + \bar{c}_C/K_1] \tag{2.19}$$

From the mass law

$$\bar{c}_C = \bar{c}_A K_1^{1/2} \tag{2.20}$$

and the mass balance

$$\bar{c}_C = [A]_0 - \bar{c}_A \tag{2.21}$$

one obtains $\bar{c}_A$ and $\bar{c}_C$ as functions of $[A]_0$ and 2.19 becomes

$$\frac{1}{\tau} = 2k_1 \frac{1 + 1/K_1^{1/2}}{1 + K_1^{1/2}} [A]_0 \tag{2.22}$$

Here the equilibrium constant must be known in order to evaluate k_1 and k_{-1}.

System $A + B + C \rightleftharpoons D$. The author is not aware of any example in the literature where a termolecular reaction could not be conducted by keeping at least one of the reactants quasi constant. Thus, the problem is reduced to a system $A + B \rightleftharpoons D$ with the relaxation time given by

$$1/\tau = k_x(\bar{c}_A + \bar{c}_B) + k_{-1} \tag{2.23}$$

with $k_x = k_1[C]_0$.

System A + B ⇄ C (several parallel pathways). When there are several parallel pathways interconnecting reactants and products this does not lead to several relaxation times as in the case for consecutive reactions, but simply to a summation of terms in the reciprocal relaxation time. This is true for any type of single-step equilibrium.

Problems

1. Derive expressions for $1/\tau$ for the following systems:

(a) $A \underset{k_{-1}}{\overset{k_1}{\rightleftharpoons}} 3B$

(b) $2A \underset{k_{-1}}{\overset{k_1}{\rightleftharpoons}} B + C$

(c) $A + B + C \underset{k_{-1}}{\overset{k_1}{\rightleftharpoons}} D + E + C$ (C = catalyst)

Discuss for each case how you would determine the rate constants from the experimentally determined $1/\tau$ values.

2. Derive Eq. 2.4.

3. Assume you have collected the following data for the system $A + B \rightleftharpoons C$. Determine k_1 and k_{-1}.

$10^4 \times [A]_0$ (M)	$10^4 \times [B]_0$ (M)	$10^{-2} \times 1/\tau$ (sec^{-1})
1.0	1.0	1.18
2.5	2.5	1.42
10	10	2.24
25	25	3.32

4. The kinetics of the dimerization of thiopental has been studied with an ultrasonic relaxation technique by Hammes and Park (*1*). The system can be described by

$$2A \underset{k_{-1}}{\overset{k_1}{\rightleftharpoons}} A_2$$

The following relaxation times were reported as a function of the stoichiometric concentration of A.

$[A]_0$ (M)	$10^9\tau$ (sec)
0.161	2.1
0.130	2.3
0.103	2.5
0.057	2.9

Calculate k_1 and k_{-1} and compare your result with that of Hammes and Park.

5. Geier (*2*) reports the following data from a temperature-jump study of the complex formation between Ni^{2+} and murexide anion (L^-) at 10°C, pH 4.

$[Ni^{2+}]_0$ (M)	$[L^-]_0$ (M)	τ (sec)
5×10^{-3}	1×10^{-4}	0.20
4×10^{-3}	1×10^{-4}	0.24
3×10^{-3}	1×10^{-4}	0.30
2×10^{-3}	1×10^{-4}	0.45
1×10^{-3}	1×10^{-4}	0.70

Assuming the reaction can be written as

$$Ni^{2+} + L^- \underset{k_{-1}}{\overset{k_1}{\rightleftharpoons}} NiL^+$$

evaluate k_1 and k_{-1}.

6. In an A + B $\rightleftharpoons$ C system, where K_1 is unknown and $\bar{c}_A$ and $\bar{c}_B$ cannot be measured directly, the following data were obtained:

$10^4 \times [A]_0$ (M)	$10^4 \times [B]_0$ (M)	τ^{-1} (sec^{-1})
5	10	206
10	10	224
20	10	288
30	10	360

Use the iteration procedure outlined in Section 2.1 to determine k_1 and k_{-1}.

References

1. G. G. Hammes and A. C. Park, *J. Amer. Chem. Soc.* **91,** 956 (1969).
2. G. Geier, *Helv. Chim. Acta* **51,** 94 (1968).

Chapter 3 | Relaxation Times in Two-Step Systems

3.1 General Considerations Regarding Multistep Systems

Most reaction mechanisms involve several steps and are characterized by more than one relaxation time (relaxation spectrum). If the intermediate states accumulate to measurable concentrations, several or all the relaxation times may in fact be observed experimentally.

The number of relaxation times is always equal to the number of *independent* rate equations that can be written for the system (equivalent to the number of concentration variables minus the number of mass conservation relationships). In linear systems such as

$$A + B \rightleftharpoons C \rightleftharpoons D \rightleftharpoons E + F \tag{3.1a}$$

this is equal to the number of reaction steps (three steps in 3.1a); in cyclic systems such as

$$\begin{array}{ccccc} A + B & \rightleftharpoons & C & \rightleftharpoons & D + E \\ \updownarrow & & \updownarrow & & \updownarrow \\ G + H & \rightleftharpoons & I & \rightleftharpoons & K + L \end{array} \tag{3.1b}$$

this number is smaller than the number of steps. Also, parallel reactions such as case 9 in Table 2.1 do not increase the number of independent rate equations.

Hence a simpler and completely general procedure for the determination of the number of relaxation times is to count the number of *states*. It turns out that *the number of relaxation times is always equal to the number of states minus one*. There are four states in scheme 3.1a and hence three relaxation times; there are six states in scheme 3.1b and hence five relaxation times, although there are seven reaction steps in this latter system.

When some of the states are present at a concentration below the detectability level of the available experimental method, some of the relaxation times become undetectable. However, such unstable (under equilibrium conditions) intermediates can sometimes be generated as transients by larger perturbations—e.g., mixing experiments or flash photolysis (1,2)—and thus the missing relaxation effect be made visible.

In exceptional cases two or more relaxation times may be identical (degeneracy) which leads to a further reduction of observable relaxation times. More often two or more relaxation times may be so similar as to make it difficult to differentiate between one or several relaxation times (see Chapter 9). Thus it is apparent that the observation of several relaxation times must always be attributed to a multiple-step mechanism; however, observation of only one relaxation time does not necessarily mean that one deals with a single-step equilibrium.

Two-step systems are among the most frequently encountered multistep mechanisms in chemistry. They are relatively simple to treat mathematically and still allow the illustration of most of the principles of chemical relaxation in multistep systems. We therefore discuss these first. In Chapter 4 the problem of analyzing more complex relaxation spectra is treated in a more general way.

3.2 The $A + B \rightleftharpoons C \rightleftharpoons D$ System

The system

$$\underset{①}{A + B} \underset{k_{-1}}{\overset{k_1}{\rightleftharpoons}} \underset{②}{C} \underset{k_{-2}}{\overset{k_2}{\rightleftharpoons}} \underset{③}{D} \tag{3.2}$$

is a common scheme in many branches of chemistry; it may represent a reaction proceeding to a product (D) via an intermediate (C), enzyme–substrate complex formation followed by chemical transformation of the enzyme–substrate complex, a reaction conducted in a buffered solution leading to an acidic (basic) product C which loses (accepts) a proton in a subsequent step, and many other situations too numerous to mention.

Scheme 3.2 is characterized by two relaxation times. In general τ_1 and τ_2 will *not* just be given by the respective expressions for the isolated reactions ① $\rightleftharpoons$ ② $[\tau^{-1} = k_1(\bar{c}_A + \bar{c}_B) + k_{-1}]$ and ② $\rightleftharpoons$ ③ $(\tau^{-1} = k_2 + k_{-2})$. At least one of the relaxation times but frequently both are complex functions of all rate processes characterizing the reaction system. This is because the two reactions are coupled to each other by a common reactant (C), very much in the same way as individual oscillators, for instance the vibrating bonds in a multiatomic molecule, are coupled in a multioscillator system. Thus, one talks about normal modes of reactions (Section 7.2) in the same way as about normal modes of vibration.

3.2.1 Rapid Equilibration of the First Step

It is very common for one of the steps in Eq. 3.2 to equilibrate much more rapidly than the other. The derivation of the relaxation times is then quite simple.

Let us assume that the first step equilibrates much faster than the second step, i.e., $k_1(\bar{c}_A + \bar{c}_B) + k_{-1} \gg k_2, k_{-2}$. This will be true in many cases where the first step involves a diffusion-controlled process such as a proton transfer followed by any kind of chemical transformation of C to D; or where the first step is the formation of an ion pair between a solvated metal ion and a ligand, the second the formation of an inner sphere complex; or where the first step is the formation of an enzyme–substrate complex followed by the actual enzyme-catalyzed transformation, etc.

Let us consider the events following a perturbation. In general a perturbation affects the equilibrium conditions in such a manner that an adjustment of the concentrations of all species partaking in the system becomes necessary. The new equilibrium position for the step $A + B \rightleftharpoons C$ is reached very rapidly, in fact before equilibration of the step $C \rightleftharpoons D$ has made any significant progress. Thus, the relaxation of the first step proceeds as in a reaction of the type $A + B \rightleftharpoons C$ considered in isolation and τ_I^{-1} is simply given by

$$1/\tau_1 = k_1(\bar{c}_A{}^1 + \bar{c}_B{}^1) + k_{-1} \tag{3.3}$$

Note that we use the symbols $\bar{c}_A{}^1$ and $\bar{c}_B{}^1$ to indicate the "equilibrium concentration" reached after equilibration of the first step only; however, k_1 and k_{-1} refer to the final equilibrium conditions. Final equilibration (second relaxation time) of the system generally leads to a further small change in the concentration of A and B until the concentrations $\bar{c}_A{}^f$ and $\bar{c}_B{}^f$ are reached. Since the difference between $\bar{c}_A{}^1$ and $\bar{c}_A{}^f$, etc., is typically very small, one usually makes no distinction between $\bar{c}_A{}^1$ and $\bar{c}_A{}^f$. If the experiment can be conducted under pseudo-first-order conditions, the distinction is of no consequence at all.

In order to derive the second relaxation time we first set up the rate equation for the slow step. The choice of a suitable concentration variable poses an interesting problem. c_D is the most obvious choice since D is involved in the slow step only; c_C obviously is unsuitable because C is taking part in the slow *and* the fast step. However, $c_A + c_C$ (or $c_B + c_C$) is equivalent to c_D since adding the two rate equations

$$-dc_A/dt = k_1c_Ac_B - k_{-1}c_C$$
$$-dc_C/dt = -k_1c_Ac_B + k_{-1}c_C + k_2c_C - k_{-2}c_D$$

provides

$$dc_D/dt = -d(c_A + c_C)/dt = k_2c_C - k_{-2}c_D \tag{3.4}$$

Substituting $\bar{c}_D{}^f + \Delta c_D$ for c_D, and $\bar{c}_C{}^f + \Delta c_C$ for c_C (see Eq. 1.9) we obtain

$$d\,\Delta c_D/dt = k_2(\bar{c}_C{}^f + \Delta c_C) - k_{-2}(\bar{c}_D{}^f + \Delta c_D) = k_2\,\Delta c_C - k_{-2}\,\Delta c_D \tag{3.5}$$

We now wish to transform Eq. 3.5 into an equation of the form

$$d\,\Delta c_D/dt = -(1/\tau_2)\,\Delta c_D \tag{3.6}$$

in analogy to Eq. 1.20. For this we must express Δc_C in terms of Δc_D, which can be achieved as follows. From mass balance considerations we have

$$\Delta c_D = -\Delta c_A - \Delta c_C \tag{3.7}$$

$$\Delta c_B = \Delta c_A \tag{3.8}$$

A further equation relating Δc_A, Δc_C, and Δc_D can be produced by specifically considering only situations at $t \gg \tau_1$, i.e., when step ① $\rightleftharpoons$ ② is at equilibrium while the second process is relaxing. As mentioned above equilibration of the first step does not lead to the establishment of the *final* equilibrium concentrations $\bar{c}_A{}^f$, $\bar{c}_B{}^f$, and $\bar{c}_C{}^f$, but leads to a set of (time-dependent) pseudo-equilibrium concentrations $\tilde{c}_A{}^1$, $\tilde{c}_B{}^1$, and $\tilde{c}_C{}^1$ which are related through

$$\tilde{c}_C{}^1/\tilde{c}_A{}^1\tilde{c}_B{}^1 = \bar{c}_C{}^f/\bar{c}_A{}^f\bar{c}_B{}^f = K_1 = k_1/k_{-1} \tag{3.9}$$

($K_1 = K_1{}^f$ defined for final condition). Note that $\tilde{c}_A{}^1$, $\tilde{c}_B{}^1$, and $\tilde{c}_C{}^1$ change from $\bar{c}_A{}^1$, $\bar{c}_B{}^1$, and $\bar{c}_C{}^1$ (Eq. 3.3) when $\tau_2 \gg t \gg \tau_1$, to $\bar{c}_A{}^f$, $\bar{c}_B{}^f$, and $\bar{c}_C{}^f$ at $t \gg \tau_2$.

Expressing

$$\tilde{c}_A{}^1 = \bar{c}_A{}^f + \Delta c_A, \qquad \tilde{c}_B{}^1 = \bar{c}_B{}^f + \Delta c_B = \bar{c}_B{}^f + \Delta c_A, \qquad \tilde{c}_C{}^1 = \bar{c}_C{}^f + \Delta c_C$$

we can write Eq. 3.9 as

$$\frac{\bar{c}_C{}^f + \Delta c_C}{(\bar{c}_A{}^f + \Delta c_A)(\bar{c}_B{}^f + \Delta c_A)} = K_1 = \frac{\bar{c}_C{}^f}{\bar{c}_A{}^f\bar{c}_B{}^f} \tag{3.10}$$

For small displacements ($K_1\,\Delta c_A \cdot \Delta c_A \approx 0$), Eq. 3.10 leads to

$$\Delta c_C = K_1(\bar{c}_A{}^f + \bar{c}_B{}^f)\,\Delta c_A \tag{3.11}$$

It should be noted at this point that Eq. 3.11 can be obtained more directly by differentiation of the mass law expression $\bar{c}_C{}^f = K_1\bar{c}_A{}^f \cdot \bar{c}_B{}^f$, namely

$$\Delta c_C = \left(\frac{\partial \bar{c}_C{}^f}{\partial \bar{c}_A{}^f}\right)_{\bar{c}_B{}^f} \Delta c_A + \left(\frac{\partial \bar{c}_C{}^f}{\partial \bar{c}_B{}^f}\right)_{\bar{c}_A{}^f} \Delta c_B = K_1\bar{c}_B{}^f\,\Delta c_A + K_1\bar{c}_A{}^f\,\Delta c_B$$

Substituting Δc_A from Eq. 3.11 into Eq. 3.7 leads to

$$\Delta c_C = -K_1(\bar{c}_A{}^f + \bar{c}_B{}^f)(\Delta c_C + \Delta c_D) \tag{3.12}$$

and after rearrangement to

$$\Delta c_C = -\frac{K_1(\bar{c}_A{}^f + \bar{c}_B{}^f)}{1 + K_1(\bar{c}_A{}^f + \bar{c}_B{}^f)}\,\Delta c_D \tag{3.13}$$

Finally substituting Eq. 3.13 into Eq. 3.5 affords

$$\frac{d\,\Delta c_D}{dt} = -\left[k_2\frac{K_1(\bar{c}_A{}^f + \bar{c}_B{}^f)}{1 + K_1(\bar{c}_A{}^f + \bar{c}_B{}^f)} + k_{-2}\right]\Delta c_D \tag{3.14}$$

which is of the form of Eq. 3.6 with

$$\frac{1}{\tau_2} = k_2\frac{K_1(\bar{c}_A + \bar{c}_B)}{1 + K_1(\bar{c}_A + \bar{c}_B)} + k_{-2} \tag{3.15}$$

Equation 3.15 (we again omit the superscript f) is seen to differ from τ^{-1} for an isolated C $\rightleftharpoons$ D system by a concentration-dependent factor ≤ 1 in the k_2 term. The magnitude of this factor reflects the degree to which the equilibrium ① $\rightleftharpoons$ ② is shifted to the right or left. When ② is strongly favored over ①, $K_1(\bar{c}_A + \bar{c}_B) \gg 1$ and $\tau_2^{-1} = k_2 + k_{-2}$, i.e., the system behaves as if the process ② ⇌ ③ were an isolated reaction; when $K_1(\bar{c}_A + \bar{c}_B) \ll 1$, ① is strongly favored over ② and $\tau_2^{-1} = k_2 K_1(\bar{c}_A + \bar{c}_B) + k_{-2}$.

There is no equilibration factor in the k_{-2} term since D is not directly coupled to the equilibrium ① ⇌ ②.

In principle Eq. 3.15 allows k_2, k_{-2}, and K_1 to be evaluated as follows. The intercept of a plot of τ_2^{-1} versus $(\bar{c}_A + \bar{c}_B)$ provides k_{-2} as shown in Fig. 3.1. If the

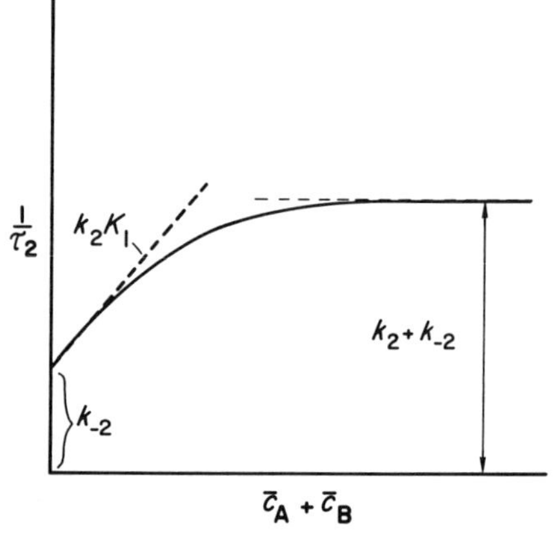

FIGURE 3.1. Concentration dependence of τ_2^{-1} for A + B ⇌ C ⇌ D system with rapid equilibration of first step.

plateau value of τ_2^{-1} can be reached, one finds $k_2 = \tau_{2(\text{plateau})}^{-1} - k_{-2}$. More frequently the plateau cannot be reached. In such a case one plots $(\tau_2^{-1} - k_{-2})^{-1}$ versus $(\bar{c}_A + \bar{c}_B)^{-1}$ which according to

$$\frac{1}{\tau_2^{-1} - k_{-2}} = \frac{1}{k_2} + \frac{1}{k_2 K_1(\bar{c}_A + \bar{c}_B)} \tag{3.16}$$

(inversion plot) gives a straight line with interecpt $1/k_2$ and slope $1/k_2K_1$ (Fig. 3.2).

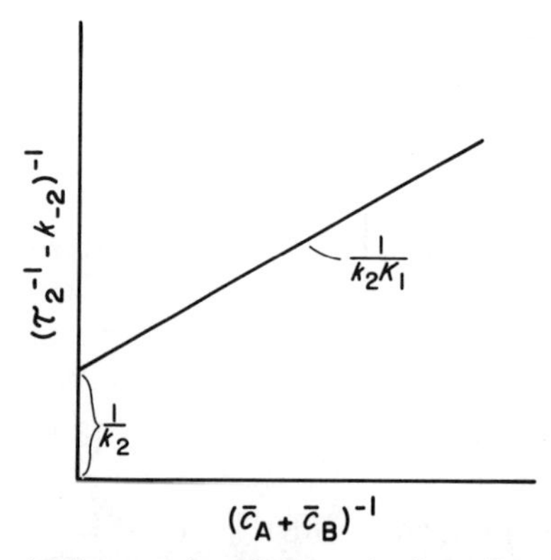

FIGURE 3.2. Inversion plot according to Eq. 3.16.

An interesting feature is that K_1 can be evaluated from the ratio of the intercept to the slope of the inversion plot. Frequently this is the only way K_1 can be obtained experimentally; an equilibrium study of the step ① ⇌ ② is usually precluded unless τ_2 is very long, and τ_1 may be too short for a determination of K_1 from k_1 and k_{-1}.

When K_1 can be obtained independently, Eq. 3.16 allows a check for consistency and may serve to confirm the assumed mechanism.

The practical problems in evaluating k_{-2}, k_2, and K_1 from τ_2^{-1} are similar to the ones with the single-step system $A + B \rightleftharpoons C$, since it is usually the stoichiometric rather than the equilibrium concentrations of A and B which are known. Whenever possible A will be kept quasi constant so that Eq. 3.15 and Eq. 3.16 simplify to

$$\frac{1}{\tau_2} = k_2 \frac{K_1[\mathrm{A}]_0}{1 + K_1[\mathrm{A}]_0} + k_{-2} \tag{3.17}$$

and

$$\frac{1}{\tau_2^{-1} - k_{-2}} = \frac{1}{k_2} + \frac{1}{k_2 K_1[\mathrm{A}]_0} \tag{3.18}$$

When this is not feasible one may use an iteration procedure similar to the one discussed for the system $A + B \rightleftharpoons C$, i.e., one first plots the expression $[\tau_2^{-1}$-intercept of plot τ_2^{-1} versus $([\mathrm{A}]_0 + [\mathrm{B}]_0)]^{-1}$ versus $([\mathrm{A}]_0 + [\mathrm{B}]_0)^{-1}$, and so forth.

Note that, just as for single-step reactions, keeping A quasi constant has the additional considerable advantage of allowing a large equilibrium perturbation.

Whether each of the three constants k_2, k_{-2}, and K_1 can be obtained from τ_2^{-1} with high precision depends on several factors. Ideally k_2 should be several-fold larger than k_{-2} and the magnitude of K_1 should be such as to allow a variation of the term $K_1(\bar{c}_A + \bar{c}_B)$ or $K_1[\mathrm{A}]_0$, respectively, from $\ll 1$ to $\gg 1$. Many real systems do not approach this ideal.

Let us consider two specific situations.

1. When k_2 is relatively small compared to k_{-2}, the expression $\tau_2^{-1} - k_{-2}$ may become comparable to the experimental uncertainty in τ_2^{-1} for a great number of experimental points and thus a plot according to Eq. 3.16 or Eq. 3.18 may be useless. Only k_{-2} can be determined with precision.
2. When K_1 is so small that $K_1(\bar{c}_A + \bar{c}_B)$ barely approaches the value of unity at the highest concentrations, the inversion plot (Eqs. 3.16 and 3.18) has an intercept that may not be distinguishable from zero within experimental error. Thus only k_{-2} and the quantity k_2K_1 can be obtained.

In the extreme case where $K_1(\bar{c}_A + \bar{c}_B) \ll 1$ throughout, no curvature in the plot of τ_2^{-1} versus $(\bar{c}_A + \bar{c}_B)$ is apparent and in the absence of independent evidence (e.g., detection of τ_1) the investigator may remain unaware of the presence of an additional equilibrium and assume that the slope of τ_2^{-1} versus $(\bar{c}_A + \bar{c}_B)$ is a true rate constant.

3.2.2 General Case

When the two steps in reaction 3.2 equilibrate at comparable rates there is strong mutual coupling between the two steps. As a consequence both relaxation times depend in a complex way on all four rate processes of the system.

Let us write the three rate equations characterizing reaction 3.2:

$$dc_A/dt = dc_B/dt = -k_1c_Ac_B + k_{-1}c_C \tag{3.19}$$

$$dc_C/dt = k_1c_Ac_B - (k_{-1} + k_2)c_C + k_{-2}c_D \tag{3.20}$$

$$dc_D/dt = k_2c_C - k_{-2}c_D \tag{3.21}$$

Linearization affords

$$d\,\Delta c_A/dt = d\,\Delta c_B/dt = -k_1(\bar{c}_A + \bar{c}_B)\,\Delta c_A + k_{-1}\,\Delta c_C \tag{3.22}$$

$$d\,\Delta c_C/dt = k_1(\bar{c}_A + \bar{c}_B)\,\Delta c_A - (k_{-1} + k_2)\,\Delta c_C + k_{-2}\,\Delta c_D \tag{3.23}$$

$$d\,\Delta c_D/dt = k_2\,\Delta c_C - k_{-2}\,\Delta c_D \tag{3.24}$$

The differential equations 3.19–3.21 or 3.22–3.24 are not all independent from each other; for example, Eq. 3.22 = −Eq. 3.23 − Eq. 3.24. This is a consequence of the stoichiometric relations 3.7 and 3.8.

Thus we may choose any two of the three equations to fully describe the system, for example Eqs. 3.22 and 3.24, which after substituting $-(\Delta c_A + \Delta c_D)$ for Δc_C become

$$d\,\Delta c_A/dt = -[k_1(\bar{c}_A + \bar{c}_B) + k_{-1}]\,\Delta c_A - k_{-1}\,\Delta c_D \tag{3.25}$$

$$d\,\Delta c_D/dt = -k_2\,\Delta c_A - [k_2 + k_{-2}]\,\Delta c_D \tag{3.26}$$

Equations 3.25 and 3.26 are of the general form

$$dx_1/dt + a_{11}x_1 + a_{12}x_2 = 0 \tag{3.27}$$

$$dx_2/dt + a_{21}x_1 + a_{22}x_2 = 0 \tag{3.28}$$

with

$$a_{11} = k_1(\bar{c}_A + \bar{c}_B) + k_{-1} \tag{3.29}$$

$$a_{12} = k_{-1} \tag{3.30}$$

$$a_{21} = k_2 \tag{3.31}$$

$$a_{22} = k_2 + k_{-2} \tag{3.32}$$

$$x_1 = \Delta c_A \tag{3.33}$$

$$x_2 = \Delta c_D \tag{3.34}$$

Simultaneous differential equations of the form of Eqs. 3.27 and 3.28 have been integrated by a number of authors (*3–5*) and the solutions have been shown to be of the general form

$$x_1 = x_1^{01}\exp(-t/\tau_1) + x_1^{02}\exp(-t/\tau_2) \tag{3.35}$$

$$x_2 = x_2^{01}\exp(-t/\tau_1) + x_2^{02}\exp(-t/\tau_2) \tag{3.36}$$

τ_1 and τ_2 are the two relaxation times whereas the x_1^{01} and x_1^{02} represent the amount of change in c_A associated with τ_1 and τ_2, respectively, x_2^{01} and x_2^{02} the change in

c_D associated with τ_1 and τ_2, respectively; they are the quantities that determine the relaxation amplitudes (Chapter 7). Though they are often relatively complex functions of the *total* change in concentration ($x_1{}^0$ and $x_2{}^0$), it is to be noted that the following equations hold:

$$x_1{}^0 = x_1^{01} + x_1^{02} \tag{3.37}$$

$$x_2{}^0 = x_2^{01} + x_2^{02} \tag{3.38}$$

The mathematical problem of finding τ_1 and τ_2 is much easier than that of finding the relaxation amplitudes. As long as one is only interested in the rate coefficients it is not necessary to find these amplitudes; in Chapters 6 and 7 we address ourselves to the problem of calculating the x_1^{01}, x_1^{02}, x_2^{01}, and x_2^{02} and discuss their relation to the experimental relaxation amplitudes.

We can find the relaxation times by assuming two particular solutions of the system of differential equations to be of the form

$$x_1 = x_1{}^0 \exp(-t/\tau) \tag{3.39}$$

$$x_2 = x_2{}^0 \exp(-t/\tau) \tag{3.40}$$

It follows that

$$\frac{dx_1}{dt} = -\frac{1}{\tau}x_1{}^0 \exp\left(\frac{-t}{\tau}\right) = -\frac{1}{\tau}x_1 \tag{3.41}$$

$$\frac{dx_2}{dt} = -\frac{1}{\tau}x_2{}^0 \exp\left(\frac{-t}{\tau}\right) = -\frac{1}{\tau}x_2 \tag{3.42}$$

Substituting back into Eqs. 3.27 and 3.28 one obtains

$$(a_{11} - 1/\tau)x_1 + a_{12}x_2 = 0 \tag{3.43}$$

$$a_{21}x_1 + (a_{22} - 1/\tau)x_2 = 0 \tag{3.44}$$

This system of simultaneous equations can easily be solved for τ^{-1} by eliminating x_1 and x_2. Those versed in algebra will recognize that the most elegant way to do this is to write the determinantal equation

$$\begin{vmatrix} a_{11} - 1/\tau & a_{12} \\ a_{21} & a_{22} - 1/\tau \end{vmatrix} = 0 \tag{3.45}$$

In mathematical terminology 3.45 is known as the *characteristic* or *secular equation* and its solutions (the reciprocal relaxation times) are referred to as the *eigenvalues* of the characteristic equation.

Equation 3.45 is equivalent to the quadratic equation

$$(a_{11} - \tau^{-1})(a_{22} - \tau^{-1}) - a_{12}a_{21} = \tau^{-2} - (a_{11} + a_{22})\tau^{-1} + a_{11}a_{22} - a_{12}a_{21} = 0 \tag{3.46}$$

with the two solutions

$$1/\tau_1 = \tfrac{1}{2}(a_{11} + a_{22}) + \{[\tfrac{1}{2}(a_{11} + a_{22})]^2 + a_{12}a_{21} - a_{11}a_{22}\}^{1/2} \tag{3.47}$$

$$1/\tau_2 = \tfrac{1}{2}(a_{11} + a_{22}) - \{[\tfrac{1}{2}(a_{11} + a_{22})]^2 + a_{12}a_{21} - a_{11}a_{22}\}^{1/2} \tag{3.48}$$

where τ_1 refers to the shorter of the two relaxation times. For reaction scheme 3.2 these are, after combining with Eqs. 3.29–3.32, equivalent to

$$1/\tau_1 = \tfrac{1}{2}\sum k + \left[\left(\tfrac{1}{2}\sum k\right)^2 - \prod k\right]^{1/2} \tag{3.49}$$

$$1/\tau_2 = \tfrac{1}{2}\sum k - \left[\left(\tfrac{1}{2}\sum k\right)^2 - \prod k\right]^{1/2} \tag{3.50}$$

with

$$\sum k = k_1(\bar{c}_A + \bar{c}_B) + k_{-1} + k_2 + k_{-2} \tag{3.51}$$

$$\prod k = k_1[k_2 + k_{-2}](\bar{c}_A + \bar{c}_B) + k_{-1}k_{-2} \tag{3.52}$$

Equations such as 3.47–3.50 are complex expressions and illsuited for the determination of the rate coefficients, which after all is what the kineticist is primarily interested in. In our specific case the problem can easily be solved, however, by considering the quantities

$$1/\tau_1 + 1/\tau_2 = (a_{11} + a_{22}) = \sum k = k_1(\bar{c}_A + \bar{c}_B) + k_{-1} + k_2 + k_{-2} \tag{3.53}$$

$$(1/\tau_1)(1/\tau_2) = a_{11}a_{22} - a_{12}a_{21}$$
$$= \prod k = k_1[k_2 + k_{-2}](\bar{c}_A + \bar{c}_B) + k_{-1}k_{-2} \tag{3.54}$$

By plotting both $\tau_1^{-1} + \tau_2^{-1}$ and $\tau_1^{-1}\tau_2^{-1}$ versus $\bar{c}_A + \bar{c}_B$ as indicated in Fig. 3.3

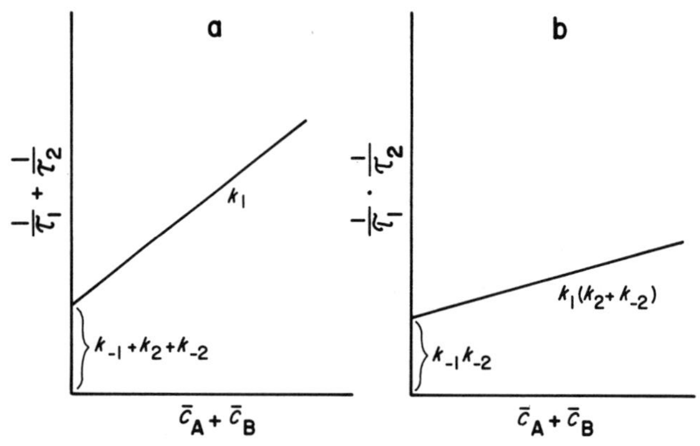

FIGURE 3.3. Evaluation of kinetic parameters for A + B ⇌ C ⇌ D system when both steps equilibrate at similar rates.

one obtains two straight lines. The four rate constants are determined as follows:

$$k_1 = \text{slope (a)} \qquad k_{-2} = \frac{\text{intercept (b)}}{k_{-1}}$$

$$k_{-1} = \text{intercept (a)} - \frac{\text{slope (b)}}{\text{slope (a)}} \qquad k_2 = \text{intercept (a)} - k_{-1} - k_{-2}$$

In principle this appears quite straightforward. However, we should briefly mention one problem which is discussed at some length in Section 9.2. When the rates of all elementary steps are comparable, τ_1 and τ_2 are similar and it may be difficult to measure them with good precision or even to recognize them as separate processes. One possible way out of this difficulty is to change the concentrations or modify the reaction conditions such as temperature and solvent, which may affect the relative rates of the various processes and thereby separate τ_1 and τ_2 sufficiently.

When this is not desirable or possible, the choice of a particular perturbation method in combination with a particular method of monitoring relaxation can sometimes render one of the two processes invisible (zero amplitude) or at least negligible so that the relaxation time of the other process is easily measured. This is illustrated in Section 7.3. For a computer fitting procedure based on only one of the equations 3.47 or 3.48 see Kirschenbaum and Kustin (*6*). Another possibility is to evaluate the "mean relaxation" times, as discussed in Section 9.3.

3.2.3 Some Useful Generalizations

It can easily be shown that for any reaction system comprising three independent states one can write two independent linearized rate equations of the general form of Eqs. 3.27 and 3.28. The solutions are always given by Eqs. 3.47 and 3.48 with the a_{ij} values being characteristic of the particular reaction mechanism.

Two features are particularly noteworthy.

1. The sum $\tau_1^{-1} + \tau_2^{-1}$ is always equal to $(a_{11} + a_{22})$, which in turn is always equal to the sum of all rate terms of the mechanism, symbolized as $\sum k$ (e.g., see Eq. 3.53).
2. The product $\tau_1^{-1}\tau_2^{-1}$ is always equal to $a_{11}a_{22} - a_{12}a_{21}$, which is always equal to a somewhat more complex function ($\prod k$) of the various rate terms (e.g., see Eq. 3.54).

As we shall see in the following section these relations can sometimes be taken advantage of in two ways. If the analytical expression for one of the relaxation times, say τ_1, is known, the other can be found immediately from

$$1/\tau_2 = \sum k - 1/\tau_1 \tag{3.55}$$

This process works only for *exact* expressions of τ_1^{-1}, however, not for approximations. If only an approximate analytical expression of τ_1^{-1} is available, τ_2^{-1} can be found from

$$1/\tau_2 = \tau_1 \prod k \tag{3.56}$$

as illustrated in the next section.

3.2.4. Special Cases Derived from the General Case

A. $k_1(\bar{c}_A + \bar{c}_B) + k_{-1} \gg k_2, k_{-2}$

As shown in Section 3.2.1, it is not necessary to solve the characteristic equation 3.45 in order to find τ_1^{-1} and τ_2^{-1} when $k_1(\bar{c}_A + \bar{c}_B) + k_{-1} \gg k_2, k_{-2}$. Nevertheless,

Eqs. 3.3 and 3.15 must also follow from Eqs. 3.47 and 3.48 (or Eqs. 3.49 and 3.50, respectively).

Equation 3.3 (though with $\bar{c}_A{}^f$ and $\bar{c}_B{}^f$ instead of $\bar{c}_A{}^1$ and $\bar{c}_B{}^1$) follows immediately from Eq. 3.49 since $(\sum k/2)^2 \gg \prod k$ so that the square root term is approximately equal to $\sum k/2$ and thus $\tau_1^{-1} \simeq \sum k$.

Since this is only an approximation we cannot use Eq. 3.55 to find τ_2^{-1}; it would yield $\tau_2^{-1} = k_2 + k_{-2}$, which we know to be incorrect (cf. Eq. 3.15). However, Eq. 3.56 provides

$$\frac{1}{\tau_2} = \frac{\prod k}{k_1(\bar{c}_A + \bar{c}_B) + k_{-1}} = \frac{k_1 k_2(\bar{c}_A + \bar{c}_B)}{k_{-1} + k_1(\bar{c}_A + \bar{c}_B)} + k_{-2}$$

which is equivalent to Eq. 3.15.

B. $k_2 + k_{-2} \gg k_1(\bar{c}_A + \bar{c}_B), k_{-1}$

In analogy to case *A* we use Eq. 3.49 to obtain

$$1/\tau_1 = k_2 + k_{-2} \tag{3.57}$$

(note that we use the lower numbered subscript to denote the shorter relaxation time, $\tau_1^{-1} \gg \tau_2^{-1}$) while dividing $\prod k$ through τ_1^{-1} affords

$$\tau_2^{-1} = k_1(\bar{c}_A + \bar{c}_B) + \frac{k_{-1}k_{-2}}{k_2 + k_{-2}} \tag{3.58}$$

Needless to say Eqs. 3.57 and 3.58 can also be derived following the principles outlined in Section 3.2.1.

C. $k_{-1} + k_2 \gg k_1(\bar{c}_A + \bar{c}_B), k_{-2}$

In this common situation *C* is frequently referred to as a "steady state" or "stationary state" intermediate. Equation 3.49 simplifies to

$$1/\tau_1 = k_{-1} + k_2 \tag{3.59}$$

Again applying Eq. 3.56 we obtain

$$\frac{1}{\tau_2} = \frac{k_1 k_2(\bar{c}_A + \bar{c}_B)}{k_{-1} + k_2} + \frac{k_{-1}k_{-2}}{k_{-1} + k_2} \tag{3.60}$$

which can also be obtained directly by means of the steady state assumption. Steady state is reached for $t \gg \tau_1$. We define

$$\Delta c_A = \bar{c}_A{}^1 - \bar{c}_A, \qquad \Delta c_B = \bar{c}_B{}^1 - \bar{c}_B, \qquad \Delta c_C = \bar{c}_C{}^1 = \bar{c}_C$$

where $\bar{c}_A{}^1$, $\bar{c}_B{}^1$, and $\bar{c}_C{}^1$ are the concentrations at $t \gg \tau_1$. At steady state $\Delta c_C \ll \Delta c_A$, Δc_D, and thus $\Delta c_A \simeq -\Delta c_D$. Solving for Δc_C after setting $d\,\Delta c_C/dt = 0$ in Eq. 3.23 provides

$$\Delta c_C = \frac{-k_1(\bar{c}_A + \bar{c}_B) + k_{-2}}{k_{-1} + k_2}\Delta c_D \tag{3.61}$$

Thus Eq. 3.24 becomes

$$\frac{d\,\Delta c_{\mathrm{D}}}{dt} = -\left[\frac{k_1 k_2(\bar{c}_{\mathrm{A}} + \bar{c}_{\mathrm{B}})}{k_{-1} + k_2} + \frac{k_{-1}k_{-2}}{k_{-1} + k_2}\right]\Delta c_{\mathrm{D}} \tag{3.62}$$

If k_{-1} and k_2 are very much larger than $k_1(\bar{c}_{\mathrm{A}} + \bar{c}_{\mathrm{B}})$ and k_{-2}, the rapid first relaxation process may be undetectable (too small concentration changes or "amplitude") with most perturbation and detection techniques and the experimenter may be led to believe he is dealing with the simple system

$$\mathrm{A} + \mathrm{B} \underset{k_{\mathrm{r}}}{\overset{k_{\mathrm{f}}}{\rightleftharpoons}} \mathrm{D} \tag{3.63}$$

with

$$1/\tau = k_{\mathrm{f}}(\bar{c}_{\mathrm{A}} + \bar{c}_{\mathrm{B}}) + k_{\mathrm{r}} \tag{3.64}$$

where in fact $k_{\mathrm{f}} = k_1 k_2/(k_{-1} + k_2)$ and $k_{\mathrm{r}} = k_{-1}k_{-2}/(k_{-1} + k_2)$ are steady state constants.

In Section 7.3 possibilities for detecting τ_1 will be discussed along with some other special cases of two-step equilibria.

3.3 Some Other Two-Step Systems

3.3.1 Rapid Equilibration of One Step ($1/\tau_1 \gg 1/\tau_2$)

The reciprocal relaxation times for the common situation where one step equilibrates much more rapidly than the other are summarized for a number of examples in Table 3.1. Note that for simplicity no distinction has been made between final equilibrium concentrations and those reached when $\tau_2 \gg t \gg \tau_1$. These examples suffice to illustrate some characteristic patterns which allow one to derive τ_1^{-1} and τ_2^{-1} by mere inspection of the reaction scheme in most cases.

Since the rapid step can always be treated in isolation, τ_1^{-1} is found by the principles set forth in Chapter 2. The expression for τ_2^{-1} is usually directly based on the one for the reaction in isolation, but with the additional feature that the term arising from the reaction of the species coupled to the rapid equilibrium step is modified by an equilibration factor. In the examples of Table 3.1 it is always the k_2 or the k_{-1} term which is thus modified.

When the two reactions of the middle state toward the outer states (k_{-1} and k_2) are both monomolecular (cases 1–3), or in case of higher molecular reactions when they both involve the same species (case 4), the equilibration factor always has the same basic structure. It is equal to that component of τ_1^{-1} which arises from the reaction *toward* the middle state [$k_1(\bar{c}_{\mathrm{A}} + \bar{c}_{\mathrm{B}})$ for case 1], divided by τ_1^{-1} [$k_1(\bar{c}_{\mathrm{A}} + \bar{c}_{\mathrm{B}}) + k_{-1}$ for case 1].

When the reactions of the middle state toward the outer states are of different molecularity (case 5), or are of the same molecularity but involve different species (case 6), the expressions for τ_2^{-1} are somewhat more complex.

Examples conforming to case 6 are of particular interest and occur frequently.

Table 3.1. Relaxation Times for Various Two-Step Systems for $\tau_1^{-1} \gg \tau_2^{-1}$

System	τ_1^{-1} (short)	τ_2^{-1} (long)
(1) $A \underset{k_{-1}}{\overset{k_1}{\rightleftharpoons}} B \underset{k_{-2}}{\overset{k_2}{\rightleftharpoons}} C$		
$k_1 + k_{-1} \gg k_2, k_{-2}$	$k_1 + k_{-1}$	$k_2 \dfrac{k_1}{k_1 + k_{-1}} + k_{-2}$
(2) $A + B \underset{k_{-1}}{\overset{k_1}{\rightleftharpoons}} C \underset{k_{-2}}{\overset{k_2}{\rightleftharpoons}} D$		
(a) $k_1(\bar{c}_A + \bar{c}_B) + k_{-1} \gg k_2, k_{-2}$	$k_1(\bar{c}_A + \bar{c}_B) + k_{-1}$	$k_2 \dfrac{k_1(\bar{c}_A + \bar{c}_B)}{k_1(\bar{c}_A + \bar{c}_B) + k_{-1}} + k_{-2}$
(b) $k_1(\bar{c}_A + \bar{c}_B), k_{-1} \ll k_2 + k_{-2}$	$k_2 + k_{-2}$	$k_1(\bar{c}_A + \bar{c}_B) + k_{-1} \dfrac{k_{-2}}{k_2 + k_{-2}}$
(3) $A + B \underset{k_{-1}}{\overset{k_1}{\rightleftharpoons}} C \underset{k_{-2}}{\overset{k_2}{\rightleftharpoons}} D + E$		
$k_1(\bar{c}_A + \bar{c}_B) + k_{-1} \gg k_2, k_{-2}(\bar{c}_D + \bar{c}_E)$	$k_1(\bar{c}_A + \bar{c}_B) + k_{-1}$	$k_2 \dfrac{k_1(\bar{c}_A + \bar{c}_B)}{k_1(\bar{c}_A + \bar{c}_B) + k_{-1}} + k_{-2}(\bar{c}_D + \bar{c}_E)$
(4) $A \underset{k_{-1}}{\overset{k_1}{\rightleftharpoons}} B + C \underset{k_{-2}}{\overset{k_2}{\rightleftharpoons}} D$		
$k_1 + k_{-1}(\bar{c}_B + \bar{c}_C) \gg k_2(\bar{c}_B + \bar{c}_C), k_{-2}$	$k_1 + k_{-1}(\bar{c}_B + \bar{c}_C)$	$k_2(\bar{c}_B + \bar{c}_C) \dfrac{k_1}{k_1 + k_{-1}(\bar{c}_B + \bar{c}_C)} + k_{-2}$
(5) $A + B \underset{k_{-1}}{\overset{k_1}{\rightleftharpoons}} C$ $C + D \underset{k_{-2}}{\overset{k_2}{\rightleftharpoons}} E$		
$k_1(\bar{c}_A + \bar{c}_B) + k_{-1} \gg k_2(\bar{c}_C + \bar{c}_D), k_{-2}$	$k_1(\bar{c}_A + \bar{c}_B) + k_{-1}$	$k_2\bar{c}_C + k_2\bar{c}_D \dfrac{k_1(\bar{c}_A + \bar{c}_B)}{k_1(\bar{c}_A + \bar{c}_B) + k_{-1}} + k_{-2}$
(6) $A \underset{k_{-1}}{\overset{k_1}{\rightleftharpoons}} B + C$ $C + D \underset{k_{-2}}{\overset{k_2}{\rightleftharpoons}} E$		
$k_1 + k_{-1}(\bar{c}_B + \bar{c}_C) \gg k_2(\bar{c}_C + \bar{c}_D), k_{-2}$	$k_1 + k_{-1}(\bar{c}_B + \bar{c}_C)$	$k_2\left(\bar{c}_C + \bar{c}_D \dfrac{k_1 + k_{-1}\bar{c}_C}{k_1 + k_{-1}(\bar{c}_B + \bar{c}_C)}\right) + k_{-2}$ $= k_2\left(\bar{c}_C + \bar{c}_D \dfrac{K_1 + \bar{c}_C}{K_1 + \bar{c}_B + \bar{c}_C}\right) + k_{-2}$

They include metal complex formation reactions with a ligand which is in an acid–base equilibrium (C, ligand; B, proton; A, protonated ligand; D, metal ion; E, complex) as discussed in Section 11.5.2; or the common situation where the rates of relatively slow proton transfer reactions (e.g., in carbon acids) are to be measured in the presence of small amounts of a strongly absorbing acid–base indicator which is used to monitor the relaxation process.

Let us rewrite the reaction scheme for such a case as

$$\mathrm{InH^+} \underset{k_{-1}}{\overset{k_1}{\rightleftharpoons}} \mathrm{In + H^+} \tag{3.65a}$$

$$\mathrm{A^- + H^+} \underset{k_{-2}}{\overset{k_2}{\rightleftharpoons}} \mathrm{AH} \tag{3.65b}$$

In Eqs. 3.65, AH and A^- comprise the acid–base pair for which the rate of proton transfer is to be measured, whereas InH^+ and In are the two forms of the indicator for which the proton transfer equilibrium is very rapid. Note that Eqs. 3.65 are a simplified scheme; the full scheme is discussed in Section 4.4.3.

In deriving the long relaxation time τ_2, we begin with the linearized rate equation for one of the species that partake in slow reactions only:

$$d\,\Delta c_{\mathrm{AH}}/dt = k_2\bar{c}_{\mathrm{H}}\,\Delta c_{\mathrm{A}} + k_2\bar{c}_{\mathrm{A}}\,\Delta c_{\mathrm{H}} - k_{-2}\,\Delta c_{\mathrm{AH}} \tag{3.66}$$

From stoichiometric relations one obtains

$$\Delta c_{\mathrm{In}} = -\Delta c_{\mathrm{InH}} \tag{3.67}$$

$$\Delta c_{\mathrm{A}} = -\Delta c_{\mathrm{AH}} \tag{3.68}$$

$$\Delta c_{\mathrm{H}} + \Delta c_{\mathrm{InH}} + \Delta c_{\mathrm{AH}} = 0 \tag{3.69}$$

At $t \gg \tau_1$, the following relation is fulfilled as can be shown by arguments similar to those that led to Eq. 3.10:

$$\frac{(\bar{c}_{\mathrm{In}} + \Delta c_{\mathrm{In}})(\bar{c}_{\mathrm{H}} + \Delta c_{\mathrm{H}})}{\bar{c}_{\mathrm{InH}} + \Delta c_{\mathrm{InH}}} = \frac{\bar{c}_{\mathrm{In}}\bar{c}_{\mathrm{H}}}{\bar{c}_{\mathrm{InH}}} = K_1 \tag{3.70}$$

After multiplication, substitution of $-\Delta c_{\mathrm{InH}}$ for Δc_{In}, and rearrangement, one obtains

$$\Delta c_{\mathrm{InH}} = \Delta c_{\mathrm{H}} \frac{\bar{c}_{\mathrm{In}}}{K_1 + \bar{c}_{\mathrm{H}}} \tag{3.71}$$

Substituting Eq. 3.71 for Δc_{InH} into Eq. 3.69 and solving for Δc_{H} affords

$$\Delta c_{\mathrm{H}} = -\Delta c_{\mathrm{AH}} \frac{K_1 + \bar{c}_{\mathrm{H}}}{K_1 + \bar{c}_{\mathrm{H}} + \bar{c}_{\mathrm{In}}} \tag{3.72}$$

Combining Eq. 3.66 with Eqs. 3.68 and 3.72 finally leads to

$$\frac{d\,\Delta c_{\mathrm{AH}}}{dt} = -\left[k_2\bar{c}_{\mathrm{H}} + k_2\bar{c}_{\mathrm{A}} \frac{K_1 + \bar{c}_{\mathrm{H}}}{K_1 + \bar{c}_{\mathrm{A}} + \bar{c}_{\mathrm{In}}} + k_{-2}\right] \Delta c_{\mathrm{AH}} \tag{3.73}$$

with

$$\frac{1}{\tau_2} = k_2\bar{c}_{\mathrm{H}} + k_2\bar{c}_{\mathrm{A}} \frac{K_1 + \bar{c}_{\mathrm{H}}}{K_1 + \bar{c}_{\mathrm{H}} + \bar{c}_{\mathrm{In}}} + k_{-2} \tag{3.74}$$

There is a limiting case which is of special interest. If the indicator concentration can be chosen low enough as to make $\bar{c}_{In} \ll (K_1 + \bar{c}_H)$, which is equivalent to $[In]_0 \ll (K_1 + \bar{c}_H)^2/K_1$ where K_1 is the ionization constant of the indicator and $[In]_0$ its total stoichiometric concentration, Eq. 3.74 simplifies to

$$1/\tau_2 = k_2(\bar{c}_H + \bar{c}_A) + k_{-2} \tag{3.75}$$

i.e., chemical relaxation due to the proton transfer between A^- and AH behaves as if it were an isolated reaction despite the fact that it is the change in absorbance of the indicator which renders the relaxation process detectable. This is a remarkable and very useful result which frequently can be achieved in practice since the extinction coefficient of indicators is usually very high.

When this simplification does not apply, a plot of τ_2^{-1} versus $\bar{c}_H + \bar{c}_A(K_1 + \bar{c}_H)/(K_1 + \bar{c}_H + \bar{c}_{In})$ affords a straight line with slope k_2 and intercept k_{-2}.

Other typical situations pertaining to acid–base equilibria are dealt with in Sections 4.4.2 and 4.4.3.

3.3.2 Degenerate Cases

Table 3.2 summarizes a few special cases which are of both theoretical and practical interest. The expressions for τ_1^{-1} and τ_2^{-1} for cases 1a, 1c, 1d, and 2 are

Table 3.2. Relaxation Times for Degenerate Two-Step Systems

System	$1/\tau_1$	$1/\tau_2$
(1) $A \underset{k_{-1}}{\overset{k_1}{\rightleftharpoons}} B \underset{k_{-2}}{\overset{k_2}{\rightleftharpoons}} C$		
(a) $k_{-2} = k_1$	$k_1 + k_{-1} + k_2$	k_1
(b) $k_2 = k_{-1}; k_{-1} \gg k_1, k_{-2}$	$2k_{-1}$	$\frac{1}{2}(k_1 + k_{-2})$
(c) $k_2 = k_1; k_{-2} = k_{-1}$	$k_1 + k_{-1} + (k_1k_{-1})^{1/2}$	$k_1 + k_{-1} - (k_1k_{-1})^{1/2}$
(d) $k_1 = k_{-1} = k_2 = k_{-2} = k$	$3k$	k
(2) $A + S \underset{k_{-1}}{\overset{2k_1}{\rightleftharpoons}} AS$; $AS + S \underset{2k_{-1}}{\overset{k_1}{\rightleftharpoons}} AS_2$	$k_1(2\bar{c}_A + \bar{c}_{AS} + \bar{c}_S) + k_{-1} = k_1(\bar{c}_{\otimes} + \bar{c}_S) + k_{-1}$	$2(k_1\bar{c}_S + k_{-1})$

easily obtained by setting up the appropriate linearized rate equations in the manner discussed in Section 3.2.2. These provide the coefficients a_{ij} of the general equations 3.27 and 3.28. Solving Eq. 3.47 and 3.48 (no approximations necessary) provides the expressions in the table. The procedure is carried out for case 2 below where an alternate simpler derivation method is also discussed.

Case 1b is similar to the steady state case discussed in Section 3.2.4. Thus, τ_1^{-1} and τ_2^{-2} can be obtained directly from Eqs. 3.59 and 3.60 after replacing $k_1(\bar{c}_A + \bar{c}_B)$ with k_1.

Case 1a may represent the reaction of some species (B) to form two isomeric products (A and C) that are chemically quite similar and thus may revert back to B at the same rate, $k_{-2} = k_1$. In such a situation it is also quite possible that $k_2 = k_{-1}$ so that τ_1^{-1} further simplifies to $k_1 + 2k_{-1}$. An example where this occurs is in the formation of 1:2 σ complexes by the attack of nucleophiles on nitro-activated aromatic compounds (*7*); A and C would typically represent cis and trans isomers of the 1:2 complex. In connection with a discussion of relaxation amplitudes and the detectability of relaxation processes (Section 7.3) this interesting case will be treated in some detail.

Case 1b can occur in similar systems as just mentioned, only that now A and C are much more stable than B, e.g., in the presence of a high concentration of nucleophile (k_{-1} and k_2 steps pseudounimolecular). If also $k_{-2} = k_1$, we obtain $\tau_2^{-1} = k_1$ as in case 1a.

Case 1c is a rudimentary version of a multistep catenary mechanism as has been used to describe certain helix–coil transitions in biopolymers (*8*). Case 1d is a further degeneration of the latter.

Case 2 describes the sequential addition of two ligands to a species that has two chemically equivalent binding sites. The factors of 2 for the addition of the first ligand to A ($2k_1$) and for the dissociation of the first ligand from AS_2 ($2k_{-1}$) are statistical factors.

The derivation of τ_1^{-1} and τ_2^{-1} is as follows. Choosing Δc_A and Δc_{AS_2} as the independent concentration variables we can write the two linearized rate equations

$$d\,\Delta c_A/dt = -2k_1\bar{c}_A\,\Delta c_S - 2k_1\bar{c}_S\,\Delta c_A + k_{-1}\,\Delta c_{AS} \tag{3.76}$$

$$d\,\Delta c_{AS_2}/dt = k_1\bar{c}_{AS}\,\Delta c_S + k_1\bar{c}_S\,\Delta c_{AS} - 2k_{-1}\,\Delta c_{AS_2} \tag{3.77}$$

Δc_S and Δc_{AS} can be substituted by functions of Δc_A and Δc_{AS_2} by virtue of the two stoichiometric relations

$$\Delta c_S + \Delta c_{AS} + 2\,\Delta c_{AS_2} = 0 \tag{3.78}$$

$$\Delta c_A + \Delta c_{AS} + \Delta c_{AS_2} = 0 \tag{3.79}$$

which lead to

$$\Delta c_{AS} = -\Delta c_A - \Delta c_{AS_2} \tag{3.80}$$

$$\Delta c_S = \Delta c_A - \Delta c_{AS_2} \tag{3.81}$$

Equations 3.76 and 3.77 then become of the form of Eqs. 3.27 and 3.28 with $x_1 = \Delta c_A$, $x_2 = \Delta c_{AS_2}$, and

$$\begin{aligned} a_{11} &= 2k_1(\bar{c}_A + \bar{c}_S) + k_{-1}, & a_{12} &= -2k_1\bar{c}_A + k_{-1} \\ a_{21} &= -k_1\bar{c}_{AS} + k_1\bar{c}_S, & a_{22} &= k_1(\bar{c}_{AS} + \bar{c}_S) + 2k_{-1} \end{aligned} \tag{3.82}$$

Solving Eqs. 3.47 and 3.48 provides

$$\tau_1^{-1} = k_1(2\bar{c}_A + \bar{c}_{AS} + \bar{c}_S) + k_{-1} = k_1(\bar{c}_\Phi + \bar{c}_S) + k_{-1} \tag{3.83}$$

$$\tau_2^{-1} = 2(k_1\bar{c}_S + k_{-1}) \tag{3.84}$$

The sum $\bar{c}_\Phi = 2\bar{c}_A + \bar{c}_{AS}$ corresponds to the concentration of "free binding sites."

From the form of Eqs. 3.83 and 3.84, one suspects that only τ_1 is associated with the actual binding process. In fact Eq. 3.83 can be derived much more easily if one writes

$$\Phi + S \underset{k_{-1}}{\overset{k_1}{\rightleftharpoons}} \chi \tag{3.85}$$

to represent all the binding processes, where Φ and χ stand for free and occupied binding sites, respectively (with $c_\chi = 2c_{AS_2} + c_{AS}$). Since Eq. 3.85 includes all possible binding reactions, the second relaxation time must be associated with a process *not* involving the binding of S. As is discussed in more detail in Section 7.3.3, this process is essentially a redistribution of S among the binding sites. It can be formally represented by

$$AS_2 + A \rightleftharpoons (AS + S + A) \rightleftharpoons AS + AS \tag{3.86}$$

For equal intrinsic affinities of A and AS, reaction 3.86 does not depend on this affinity, the equilibrium constant being equal to the statistical factor 4.

Since Eq. 3.83 is not an approximation but an exact solution of 3.47, τ_2^{-1} can immediately be found from Eq. 3.55 with

$$\sum k = k_1(2\bar{c}_A + 3\bar{c}_S + \bar{c}_{AS}) + 3k_{-1} \tag{3.87}$$

An example of considerable importance is the binding of small molecules to proteins consisting of two equivalent subunits (allosteric enzymes) (*9*); this scheme, and particularly its extended version for four subunit enzymes, has been treated in great detail, and is discussed in Section 10.1.

In principle the reaction scheme of case 2 applies also to diprotic acids ($S = H^+$, $AS_2 = AH_2$, etc.) for the limiting situation when the two functional groups are far enough apart so that there is no interaction between them. However, for "normal" acids and bases (*10*) the recombination steps $A^{2-} + H^+ \rightarrow AH^-$ and $AH^- + H^+ \rightarrow AH_2$ would both be diffusion controlled and thus have about the same rate coefficient, k_1; the statistical factor for the first equilibrium would show up in the rate constant of the reverse step ($AH^- \rightarrow A^{2-} + H^+$, $\frac{1}{2}k_{-1}$) instead of the forward step. In this case the analytical expressions for τ_1^{-1} and τ_2^{-2} are no longer simple.

Problems

1. Derive the expressions for τ_2^{-1} for cases 2b and 3–5 in Table 3.1.

2. Derive the expressions for τ_1^{-1} and τ_2^{-1} for cases 1a–1d in Table 3.2.

3. The binding of adenosine diphosphate (ADP) to the enzyme creatine phosphotransferase was investigated by the temperature-jump method (*11*). One relaxation time in the range of 60–150 μsec was observed; some of the (approximate) data as functions of the sum of the equilibrium concentrations of enzyme (E) and ADP are given below.

$(\bar{c}_E + \bar{c}_{ADP})$ (mM)	$10^{-3} \times \tau^{-1}$ (sec^{-1})
0.05	3.7
0.10	5.9
0.30	8.3
0.50	10.0
0.75	11.3
1.0	13.1
2.0	14.2
3.0	15.8
5.0	16.2

The relaxation effect was assumed to arise from a conformational change in the enzyme–ADP complex

$$\text{E} + \text{ADP} \underset{}{\overset{K_1}{\rightleftharpoons}} (\text{EADP})_1 \underset{k_{-2}}{\overset{k_2}{\rightleftharpoons}} (\text{EADP})_2$$

which follows the rapid equilibrium of enzyme–ADP complex formation. Evaluate k_2, k_{-2}, and K_1 from the data and compare your results with that of Hammes and Hurst (*11*).

4. Similarly to the situation in Problem 3, the interaction of proflavin (PF) with the enzyme chymotrypsin has been described by a binding step followed by a conformational change (*12*):

$$\text{E} + \text{PF} \underset{k_{-1}}{\overset{k_1}{\rightleftharpoons}} (\text{EPF})_1 \underset{k_{-2}}{\overset{k_2}{\rightleftharpoons}} (\text{EPF})_2$$

However, here both relaxation times could be measured by the temperature-jump method (*12*). Some of the (approximate) data at pH 8.42 are given below.

$10^5 \times (\bar{c}_E + \bar{c}_{PF})$ (M)	$10^{-3} \times \tau_1^{-1}$ (sec^{-1})	$10^{-3} \times \tau_2^{-1}$ (sec^{-1})
4	7.7	3.6
8	7.9	5.8
10	8.35	6.75
20	13.9	7.5
25	15.7	8.8

Evaluate k_1, k_2, k_{-1}, and k_{-2} from the data and compare your results with those of Havsteen (*12*).

5. The two relaxation times associated with the mono- and diadduct formation between *N*-methyl-*N*-β-hydroxyethyl picramide and sulfite ion have been measured by the stopped-flow

H_3C CH_2CH_2OH N O_2N NO_2 NO_2 $\underset{k_{-1}}{\overset{k_1[SO_3^{2-}]}{\rightleftharpoons}}$ H_3C CH_2CH_2OH N O_2N NO_2 − SO_3^- H NO_2 $\underset{k_{-2}}{\overset{k_2[SO_3^{2-}]}{\rightleftharpoons}}$

H_3C CH_2CH_2OH N O_2N NO_2 − ^-O_3S SO_3^- H H NO_2^-

technique with a large excess of SO_3^{2-} over the substrate (*13*). Some of the data are summarized in the table below.

$10^5 \times [SO_3^{2-}]_0$ (*M*)	τ_1^{-1} (sec^{-1})	τ_2^{-1} (sec^{-1})
4.64	0.30	0.039
7.12	0.33	0.067
9.94	0.38	0.110
13.0	0.41	0.186
19.0	0.64	0.237
26.2	0.80	0.366

From initial rate measurements (acidifying a solution containing only the diadduct) $k_{-2} = 4.2 \times 10^{-3}$ sec^{-1} was determined. Evaluate the other three rate constants. Treat your data (a) assuming that the first step equilibrates much more rapidly than the second; (b) making no such assumption. Compare your findings from both treatments.

References

1. G. Porter, *in* "Technique of Organic Chemistry" (S. L. Friess, E. S. Lewis, and A. Weissberger, ed.), Vol. VIII, part 2, pp. 1055–1106. Wiley (Interscience), New York, 1963.
2. G. Porter and M. A. West, *in* "Techniques of Chemistry" (G. G. Hammes, ed.), Vol. VI, part 2, pp. 367–462. Wiley (Interscience), New York, 1973.
3. E. Picard, "Traité d'Analyse." Gauthier-Villars, Paris, 1928.
4. B. J. Zwolinski and H. Eyring, *J. Amer. Chem. Soc.* **69**, 2702 (1947).

5. K. G. Denbigh, M. Hicks, and F. M. Page, *Trans. Faraday Soc.* **44**, 479 (1948).
6. L. J. Kirschenbaum and K. Kustin, *J. Chem. Soc. A* 684 (1970).
7. C. F. Bernasconi and R. G. Bergstrom, *J. Amer. Chem. Soc.* **96**, 2397 (1974).
8. R. Lumry, R. Legare, and W. Miller, *Biopolymers* **2**, 489 (1964).
9. G. G. Hammes and C.-W. Wu, *Ann. Rev. Biophys. Bioeng.* **3**, 1 (1974).
10. M. Eigen, *Angew. Chem. Int. Ed.* **3**, 1 (1964).
11. G. G. Hammes and J. K. Hurst, *Biochemistry* **8**, 1083 (1969).
12. B. H. Havsteen, *J. Biol Chem.* **242**, 769 (1967).
13. C. F. Bernasconi and H.-C. Wang, unpublished results.

Chapter 4 | Relaxation Times in Common Multistep Systems

4.1 General System. Castellan's Treatment

The kinetics of a general multistep system consisting of $n + 1$ states can be described by the following n linearized rate equations:

$$
\begin{aligned}
\frac{dx_1}{dt} + \sum_{j=1}^{n} a_{1j}x_j &= 0 \\
\frac{dx_2}{dt} + \sum_{j=1}^{n} a_{2j}x_j &= 0 \\
&\vdots \\
\frac{dx_i}{dt} + \sum_{j=1}^{n} a_{ij}x_j &= 0 \\
&\vdots \\
\frac{dx_n}{dt} + \sum_{j=1}^{n} a_{nj}x_j &= 0
\end{aligned}
\tag{4.1}
$$

where the x_j terms are the concentration variables defined as $c_j - \bar{c}_j$ and the a_{ij} terms are functions of rate coefficients and equilibrium concentrations characteristic of the specific reaction scheme. The solution of these n differential equations are n sums of exponentials given by

$$
\begin{aligned}
x_1 &= \sum_{j=1}^{n} x_1^{0j} \exp(-t/\tau_j) \\
x_2 &= \sum_{j=1}^{n} x_2^{0j} \exp(-t/\tau_j) \\
&\vdots \\
x_i &= \sum_{j=1}^{n} x_i^{0j} \exp(-t/\tau_j) \\
&\vdots \\
x_n &= \sum_{j=1}^{n} x_n^{0j} \exp(-t/\tau_j)
\end{aligned}
\tag{4.2}
$$

The reciprocal relaxation times $1/\tau_j$ are the eigenvalues of the determinantal equation

$$\begin{vmatrix} a_{11} - 1/\tau & a_{12} & \cdots & a_{1j} & \cdots & a_{1n} \\ a_{21} & a_{22} - 1/\tau & \cdots & a_{2j} & \cdots & a_{2n} \\ \vdots & \vdots & \ddots & & & \vdots \\ a_{n1} & a_{n2} & \cdots & a_{nj} & \cdots & a_{nn} - 1/\tau \end{vmatrix} = 0 \qquad (4.3)$$

For $n = 2$, Eq. 4.3 affords a quadratic equation (3.46); for $n \geq 3$, the resulting polynomial cannot be solved for the completely general case, i.e., in the situation where all reactions have comparable rates so that they are all strongly coupled to each other. However, it should be pointed out that in most chemical systems some reaction steps are either much faster or much slower than others, which leads to great mathematical simplifications and thus allows explicit expressions to be found for at least some or even all relaxation times of the system. Also, the relaxation spectrum of certain mechanisms is easier to analyze than that of others. For example, linear mechanisms such as

$$A + B \underset{k_{-1}}{\overset{k_1}{\rightleftharpoons}} C \underset{k_{-2}}{\overset{k_2}{\rightleftharpoons}} D \underset{k_{-3}}{\overset{k_3}{\rightleftharpoons}} E \rightleftharpoons \cdots \underset{k_{-n}}{\overset{k_n}{\rightleftharpoons}} X \qquad (4.4)$$

contain, for a given number of states, a minimum number of coupled reactions. In comparison a mechanism such as Eq. 3.1b contains loops and thus more processes coupled to each other. For linear mechanisms, procedures are known which in principle allow all rate constants to be evaluated even if they are all of similar magnitude, provided all relaxation times can be measured. These procedures bypass the problem of finding explicit solutions of Eq. 4.3. This is illustrated with a three-step system in Section 4.2.2. Another approach to the analysis of complex relaxation spectra is to introduce the concept of the mean relaxation time, as discussed in Section 9.3.

The mathematical problems of analyzing complex relaxation spectra have been approached in a very general way by a number of authors (*1–3b*). The derivations are relatively sophisticated as far as the mathematical background of an organic chemist or a biochemist is concerned and therefore no attempt is made here to reproduce them. However, we shall use some of the resulting equations.

A particularly useful approach is that of Castellan (*1*). Its starting point is a formulation of the kinetic problem in the language of nonequilibrium thermodynamics. Just as the traditional approach (formulating the linearized rate equations 4.1) it leads to a secular determinant:

$$\begin{vmatrix} b_{11} - 1/\tau & b_{12} & \cdots & b_{1n} \\ b_{21} & b_{22} - 1/\tau & \cdots & b_{2n} \\ \vdots & \vdots & \ddots & \vdots \\ b_{n1} & b_{n2} & \cdots & b_{nn} - 1/\tau \end{vmatrix} = 0 \qquad (4.5)$$

the eigenvalues of which are again the reciprocal relaxation times $1/\tau_j$. In fact, developing Eq. 4.5 must of course lead to the same polynomial as obtained in developing Eq. 4.3. The advantages of Eq. 4.5 over Eq. 4.3 are twofold.

1. In contrast to Eq. 4.3, many of the elements, b_{ij}, of the determinant in Eq. 4.5 are zero, thus greatly reducing the computational labor in developing the determinant. For $n > 2$ this saving in labor is very significant indeed.

2. There is a certain symmetry in Castellan's determinant which allows us to write it down upon mere inspection of the mechanism, thus saving the labor of formulating the linearized rate equations. For example, for the linear mechanism 4.4, Eq. 4.5 becomes

$$\begin{vmatrix} k_1'+k_{-1}-1/\tau & -k_{-1} & 0 & \cdots & 0 & 0 & 0 \\ -k_2 & k_2+k_{-2}-1/\tau & -k_{-2} & \cdots & 0 & 0 & 0 \\ \vdots & \vdots & \vdots & \ddots & \vdots & \vdots & \vdots \\ 0 & 0 & 0 & \cdots & -k_{n-1} & k_{n-1}+k_{-(n-1)}-1/\tau & -k_{-(n-1)} \\ 0 & 0 & 0 & \cdots & 0 & -k_n & k_n+k_{-n}-1/\tau \end{vmatrix} = 0 \tag{4.6}$$

with $k_1' = k_1(\bar{c}_A + \bar{c}_B)$. Note that for a two-step reaction the determinantal equation 4.3 (or 3.45) and Castellan's determinant

$$\begin{vmatrix} k_1' + k_{-1} - 1/\tau & -k_{-1} \\ -k_2 & k_2 + k_{-2} - 1/\tau \end{vmatrix} = 0 \tag{4.7}$$

become identical except for a minus sign in the k_{-1} and k_2 terms which is of no consequence since in developing the determinant these minus signs cancel each other.

For more complex mechanisms, Castellan's determinant may contain a few more nonzero elements but is still much simpler than Eq. 4.3 (*1–3*). By using Castellan's determinant as the starting point it is also relatively easy to obtain explicit expressions for the $1/\tau$ values if some steps equilibrate much more rapidly or much more slowly than others. The mathematically unsophisticated reader may, however, feel more comfortable with the procedures outlined in the following section, which are based directly on the principles outlined in Chapter 3.

4.2 The A + B $\rightleftharpoons$ C $\rightleftharpoons$ D $\rightleftharpoons$ E (+F) System

The scheme

$$\underset{①}{A + B} \underset{k_{-1}}{\overset{k_1}{\rightleftharpoons}} \underset{②}{C} \underset{k_{-2}}{\overset{k_2}{\rightleftharpoons}} \underset{③}{D} \underset{k_{-3}}{\overset{k_3}{\rightleftharpoons}} \underset{④}{E + F} \tag{4.8}$$

serves well to illustrate how the principles discussed in Chapter 3 can be applied and extended to systems comprising more than three states. Equation 4.8 is representative of several types of chemical systems. For example, it could be a reaction where the principal process, ② $\rightleftharpoons$ ③, is both preceded and followed by acid–base equilibria; or if state ④ contains only one species, the scheme corresponds to the general mechanism of metal complex formation

$$M^+(\text{sol}) + L^-(\text{sol}) \rightleftharpoons M^+(\text{sol, sol})L^- \rightleftharpoons M^+(\text{sol})L^- \rightleftharpoons ML \quad (4.9)$$

formulated by Eigen (*4*) (see also Section 11.5.2).

For the special case where B = E, Eq. 4.8 refers to a typical scheme of an enzyme-catalyzed reaction (where E is an enzyme). This situation can lead to complications not present in the case B ≠ E because of the coupling of steps ① ⇌ ② and ③ ⇌ ④ through the common species E. This situation is therefore treated separately in Section 4.3. However, enzyme reactions proceeding through a large number of intermediates can sometimes be separated into a relatively fast sequence of initial steps, followed by relatively slow product-forming step(s). For the initial sequence one may then have

$$E + S \rightleftharpoons X_1 \rightleftharpoons X_2 \rightleftharpoons X_3 \quad (4.10)$$

Depending on the relative rates of the various steps in reaction 4.8 one can distinguish many special cases, a number of which are summarized in Table 4.1;

Table 4.1. Relaxation Times for the System $A + B \underset{k_{-1}}{\overset{k_1}{\rightleftharpoons}} C \underset{k_{-2}}{\overset{k_2}{\rightleftharpoons}} D \underset{k_{-3}}{\overset{k_3}{\rightleftharpoons}} E + F$

$k_1' = k_1(\bar{c}_A + \bar{c}_B)$, $K_1' = K_1(\bar{c}_A + \bar{c}_B)$, $k'_{-3} = k_{-3}(\bar{c}_E + \bar{c}_F)$, $K_3' = K_3(\bar{c}_E + \bar{c}_F)^{-1}$

	$1/\tau_1$ (short)	$1/\tau_2$ (middle)	$1/\tau_3$ (long)
(1) $k_1', k_{-1} \gg k_2, k_{-2} \gg k_3, k'_{-3}$	$k_1' + k_{-1}$	$k_2 \frac{K_1'}{1 + K_1'} + k_{-2}$	$k_3 \frac{K_1'K_2}{1 + K_1' + K_1'K_2} + k'_{-3}$
(2) $k_1', k_{-1}, k_2, k_{-2} \gg k_3, k'_{-3}$	Eq. 3.49	Eq. 3.50	$k_3 \frac{K_1'K_2}{1 + K_1' + K_1'K_2} + k'_{-3}$
(3) $k_1', k_{-1} \gg k_2, k_{-2} \ll k_3, k'_{-3}$	$k_1' + k_{-1}$	$k_3 + k'_{-3}$	$k_2 \frac{K_1'}{1 + K_1'} + k_{-2} \frac{1}{1 + K_3'}$
(4) $k_1', k_{-1} \gg k_2, k_{-2}, k_3, k'_{-3}$	$k_1' + k_{-1}$	Eq. 4.34[a]	Eq. 4.35[a]
(5) $k_2, k_{-2} \gg k_1', k_{-1}, k_3, k'_{-3}$	$k_2 + k_{-2}$	Eq. 4.34[b]	Eq. 4.35[b]

[a] With the a_{ij} defined on page 46.
[b] With

$$a_{11} = k_1' + k_{-1} \frac{1}{1 + K_2}, \qquad a_{12} = k_{-1} \frac{1}{1 + K_2}$$

$$a_{21} = k_3 \frac{K_2}{1 + K_2}, \qquad a_{22} = k_3 \frac{K_2}{1 + K_2} + k'_{-3}$$

a few are now discussed in some detail.

4.2.1 Special Cases

A. $k_1(\bar{c}_A + \bar{c}_B), k_{-1} \gg k_2, k_{-2} \gg k_3, k_{-3}(\bar{c}_E + \bar{c}_F)$

τ_1^{-1} and τ_2^{-1} are the same as for the system $A + B \rightleftharpoons C \rightleftharpoons D$ with the first step equilibrating much more rapidly than the second. The slowest relaxation time is the interesting one and is now derived.

The linearized rate equation for one of the species that participate only in the slowest reactions is given by

$$d\,\Delta c_E/dt = k_3\,\Delta c_D - k'_{-3}\,\Delta c_E \tag{4.11}$$

$[k'_{-3} = k_{-3}(\bar{c}_E + \bar{c}_F)]$, whereas the stoichiometric relations are given by

$$\Delta c_A = \Delta c_B \tag{4.12}$$

$$\Delta c_E = \Delta c_F \tag{4.13}$$

$$\Delta c_A + \Delta c_C + \Delta c_D + \Delta c_E = 0 \tag{4.14}$$

At $t \gg \tau_2$ the steps ① $\rightleftharpoons$ ② $\rightleftharpoons$ ③ have reached equilibrium and we can write the two relations

$$\frac{(\bar{c}_D + \Delta c_D)}{(\bar{c}_C + \Delta c_C)} = \frac{\bar{c}_D}{\bar{c}_C} = K_2 \tag{4.15}$$

$$\frac{(\bar{c}_C + \Delta c_C)}{(\bar{c}_A + \Delta c_A)(\bar{c}_B + \Delta c_A)} = \frac{\bar{c}_C}{\bar{c}_A \cdot \bar{c}_B} = K_1 \tag{4.16}$$

which are obtained in a way similar to that for Eq. 3.10 and which lead to

$$\Delta c_C = \frac{\Delta c_D}{K_2} \tag{4.17}$$

$$\Delta c_A = \frac{\Delta c_C}{K_1'} = \frac{\Delta c_D}{K_1' K_2} \tag{4.18}$$

with $K_1' = K_1(\bar{c}_A + \bar{c}_B)$. Combining Eqs. 4.17 and 4.18 with Eq. 4.14 gives

$$\Delta c_D = -\Delta c_E \frac{K_1' K_2}{1 + K_1' + K_1' K_2} \tag{4.19}$$

Substituting Δc_D from Eq. 4.19 into Eq. 4.11 leads to

$$\frac{d\,\Delta c_E}{dt} = -\frac{1}{\tau_3}\,\Delta c_E \tag{4.20}$$

with

$$\frac{1}{\tau_3} = k_3 \frac{K_1' K_2}{1 + K_1' + K_1' K_2} + k'_{-3} \tag{4.21}$$

Two features about Eq. 4.21 deserve to be mentioned.

1. Since the only condition for Eq. 4.21 to be valid is that the relations 4.15 and 4.16 hold, one gets the same expression for τ_3^{-1} regardless of whether the step ① ⇌ ② or ② ⇌ ③ equilibrates more rapidly than the other as long as the slower of the two still equilibrates much more rapidly than step ③ ⇌ ④.

2. It is again only the rate term coming from the species coupled to the rapid equilibria which is modified by an equilibration factor. This would still be true for an infinite number of rapid equilibria preceding the slow step. It can easily be shown that the equilibration factor for a system with n rapid preequilibria

$$\mathrm{A + B} \underset{}{\overset{K_1}{\rightleftharpoons}} \mathrm{C} \overset{K_2}{\rightleftharpoons} \mathrm{D} \overset{K_i}{\rightleftharpoons} \cdots \overset{K_n}{\rightleftharpoons} \mathrm{X} \rightleftharpoons \mathrm{Y + Z}$$

is of the form

$$\frac{K_1'K_2\cdots K_i\cdots K_n}{1 + K_1' + K_1'K_2 + \cdots + K_1'K_2\cdots K_i + \cdots + K_1'K_2\cdots K_i\cdots K_n}$$

B. $k_1(\bar{c}_A + \bar{c}_B) + k_{-1} \gg k_2, k_{-2} \ll k_3 + k_{-3}(\bar{c}_E + \bar{c}_F)$

The slow equilibration of step ② ⇌ ③ acts like a barrier; hence, steps ① ⇌ ② and ③ ⇌ ④ equilibrate independently and τ_1^{-1} and τ_2^{-1} are the same as if each were an isolated system.

The derivation of τ_3 involves some features not yet encountered in previous systems and is therefore briefly discussed. Setting up a linearized rate equation is not as straightforward as in earlier examples because *both* species participating in the slow process are also coupled to a rapid equilibrium. In such a case one treats the *sum* of states ① and ②, or of states ③ and ④, as one state, i.e., one adds the two respective linearized rate equations, e.g.,

$$d\,\Delta c_D/dt = k_2\,\Delta c_C - k_{-2}\,\Delta c_D + k'_{-3}\,\Delta c_E - k_3\,\Delta c_D$$
$$d\,\Delta c_E/dt = k_3\,\Delta c_D - k'_{-3}\,\Delta c_E$$

which leads to

$$d(\Delta c_D + \Delta c_E)/dt = k_2\,\Delta c_C - k_{-2}\,\Delta c_D \tag{4.22}$$

For $t \gg \tau_1, \tau_2$ equilibrium is established for steps ① ⇌ ② and ③ ⇌ ④. Thus Eq. 4.16 and

$$\frac{(\bar{c}_E + \Delta c_E)(\bar{c}_F + \Delta \bar{c}_E)}{\bar{c}_D + \Delta c_D} = \frac{\bar{c}_E\bar{c}_F}{\bar{c}_D} = K_3 \tag{4.23}$$

are valid. Equation 4.16 leads to

$$\Delta c_A = \frac{\Delta c_C}{K_1(\bar{c}_A + \bar{c}_B)} \tag{4.24}$$

while Eq. 4.23 leads to

$$\Delta c_E = \frac{\Delta c_D K_3}{\bar{c}_E + \bar{c}_F} \tag{4.25}$$

Combining Eqs. 4.24 and 4.25 with Eq. 4.14 leads to

$$\Delta c_{\mathrm{C}} = -\Delta c_{\mathrm{D}} \frac{(K_3 + \bar{c}_{\mathrm{E}} + \bar{c}_{\mathrm{F}})K_1(\bar{c}_{\mathrm{A}} + \bar{c}_{\mathrm{B}})}{(\bar{c}_{\mathrm{E}} + \bar{c}_{\mathrm{F}})(1 + K_1(\bar{c}_{\mathrm{A}} + \bar{c}_{\mathrm{B}}))} \tag{4.26}$$

Substituting Eq. 4.26 for Δc_{C} and Eq. 4.25 for Δc_{E} into Eq. 4.22 gives

$$\frac{d\,\Delta c_{\mathrm{D}}}{dt} = -\frac{1}{\tau_3}\Delta c_{\mathrm{D}} \tag{4.27}$$

with

$$\frac{1}{\tau_3} = k_2 \frac{K_1(\bar{c}_{\mathrm{A}} + \bar{c}_{\mathrm{B}})}{K_1(\bar{c}_{\mathrm{A}} + \bar{c}_{\mathrm{B}}) + 1} + k_{-2} \frac{\bar{c}_{\mathrm{E}} + \bar{c}_{\mathrm{F}}}{\bar{c}_{\mathrm{E}} + \bar{c}_{\mathrm{F}} + K_3} \tag{4.28}$$

By substituting k_1/k_{-1} for K_1 and k_3/k_{-3} for K_3, Eq. 4.28 can also be written as

$$\frac{1}{\tau_3} = k_2 \frac{k_1(\bar{c}_{\mathrm{A}} + \bar{c}_{\mathrm{B}})}{k_1(\bar{c}_{\mathrm{A}} + \bar{c}_{\mathrm{B}}) + k_{-1}} + k_{-2} \frac{k_{-3}(\bar{c}_{\mathrm{E}} + \bar{c}_{\mathrm{F}})}{k_{-3}(\bar{c}_{\mathrm{E}} + \bar{c}_{\mathrm{F}}) + k_3} \tag{4.29}$$

In Eq. 4.29 it can be seen more clearly that k_{-2} is multiplied with the same equilibration factor as in the part system $\mathrm{C} \rightleftharpoons \mathrm{D} \rightleftharpoons \mathrm{E} + \mathrm{F}$ with the second step equilibrating more rapidly than the first, whereas the equilibration factor for k_2 is the familiar expression for the part system $\mathrm{A} + \mathrm{B} \rightleftharpoons \mathrm{C} \rightleftharpoons \mathrm{D}$ with a rapid first equilibrium; i.e., Eq. 4.29 is completely symmetrical.

C. $k_1(\bar{c}_{\mathrm{A}} + \bar{c}_{\mathrm{B}}) + k_{-1} \gg k_2, k_{-2}, k_3, k_{-3}\,(\bar{c}_{\mathrm{E}} + \bar{c}_{\mathrm{F}})$

We set up two linearized rate equations involving two species partaking only in slow reactions, e.g.,

$$d\,\Delta c_{\mathrm{D}}/dt = k_2\,\Delta c_{\mathrm{C}} - (k_{-2} + k_3)\,\Delta c_{\mathrm{D}} + k'_{-3}\,\Delta c_{\mathrm{E}} \tag{4.30}$$

$$d\,\Delta c_{\mathrm{E}}/dt = k_3\,\Delta c_{\mathrm{D}} - k'_{-3}\,\Delta c_{\mathrm{E}} \tag{4.31}$$

Making use of Eq. 3.11 for $t \gg \tau_1$ and combining it with Eq. 4.14 gives

$$\Delta c_{\mathrm{C}} = -\frac{K_1'(\Delta c_{\mathrm{D}} + \Delta c_{\mathrm{E}})}{1 + K_1'} \tag{4.32}$$

Thus Eq. 4.30 becomes

$$\frac{d\,\Delta c_{\mathrm{D}}}{dt} = -\left(\frac{k_2 K_1'}{1 + K_1'} + k_{-2} + k_3\right)\Delta c_{\mathrm{D}} - \left(\frac{k_2 K_1'}{1 + K_1'} - k'_{-3}\right)\Delta c_{\mathrm{E}} \tag{4.33}$$

Equations 4.31 and 4.33 are of the general form of Eqs. 3.27 and 3.28 with

$$a_{11} = \frac{k_2 K_1'}{1 + K_1'} + k_{-2} + k_3, \qquad a_{12} = \frac{k_2 K_1'}{1 + K_1'} - k'_{-3}$$

$$a_{21} = -k_3, \qquad a_{22} = k'_{-3}$$

$$x_1 = \Delta c_{\mathrm{D}}, \qquad x_2 = \Delta c_{\mathrm{E}}$$

The problem has been reduced to a familiar one; τ_2^{-1} and τ_3^{-1} are given by

$$1/\tau_2 = \tfrac{1}{2}(a_{11} + a_{22}) + \{[\tfrac{1}{2}(a_{11} + a_{22})]^2 + a_{12}a_{21} - a_{11}a_{22}\}^{1/2} \quad (4.34)$$

$$1/\tau_3 = \tfrac{1}{2}(a_{11} + a_{22}) - \{[\tfrac{1}{2}(a_{11} + a_{22})]^2 + a_{12}a_{21} - a_{11}a_{22}\}^{1/2} \quad (4.35)$$

By taking the sum and the product

$$\tau_2^{-1} + \tau_3^{-1} = (a_{11} + a_{22}) = k_2 \frac{K_1(\bar{c}_A + \bar{c}_B)}{1 + K_1(\bar{c}_A + \bar{c}_B)} + k_{-2} + k_3 + k_{-3}(\bar{c}_E + \bar{c}_F) \quad (4.36)$$

$$\begin{aligned}\tau_2^{-1}\tau_3^{-1} &= a_{11}a_{22} - a_{12}a_{21} \\ &= k_2 \frac{K_1(\bar{c}_A + \bar{c}_B)}{1 + K_1(\bar{c}_A + \bar{c}_B)} [k_3 + k_{-3}(\bar{c}_E + \bar{c}_F)] + k_{-2}k_{-3}(\bar{c}_E + \bar{c}_F) \quad (4.37)\end{aligned}$$

one obtains expressions that allow the determination of the various rate coefficients. In most practical situations it will be possible to keep the concentration of at least one reagent quasi constant and preferably of two, which simplifies the problem considerably. If A and E are kept quasi constant, Eqs. 4.36 and 4.37 reduce to

$$\tau_2^{-1} + \tau_3^{-1} = k_2 \frac{K_1[A]_0}{1 + K_1[A]_0} + k_{-2} + k_3 + k_{-3}[E]_0 \quad (4.38)$$

$$\tau_2^{-1}\tau_3^{-1} = k_2 \frac{K_1[A]_0}{1 + K_1[A]_0} (k_3 + k_{-3}[E]_0) + k_{-2}k_{-3}[E]_0 \quad (4.39)$$

The rate coefficients and K_1 are found as follows:

1. A plot of $\tau_2^{-1} + \tau_3^{-1}$ versus $[E]_0$ with $[A]_0$ a constant has a slope of k_{-3}.
2. A plot of $\tau_2^{-1}\tau_3^{-1}$ versus $[A]_0$ with $[E]_0$ a constant has an intercept of $k_{-2}k_{-3}[E]_0$; thus k_{-2} = intercept/$k_{-3}[E]_0$.
3. A plot of $\tau_2^{-1} + \tau_3^{-1}$ versus $[A]_0$ with $[E]_0$ a constant has an intercept of $k_{-2} + k_3 + k_{-3}[E]_0$; thus k_3 = intercept $- k_{-2} - k_{-3}[E]_0$.
4. A plot of $(\tau_2^{-1} + \tau_3^{-1} - k_{-2} - k_3 - k_{-3}[E]_0)^{-1}$ versus $[A]_0^{-1}$ with $[E]_0$ a constant has an intercept $(k_2)^{-1}$ and a slope $(k_2K_1)^{-1}$; thus both k_2 and K_1 can be obtained.
5. A further plot, namely $(\tau_2^{-1}\tau_3^{-1} - k_{-2}k_{-3}[E]_0)^{-1}$ versus $[A]_0^{-1}$ with $[E]_0$ a constant has an intercept of $[k_2(k_3 + k_{-3}[E]_0)]^{-1}$ and a slope $[k_2K_1(k_3 + k_{-3}[E]_0)]^{-1}$; in combination with the fourth plot this provides $k_3 + k_{-3}[E]_0$.

This fifth plot can be used as an alternative to the first if the k_{-3} step is unimolecular and $k_{-3}[E]_0$ is replaced by k_{-3}. Thus from plots 4 and 5 we obtain $k_3 + k_{-3}$; subtracting $k_3 + k_{-3}$ from the intercept of plot 3 affords k_{-2}, etc.

4.2.2 General Case

It is straightforward to set up three independent linearized rate equations and thus the determinantal equation 4.3 with $n=3$. However, we shall use Castellan's (*1*) approach which immediately allows us to write

$$\begin{vmatrix} k_1' + k_{-1} - 1/\tau & -k_{-1} & 0 \\ -k_2 & k_2 + k_{-2} - 1/\tau & -k_{-2} \\ 0 & -k_3 & k_3 + k'_{-3} - 1/\tau \end{vmatrix} = 0 \tag{4.40}$$

Developing the determinant

$$(k_1' + k_{-1} - 1/\tau)\begin{vmatrix} k_2 + k_{-2} - 1/\tau & -k_{-2} \\ -k_3 & k_3 + k'_{-3} - 1/\tau \end{vmatrix}$$
$$- (-k_{-1})\begin{vmatrix} -k_2 & -k_{-2} \\ 0 & k_3 + k'_{-3} - 1/\tau \end{vmatrix} + 0\begin{vmatrix} -k_2 & k_2 + k_{-2} - 1/\tau \\ 0 & -k_3 \end{vmatrix} = 0$$

leads to the cubic polynomial

$$-(1/\tau)^3 + A_1(1/\tau)^2 - A_2(1/\tau) + A_3 = 0 \tag{4.41}$$

with

$$A_1 = k_1' + k_{-1} + k_2 + k_{-2} + k_3 + k'_{-3} \tag{4.42}$$

$$A_2 = (k_1' + k_{-1})(k_2 + k_{-2}) + (k_2 + k_{-2})(k_3 + k'_{-3}) + (k_1' + k_{-1})(k_3 + k'_{-3}) - k_{-1}k_2 - k_{-2}k_3 \tag{4.43}$$

$$A_3 = (k_1' + k_{-1})(k_2 + k_{-2})(k_3 + k'_{-3}) - (k_1' + k_{-1})k_{-2}k_3 - (k_3 + k'_{-3})k_{-1}k_2 \tag{4.44}$$

Equation 4.41 cannot be solved explicitly for the general case although numerical solutions can be found through computer analysis. However, there is a procedure that allows the evaluation of all six rate constants provided the concentration dependence of the three τ values can be determined with good precision. It is based on the relationships

$$A_1 = 1/\tau_1 + 1/\tau_2 + 1/\tau_3 \tag{4.45}$$

$$A_2 = (1/\tau_1)(1/\tau_2) + (1/\tau_1)(1/\tau_3) + (1/\tau_2)(1/\tau_3) \tag{4.46}$$

$$A_3 = (1/\tau_1)(1/\tau_2)(1/\tau_3) \tag{4.47}$$

Plotting A_1, A_2, and A_3 as defined by Eqs. 4.45–4.47, respectively, versus $(\bar{c}_A + \bar{c}_B)$ and keeping k'_{-3} constant, affords three linear plots whose slopes and intercepts are found from Eqs. 4.42–4.44, respectively. Thus, even if the k_{-3} step is unimolecular ($k'_{-3} = k_{-3}$), the three slopes and three intercepts are sufficient to solve for the six unknown rate constants (see Problem 7). Note that this procedure is simply an extension of the one used in the two-step system (Eqs. 3.53 and 3.54) and can be further extended to a linear mechanism of any number of steps (*5*). It has in fact been successfully applied to a system of five consecutive steps (*6*).

It should be pointed out that for a mechanism containing more steps than relaxation times, i.e., a mechanism with loops, the number of unknown rate constants may exceed the combined sum of obtainable slopes and intercepts and thus not all rate constants can be evaluated.

4.3 The $E + S \rightleftharpoons ES \rightleftharpoons EP \rightleftharpoons P + E$ System (Enzyme Reactions)

The following scheme,

$$\underset{①}{E + S} \underset{k_{-1}}{\overset{k_1}{\rightleftharpoons}} \underset{②}{ES} \underset{k_{-2}}{\overset{k_2}{\rightleftharpoons}} \underset{③}{EP} \underset{k_{-3}}{\overset{k_3}{\rightleftharpoons}} \underset{④}{P + E} \tag{4.48}$$

which is a special case of scheme 4.8, is the now generally accepted minimum mechanism for an enzyme-catalyzed reaction with one substrate and product (S, substrate; P, product; E, enzyme; ES, enzyme–substrate complex; EP, enzyme–product complex). Scheme 4.48 is more complicated than 4.8 because steps ① $\rightleftharpoons$ ② and ③ $\rightleftharpoons$ ④ are coupled through E.

It is common that the binding steps ① $\rightleftharpoons$ ② and ③ $\rightleftharpoons$ ④ equilibrate very rapidly compared to ② $\rightleftharpoons$ ③. In such a situation there are two short and one long relaxation times, $\tau_1^{-1}, \tau_2^{-1} \gg \tau_3^{-1}$. A very general method for the derivation of the three relaxation times has been given by Hammes and Schimmel (*3b*) but application of the principles set forth in the present book is probably easier for the novice.

The result for τ_1^{-1} and τ_2^{-1} is most conveniently expressed by

$$1/\tau_1 + 1/\tau_2 = k_1' + k_{-1} + k_3 + k'_{-3} \tag{4.49}$$

$$(1/\tau_1)(1/\tau_2) = (k_1' + k_{-1})(k_3 + k'_{-3}) - k_1 k_{-3} \bar{c}_S \bar{c}_P \tag{4.50}$$

with

$$k_1' = k_1(\bar{c}_S + \bar{c}_E), \qquad k'_{-3} = k_{-3}(\bar{c}_P + \bar{c}_E)$$

In case step ① $\rightleftharpoons$ ② equilibrates much more rapidly than ③ $\rightleftharpoons$ ④, Eq. 4.49 simplifies to

$$1/\tau_1 = k_1' + k_{-1}$$

while $1/\tau_2$ ($\ll 1/\tau_1$) is found by dividing Eq. 4.50 by $1/\tau_1$:

$$1/\tau_2 = (k_3 + k'_{-3}) - \frac{k_1 k_{-3} \bar{c}_S \bar{c}_P}{k_1' + k_{-1}} \tag{4.51}$$

Rearrangement leads to

$$\frac{1}{\tau_2} = k_3 + k'_{-3}\left[1 - \frac{k_1'}{k_1' + k_{-1}} \frac{\bar{c}_S \bar{c}_P}{(\bar{c}_S + \bar{c}_E)(\bar{c}_P + \bar{c}_E)}\right] \tag{4.52}$$

Note the minus sign in Eq. 4.52; the plus sign in Eq. 3.54 of Eigen and De Maeyer (*7*) is apparently a misprint.

It is interesting to note that for $\bar{c}_E \gg \bar{c}_S, \bar{c}_P$ the $k_1 k_{-3} \bar{c}_S \bar{c}_P$ term in Eq. 4.50 becomes negligibly small so that

$$1/\tau_1 = k_1 \bar{c}_E + k_{-1}; \qquad 1/\tau_2 = k_3 + k_{-3} \bar{c}_E$$

regardless of the relative magnitude of k_1', k_{-1}, k_3, and k'_{-3}, just as in system 4.8 (case 3 in Table 4.1, $\bar{c}_B \gg \bar{c}_A$, $\bar{c}_E \gg \bar{c}_F$). This means that at high enzyme concentration steps ① ⇌ ② and ③ ⇌ ④ become decoupled. That this must be so can be understood intuitively by recognizing that when there is plenty of E available, S and P do not need to "compete" for E in order to form the respective complexes.

For the third relaxation time one obtains

$$1/\tau_3 = \alpha k_2 + \beta k_{-2} \tag{4.53}$$

with

$$\alpha = \left[1 + \frac{k_1 k_3 \bar{c}_S + k_{-1}k_{-3}\bar{c}_E + k_{-1}k_{-3}\bar{c}_P + k_{-1}k_3}{k_1 k_3 \bar{c}_E + k_1 k_{-3}\bar{c}_E(\bar{c}_E + \bar{c}_S + \bar{c}_P)}\right]^{-1} \tag{4.54}$$

$$\beta = \left[1 + \frac{k_1 k_3 \bar{c}_S + k_1 k_3 \bar{c}_E + k_{-1}k_{-3}\bar{c}_P + k_{-1}k_3}{k_{-1}k_{-3}\bar{c}_E + k_1 k_{-3}\bar{c}_E(\bar{c}_E + \bar{c}_S + \bar{c}_P)}\right]^{-1} \tag{4.55}$$

which is the form given by Hammes and Schimmel (*3b*, *8*). Introducing the binding constants $K_{ES} = k_1/k_{-1}$ and $K_{EP} = k_{-3}/k_3$, α and β can be expressed as (*7*)

$$\alpha = \frac{K_{ES}\bar{c}_E}{1 + K_{ES}\bar{c}_E + \delta_S} \tag{4.56}$$

$$\beta = \frac{K_{EP}\bar{c}_E}{1 + K_{EP}\bar{c}_E + \delta_P} \tag{4.57}$$

with

$$\delta_S = \frac{\bar{c}_S(K_{ES} - K_{EP})}{1 + K_{EP}(\bar{c}_E + \bar{c}_S + \bar{c}_P)} \tag{4.58}$$

$$\delta_P = \frac{\bar{c}_P(K_{EP} - K_{ES})}{1 + K_{ES}(\bar{c}_E + \bar{c}_S + \bar{c}_P)} \tag{4.59}$$

It is to be noted that for the degenerate case $K_{ES} = K_{EP}$ we have $\delta_S = \delta_P = 0$; δ_S and δ_P also become negligibly small when $\bar{c}_E \gg \bar{c}_S$, $\bar{c}_P$. This simplifies Eq. 4.53 to

$$\frac{1}{\tau_3} = k_2 \frac{K_{ES}\bar{c}_E}{1 + K_{ES}\bar{c}_E} + k_{-2}\frac{K_{EP}\bar{c}_E}{1 + K_{EP}\bar{c}_E} \tag{4.60}$$

We have seen above that steps ① ⇌ ② and ③ ⇌ ④ become decoupled when $\bar{c}_E \gg \bar{c}_S$, $\bar{c}_P$ and thus we are not surprised that Eq. 4.60 is the same as for system 4.8 (case 3 in Table 4.1). It is evident that when the enzyme concentration is large enough to make $K_{ES}\bar{c}_E \gg 1$ and $K_{EP}\bar{c}_E \gg 1$, Eq. 4.60 simplifies to $\tau_3^{-1} = k_2 + k_{-2}$; here essentially all S and P are in the form of ES and EP, respectively, and the system reduces to the simple reaction ES ⇌ EP. It should be pointed out that contrary to a statement in a recent review (*8*) a similar simplification of τ_3^{-1} does *not* occur for high $\bar{c}_S$ and $\bar{c}_P$ instead of high $\bar{c}_E$.

4.4 Cyclic Reaction Schemes

4.4.1 Simple Triangular Schemes

Cyclic reaction schemes are quite common in chemistry. Though the general principles of derivation of the relaxation times are the same as for the linear systems discussed in Chapter 3, the following examples serve to illustrate some additional important aspects of chemical relaxation.

The simplest cyclic scheme is the triangle

$$\begin{array}{ccc} & \textcircled{2} & \\ & \mathrm{I} & \\ k_{12} \nearrow\swarrow k_{21} & & k_{32} \nwarrow\searrow k_{23} \\ \textcircled{1}\ \mathrm{A} & \underset{k_{31}}{\overset{k_{13}}{\rightleftharpoons}} & \mathrm{D}\ \textcircled{3} \end{array} \tag{4.61}$$

This scheme or variations thereof (some steps of higher molecularity) represent the classical situation where the reaction of some starting material (A) to form a product (D) can proceed by two competing mechanisms, one being a concerted one-step reaction, the other involving the formation of an intermediate (I). Familiar examples are S_N1 versus S_N2, E1 versus E2, or the situation where the direct route $A \rightleftharpoons D$ represents a general acid- or base-catalyzed pathway which avoids the necessity of forming a relatively unstable intermediate. Whether one can distinguish kinetically between the two pathways depends on whether the paths have different concentration dependences.

The two relaxation times are given by Eqs. 3.47–3.48 with ($x_1 = \Delta c_A$, $x_2 = \Delta c_D$)

$$a_{11} = k_{12} + k_{13} + k_{21}, \qquad a_{12} = k_{21} - k_{31}$$

$$a_{21} = k_{23} - k_{13}, \qquad a_{22} = k_{23} + k_{31} + k_{32}$$

Assume now that the k_{12} and k_{13} steps are bimolecular ($A + B \rightleftharpoons D$; $A + B \rightleftharpoons I$). For the case where $k_{12}(\bar{c}_A + \bar{c}_B) + k_{21} \gg k_{13}(\bar{c}_A + \bar{c}_B)$, k_{31}, k_{23}, k_{32}, we obtain

$$\frac{1}{\tau_1} = k_{12}(\bar{c}_A + \bar{c}_B) + k_{21} \tag{4.62}$$

$$\frac{1}{\tau_2} = \frac{(k_{23}K_{12} + k_{13})(\bar{c}_A + \bar{c}_B)}{1 + K_{12}(\bar{c}_A + \bar{c}_B)} + k_{31} + k_{32} \tag{4.63}$$

It is obvious that only $k_f = k_{23}K_{12} + k_{13}$ and $k_r = k_{31} + k_{32}$ can be evaluated from the concentration dependence of τ_2; in fact in the absence of independent evidence, there is no way to tell whether the direct pathway contributes anything (possibly $k_{13} = k_{31} = 0$: $k_f = k_{23}K_{12}$, $k_r = k_{32}$) to the overall reaction, or whether I is a true intermediate rather than a shunt (possibly $k_{23} = k_{32} = 0$: $k_f = k_{13}$,

$k_r = k_{31}$). On the other hand, if the direct pathway is of different molecularity, for example if it involves a catalyst C, Eq. 4.63 becomes

$$\frac{1}{\tau_2} = \frac{(k_{23}K_{12} + k_{13}\bar{c}_C)(\bar{c}_A + \bar{c}_B)}{1 + K_{12}(\bar{c}_A + \bar{c}_B)} + k_{31}\bar{c}_C + k_{32}$$

from which all four rate constants can be evaluated.

Protolytic reactions in hydroxylic solvents are another important class giving rise to cyclic reaction schemes. They are discussed in the following sections.

4.4.2 Protolytic Reactions: One Acid–Base Pair

The first example, scheme 4.64, deals with the reactions occurring in a simple

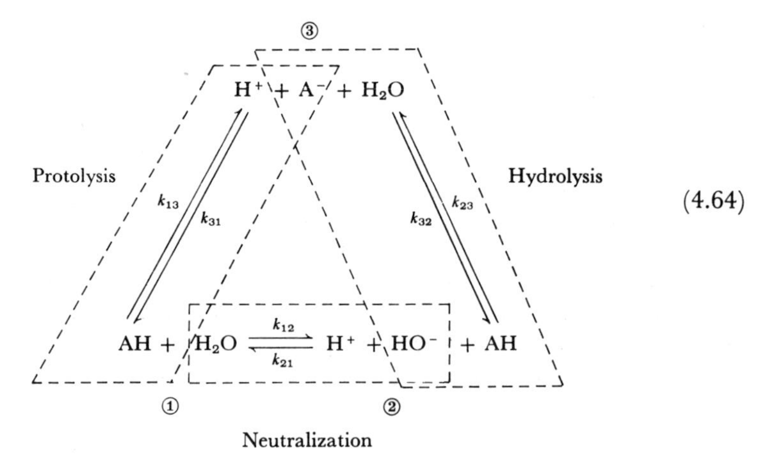

(4.64)

acid–base equilibrium in aqueous solution, whereas the second example describes all proton transfer processes occurring in aqueous solution when two different acid–base pairs are present (scheme 4.75). Both schemes are of such general interest that a detailed discussion is amply justified. The following treatment is to a large extent based on a paper by Eigen (*9*) but is somewhat more detailed and explicit where this was felt desirable for the sake of understanding and clarity.

Note that the respective reactions summarized in scheme 4.64 include only the species boxed in, e.g., the reaction ② ⇌ ③ corresponds to $AH + OH^- \rightleftharpoons A + H_2O$ only; i.e., H^+ is merely a "spectator," not a catalyst.

Scheme 4.64 is a three-state system and thus is characterized by two relaxation times. They are derived in the usual way, that is, by setting up two linearized rate equations and solving for Eqs. 3.47 and 3.48. If one chooses $\Delta c_{AH} = x_1$ and $\Delta c_H = x_2$, and considers the stoichiometric relations

$$\Delta c_A = -\Delta c_{AH} = -x_1 \tag{4.65}$$

$$\Delta c_{H_2O} = -\Delta c_{HO} = -\Delta c_{AH} - \Delta c_H = -x_1 - x_2 \tag{4.66}$$

one obtains

$$a_{11} = k_{13} + k_{31}\bar{c}_{\mathrm{H}} + k_{32} + k_{23}(\bar{c}_{\mathrm{HO}} + \bar{c}_{\mathrm{AH}})$$

$$a_{12} = -k_{31}\bar{c}_{\mathrm{A}} + k_{23}\bar{c}_{\mathrm{AH}}$$

$$a_{21} = -k_{13} - k_{31}\bar{c}_{\mathrm{H}} + k_{12} + k_{21}\bar{c}_{\mathrm{H}}$$

$$a_{22} = k_{12} + k_{21}(\bar{c}_{\mathrm{H}} + \bar{c}_{\mathrm{HO}}) + k_{31}\bar{c}_{\mathrm{A}}$$

For simplicity we have omitted the ionic charges in the concentration symbols.

As we have seen for the simpler system $A + B \rightleftharpoons C \rightleftharpoons D$ it is useful to consider the sum $\tau_{\mathrm{I}}^{-1} + \tau_{2}^{-1}$ and the product $\tau_{\mathrm{I}}^{-1}\tau_{2}^{-1}$ of the reciprocal relaxation times. The sum is given by

$$\begin{aligned} 1/\tau_1 + 1/\tau_2 &= a_{11} + a_{22} \\ &= k_{12} + k_{21}(\bar{c}_{\mathrm{H}} + \bar{c}_{\mathrm{HO}}) + k_{13} + k_{31}(\bar{c}_{\mathrm{H}} + \bar{c}_{\mathrm{A}}) + k_{32} + k_{23}(\bar{c}_{\mathrm{HO}} + \bar{c}_{\mathrm{AH}}) \\ &= 1/\tau_{12} + 1/\tau_{13} + 1/\tau_{23} \end{aligned} \tag{4.67}$$

with

$$1/\tau_{12} = k_{12} + k_{21}(\bar{c}_{\mathrm{H}} + \bar{c}_{\mathrm{HO}})$$

$$1/\tau_{13} = k_{13} + k_{31}(\bar{c}_{\mathrm{H}} + \bar{c}_{\mathrm{A}})$$

$$1/\tau_{23} = k_{32} + k_{23}(\bar{c}_{\mathrm{HO}} + \bar{c}_{\mathrm{HA}})$$

We see again that $\tau_{\mathrm{I}}^{-1} + \tau_{2}^{-1}$ is equal to the sum of all rate terms and is thus equivalent to the sum of the three hypothetical reciprocal relaxation times of the part equilibria ① ⇌ ② (neutralization), ① ⇌ ③ (protolysis), and ② ⇌ ③ (hydrolysis); these part relaxation times are those that would occur if the part equilibria were not coupled to each other but isolated reactions.

The product $\tau_{\mathrm{I}}^{-1}\tau_{2}^{-1}$ is a rather complex expression. However, it can be expressed in terms of the part relaxation times, defined above, as

$$\tau_{\mathrm{I}}^{-1}\tau_{2}^{-1} = a_{11}a_{22} - a_{12}a_{21} = \tau_{12}^{-1}\tau_{13}^{-1} + \tau_{12}^{-1}\tau_{23}^{-1} + \tau_{13}^{-1}\tau_{23}^{-1} - R \tag{4.68}$$

with

$$\begin{aligned} R = {} & k_{31}\bar{c}_{\mathrm{A}} \cdot k_{21}\bar{c}_{\mathrm{HO}} + k_{23}\bar{c}_{\mathrm{AH}}(k_{12} + k_{21}\bar{c}_{\mathrm{H}}) \\ & + (k_{23}\bar{c}_{\mathrm{HO}} + k_{32})(k_{13} + k_{31}\bar{c}_{\mathrm{H}}) \end{aligned} \tag{4.69}$$

As we shall see, the R term is often negligibly small or is compensated by one of the three product terms which leads to significant mathematical simplifications.

Let us now discuss several typical situations.

A. Solution of a Weak Base or Weak Acid

In the first we consider the behavior of a solution prepared by simply adding a weak base, say 0.1 M NH_3, to water. Since $pK_{NH_4^+}$ is 9.25, we calculate $\bar{c}_{\mathrm{AH}} \approx \bar{c}_{\mathrm{HO}} \approx 1.3 \times 10^{-3}\,M$, $\bar{c}_{\mathrm{A}} \approx 0.1\,M$, $\bar{c}_{\mathrm{H}} \approx 7.7 \times 10^{-12}\,M \approx 0$. Further, from the known values of the various rate coefficients (Table 4.2) the rate terms are

Table 4.2. Rate Coefficients for Some Typical Acid–Base Reactions[a]

Reaction		
$NH_4^+ + HO^- \underset{k_{32}}{\overset{k_{23}}{\rightleftharpoons}} NH_3 + H_2O$	$k_{23} = 3.4 \times 10^{10}$	$k_{32} = 6.0 \times 10^5$
$NH_3 + H^+ \underset{k_{13}}{\overset{k_{31}}{\rightleftharpoons}} NH_4^+$	$k_{31} = 4.3 \times 10^{10}$	$k_{13} = 24.6$
$AcOH + HO^- \underset{k_{32}}{\overset{k_{23}}{\rightleftharpoons}} AcO^- + H_2O$	$k_{23} \approx 2 \times 10^{10}$	$k_{32} \approx 10$
$AcO^- + H^+ \underset{k_{13}}{\overset{k_{31}}{\rightleftharpoons}} AcOH$	$k_{31} = 4.5 \times 10^{10}$	$k_{13} = 7.8 \times 10^5$
$H^+ + HO^- \underset{k_{12}}{\overset{k_{21}}{\rightleftharpoons}} H_2O$	$k_{21} = 1.4 \times 10^{11}$	$k_{12} = 1.4 \times 10^{-3}$

[a] Data from Eigen (*9*).

calculated to have the following numerical values (in reciprocal seconds):

$$k_{12} = 1.4 \times 10^{-3}; \qquad k_{21}(\bar{c}_{HO} + \bar{c}_{H}) \approx k_{21}\bar{c}_{HO} \approx 1.82 \times 10^8 \approx \tau_{12}^{-1}$$

$$k_{13} = 24.6; \qquad k_{31}(\bar{c}_{H} + \bar{c}_{A}) \approx k_{31}\bar{c}_{A} \approx 4.3 \times 10^9 \approx \tau_{13}^{-1}$$

$$k_{32} = 6.0 \times 10^5; \qquad k_{23}(\bar{c}_{HO} + \bar{c}_{AH}) \approx 8.82 \times 10^7 \approx \tau_{23}^{-1}$$

In this situation we have

$$R \approx \tau_{12}^{-1}\tau_{13}^{-1}$$

because all other terms in Eq. 4.69 are negligibly small compared to $k_{31}\bar{c}_A \cdot k_{21}\bar{c}_{HO}$. Consequently Eq. 4.68 becomes

$$\tau_1^{-1}\tau_2^{-1} = a_{11}a_{22} - a_{12}a_{21} \approx \tau_{12}^{-1}\tau_{23}^{-1} + \tau_{13}^{-1}\tau_{23}^{-1}$$

and therefore

$$\{[\tfrac{1}{2}(a_{11} + a_{22})]^2 + a_{12}a_{21} - a_{11}a_{22}\}^{1/2} \approx \tfrac{1}{2}(\tau_{12}^{-1} + \tau_{13}^{-1} - \tau_{23}^{-1})$$

After expressing $a_{11} + a_{22}$ as $\tau_{12}^{-1} + \tau_{13}^{-1} + \tau_{23}^{-1}$ from Eq. 4.67 it is apparent that Eqs. 3.47 and 3.48 simplify to

$$\tau_1^{-1} \approx \tau_{12}^{-1} + \tau_{13}^{-1} \tag{4.70}$$

$$\tau_2^{-1} \approx \tau_{23}^{-1} = k_{23}(\bar{c}_{HO} + \bar{c}_{AH}) + k_{32} \tag{4.71}$$

In the particular situation at hand τ_{13}^{-1} is about fivefold larger than τ_{12}^{-1}, so that τ_1^{-1} is determined mainly by $\tau_{13}^{-1} \approx k_{31}\bar{c}_A$.

A completely analogous situation exists in a solution of a weak acid, for instance in 0.1 *M* acetic acid. Here we obtain

$$\tau_1^{-1} \approx \tau_{12}^{-1} + \tau_{23}^{-1} \tag{4.72}$$

$$\tau_2^{-1} \approx \tau_{13}^{-1} = k_{31}(\bar{c}_{H} + \bar{c}_{A}) + k_{13} \tag{4.73}$$

It is important to realize that in both cases only τ_2 can be observed experimentally. The changes in concentrations associated with τ_1 are so small due to the extremely small concentration of H^+ in the case of ammonia, or of HO^- in the case of acetic acid, that they cannot be detected. This means that for all practical purposes system 4.64 behaves like the part equilibrium ② ⇌ ③ in the case of a solution of NH_3, or like the part equilibrium ① ⇌ ③ in the case of a solution of AcOH.

The rate coefficients would typically be evaluated as follows. In the solution prepared by adding a weak base (NH_3) to water the k_{32} term in $\tau_2^{-1} \approx \tau_{23}^{-1} = k_{23}(\bar{c}_{OH} + \bar{c}_{AH}) + k_{32}$, is often negligibly small for not too low base concentrations. For example, in a 0.1 M NH_3 solution $k_{23}(\bar{c}_{HO} + c_{AH}) = 8.82 \times 10^7$ sec^{-1} (see above) compared to $k_{32} = 6.0 \times 10^5$ sec^{-1}. Thus only k_{23} can be determined from a concentration dependence of τ_2^{-1}. However, k_{32} is immediately found from $k_{32} = K_{32}k_{23} = K_w k_{23}/K_a^{AH}$ where K_w is the ionic product of water and K_a^{AH} is the acid dissociation constant of AH.

Similarly for the solution of a weak acid (AcOH) we have $\tau_2^{-1} \approx \tau_{13}^{-1} = k_{31}(\bar{c}_H + \bar{c}_A) + k_{13} \approx k_{31}(\bar{c}_H + \bar{c}_A)$, etc.

B. Solution of a Weak Base at High pH or of a Weak Acid at Low pH

Increasing the pH of the ammonia solution leads to a further reduction of $\bar{c}_H$, making the system behave even more like the part equilibrium ② ⇌ ③ with the principal relaxation time given by Eq. 4.71.

Equation 4.71 eventually simplifies to $\tau_2^{-1} \approx k_{23}\bar{c}_{OH}$ because $\bar{c}_{OH}$ increases but $\bar{c}_{AH}$ decreases. At very high pH τ_2 may no longer be detectable because the concentration of AH becomes so low that Δc_{AH} cannot be measured (small "amplitude"; see Chapter 6).

Likewise, decreasing the pH of the acetic acid solution reduces the importance of state ② further and makes the system more like the one-step equilibrium ① ⇌ ③. Equation 4.73 approaches the limiting situation whose $\tau_2^{-1} \approx k_{31}\bar{c}_H$. Again τ_2 becomes undetectable at very low pH when $\bar{c}_A$ becomes so small that Δc_A can no longer be measured.

C. Solution of Weak Base or Weak Acid at pH ≈ pK

Let us assume our 0.1 M NH_3 is adjusted to a pH = pK_a = 9.25. In that case $\bar{c}_A = \bar{c}_{AH} = 0.05\ M$, $\bar{c}_{HO} = 1.8 \times 10^{-5}\ M$, $\bar{c}_H = 5.6 \times 10^{-10}\ M$, and $k_{21}\bar{c}_{HO} = 2.52 \times 10^6 \approx \tau_{12}^{-1}$, $k_{31}\bar{c}_A = 2.15 \times 10^9 \approx \tau_{13}^{-1}$, $k_{23}(\bar{c}_{HO} + \bar{c}_{AH}) \approx k_{23}\bar{c}_{AH} = 1.7 \times 10^9 \approx \tau_{23}^{-1}$.

Here $\tau_{13}^{-1}, \tau_{23}^{-1} \gg \tau_{12}^{-1}$; i.e., the part relaxation times for protolysis and hydrolysis are much shorter than the one for neutralization. This will generally be true when $\bar{c}_A, \bar{c}_{AH} \gg \bar{c}_H, \bar{c}_{HO}$.

Since no term in Eq. 4.69 is of comparable magnitude to $\tau_{13}^{-1}\tau_{23}^{-1}$, Eq. 4.68 reduces to

$$a_{11}a_{22} - a_{12}a_{21} \approx \tau_{13}^{-1}\tau_{23}^{-1}$$

Furthermore, since

$$a_{11} + a_{22} \approx \tau_{13}^{-1} + \tau_{23}^{-1}$$

Eqs. 3.47 and 3.48 become

$$\tau_1^{-1} = \tau_{13}^{-1}, \qquad \tau_2^{-1} = \tau_{23}^{-1}$$

This means that the system behaves as if the part equilibria ① ⇌ ③ and ② ⇌ ③ were isolated reactions.

The degenerate case where, as a consequence of the equality

$$k_{23}\bar{c}_{AH} = k_{31}\bar{c}_{A} \tag{4.74}$$

we have $\tau_1^{-1} = \tau_2^{-1} = \tau^{-1}$, is of particular interest since the system is characterized by one relaxation time only. It occurs usually at a pH $\approx$ pK because k_{23} and k_{31} frequently refer to diffusion-controlled processes and thus are both of similar magnitude. In our example of an ammonia solution Eq. 4.74 holds at pH $= 9.15$ (see Table 4.2 for k_{23} and k_{31}).

At this pH the rate of approach to the final equilibrium state is at a relative maximum; if the pH is raised or lowered, one of the relaxation times becomes shorter but the other, more important one, becomes longer. For example, lowering the pH increases $\bar{c}_{AH}$ and with it $\tau_{23}^{-1} = k_{23}\bar{c}_{AH}$, but decreases $\bar{c}_A$ and with it $\tau_{13}^{-1} = k_{31}\bar{c}_A$. As a consequence attainment of the final equilibrium state (at $t \gg \tau_{13}$) is slower. Or increasing the pH decreases $\bar{c}_{AH}$ and with it $\tau_{23}^{-1} = k_{23}\bar{c}_{HA}$. Note, however, that there is only a certain pH range where this happens, namely as long as $\bar{c}_A, \bar{c}_{AH} \gg \bar{c}_H, \bar{c}_{HO}$. For example, when the pH is increased strongly, τ_{23}^{-1}, which first decreases due to a decrease in $\bar{c}_{HA}$ ($\tau_{23}^{-1} = k_{23}\bar{c}_{AH}$), eventually increases again because $\bar{c}_{HO}$ is no longer negligible [$\tau_{23}^{-1} = k_{32} + k_{23}(\bar{c}_{HA} + \bar{c}_{HO})$] and the equilibration of the total system becomes faster again.

Such considerations of the rate at which the whole system reaches an equilibrium state have interesting implications for buffering and acid–base catalysis (*9*).

4.4.3 Proton Transfer between Two Acid–Base Pairs

This is a very common experimental situation, as represented in scheme 4.75.

Protolysis

①

$H^+ + A + B + H_2O$

k_{31} / k_{13} (between ③ and ①) $\qquad$ Direct transfer $\qquad$ $k_{13'}$ / $k_{3'1}$ (between ① and ③′)

$$③\ AH + B + H_2O \underset{k_{3'3}}{\overset{k_{33'}}{\rightleftharpoons}} A + BH + H_2O\ ③' \tag{4.75}$$

k_{23} / k_{32} (between ③ and ②) $\qquad$ $k_{23'}$ / $k_{3'2}$ (between ② and ③′)

$AH + BH + OH^-$

②

Hydrolysis

Reaction scheme 4.75 describes all proton transfer processes occurring in such a system. Note, however, that the neutralization reaction, $H^+ + OH^- \rightleftharpoons H_2O$, is not explicitly included; in principle it has to be considered as well but usually (for high concentrations of AH, A, BH, B) it does not play a significant role. Just as for scheme 4.64 the respective reactions involve only species that undergo a transformation, e.g., ① ⇌ ③ corresponds to $H^+ + A \rightleftharpoons AH$ only (with B and H_2O being spectators rather than catalysts).

Scheme 4.75 represents, for example, the behavior of an acid–base pair (AH/A) in the presence of a buffer (BH/B), or the process of pH indication by an indicator in the presence of an acid–base pair. In this latter case the indicator will be present in very small amounts in order to interfere as little as possible with the process to be indicated which usually leads to considerable mathematical simplifications. Similarly, in the first example the buffer will frequently be chosen in a large excess over the other acid–base pair for effective buffering action. Scheme 4.75 also pertains to many mechanisms of acid–base catalysis in a variety of organic and biochemical reactions.

Scheme 4.75 is characterized by three relaxation times since it includes four states. However, it is very rare that all three relaxation times can be observed in a given system, and we can restrict our discussion to special cases. Again we essentially elaborate Eigen's (*9*) outline; a somewhat similar discussion has been offered by Yapel and Lumry (*10*).

(*1*) *pH ≈ 4–10.* In general the concentrations of AH, A, BH, and B will be large compared to c_H and c_{HO}. In deriving the principal relaxation time one can in such cases consider ① and ② as steady states, which simplifies matters considerably. The problem is reduced to one of finding the relaxation time of a two-state system with three parallel reaction pathways, similar to case 9 in Table 2.1 (page 14).

Each pathway contributes additively to τ^{-1} according to

$$1/\tau = 1/\tau_{33'} + 1/\tau_{313'} + 1/\tau_{323'} \tag{4.76}$$

where $\tau_{33'}^{-1}$ refers to the direct proton transfer, $\tau_{313'}^{-1}$ to the protolysis, and $\tau_{323'}^{-1}$ to the hydrolysis pathway.

The principal difference between the present system and case 9 in Table 2.1 is that we have to find the steady state rate coefficients for the two processes via the steady states ① and ②, respectively.

Let us begin with the part relaxation time $\tau_{33'}$, i.e., the contribution to τ^{-1} by the direct pathway. For clarity the reaction is rewritten as

$$\underset{③}{AH + B} \underset{k_{3'3}}{\overset{k_{33'}}{\rightleftharpoons}} \underset{③'}{A + BH} \tag{4.77}$$

which is strictly analogous to case 5 in Table 2.1, and therefore

$$1/\tau_{33'} = k_{33'}(\bar{c}_{AH} + \bar{c}_B) + k_{3'3}(\bar{c}_A + \bar{c}_{BH}) \tag{4.78}$$

For the derivation of $\tau_{313'}^{-1}$ we proceed as follows. The steady state condition for our part system, which is rewritten as

$$\mathrm{B} + \left(\mathrm{AH} \underset{k_{13}}{\overset{k_{31}}{\rightleftharpoons}} \mathrm{A} + \left(\mathrm{H}^+\right) + \mathrm{B} \underset{k_{3'1}}{\overset{k_{13'}}{\rightleftharpoons}} \mathrm{BH}\right) + \mathrm{A} \tag{4.79}$$

③ ① ③'

is expressed as

$$dc_{\mathrm{H}}/dt = k_{31}c_{\mathrm{AH}} - k_{13}c_{\mathrm{A}}c_{\mathrm{H}} - k_{13'}c_{\mathrm{B}}c_{\mathrm{H}} + k_{3'1}c_{\mathrm{BH}} = 0$$

It follows for the steady state concentration of H^+

$$c_{\mathrm{H}^+} = \frac{k_{31}c_{\mathrm{AH}} + k_{3'1}c_{\mathrm{BH}}}{k_{13}c_{\mathrm{A}} + k_{13'}c_{\mathrm{B}}} \tag{4.80}$$

The rate equation for system 4.79 can now be written in terms of any of the species AH, A, BH, or B, for example

$$-dc_{\mathrm{AH}}/dt = k_{31}c_{\mathrm{AH}} - k_{13}c_{\mathrm{A}}c_{\mathrm{H}} \tag{4.81}$$

Substituting c_{H^+} from Eq. 4.80 into Eq. 4.81 leads, after some algebraic manipulations, to

$$-dc_{\mathrm{AH}}/dt = k_{313'}c_{\mathrm{AH}}c_{\mathrm{B}} - k_{3'13}c_{\mathrm{A}}c_{\mathrm{BH}}$$

with the steady state rate coefficients

$$k_{313'} = \frac{k_{31}k_{13'}}{k_{13}\bar{c}_{\mathrm{A}} + k_{13'}\bar{c}_{\mathrm{B}}} \tag{4.82}$$

$$k_{3'13} = \frac{k_{3'1}k_{13}}{k_{13}\bar{c}_{\mathrm{A}} + k_{13'}\bar{c}_{\mathrm{B}}} \tag{4.83}$$

In anticipation of what follows they have already been expressed in terms of the equilibrium concentrations of A and B. The reciprocal part relaxation time for reaction 4.79 then simply becomes

$$\tau_{313'}^{-1} = k_{313'}(\bar{c}_{\mathrm{AH}} + \bar{c}_{\mathrm{B}}) + k_{3'13}(\bar{c}_{\mathrm{A}} + \bar{c}_{\mathrm{BH}}) \tag{4.84}$$

Note that in this derivation we expressed the steady state condition by $dc_{\mathrm{H}}/dt = 0$; an alternative method would be to set up the linearized rate equation for $d\,\Delta c_{\mathrm{H}}/dt$ and set it equal to zero, in analogy to the procedure used in deriving Eq. 3.60 via Eqs. 3.61 and 3.62. Both procedures work equally well.

Similarly, we find the steady state rate coefficients for the hydrolysis pathway

$$\mathrm{AH} + \left(\mathrm{B} + \mathrm{H_2O} \underset{k_{23}}{\overset{k_{32}}{\rightleftharpoons}} \mathrm{BH} + \left(\mathrm{OH}^-\right) + \mathrm{AH} \underset{k_{3'2}}{\overset{k_{23'}}{\rightleftharpoons}} \mathrm{A} + \mathrm{H_2O}\right) + \mathrm{BH} \tag{4.85}$$

③ ② ③'

to be

$$k_{323'} = \frac{k_{32}k_{23'}}{k_{23}\bar{c}_{\mathrm{BH}} + k_{23'}\bar{c}_{\mathrm{AH}}} \tag{4.86}$$

$$k_{3'23} = \frac{k_{3'2}k_{23}}{k_{23}\bar{c}_{BH} + k_{23'}\bar{c}_{AH}} \tag{4.87}$$

Thus for $\tau_{323'}^{-1}$, we obtain

$$\tau_{323'}^{-1} = k_{323'}(\bar{c}_{AH} + \bar{c}_B) + k_{3'23}(\bar{c}_A + \bar{c}_{BH}) \tag{4.88}$$

Equation 4.76 can now be written as

$$1/\tau = k_{33'}^*(\bar{c}_{AH} + \bar{c}_B) + k_{3'3}^*(\bar{c}_A + \bar{c}_{BH}) \tag{4.89}$$

with

$$k_{33'}^* = k_{33'} + k_{313'} + k_{323'} \tag{4.90}$$

$$k_{3'3}^* = k_{3'3} + k_{3'13} + k_{3'23} \tag{4.91}$$

In most situations one or two terms in Eqs. 4.90 and 4.91 can be neglected. At pH $\approx$ 7 both c_H and c_{HO} are very low and therefore the steady state pathways are relatively unimportant. One expects that here $k_{33'} \gg k_{313'} + k_{323'}$ and $k_{3'3} \gg k_{3'13} + k_{3'23}$, and hence

$$k_{33'}^* \approx k_{33'}, \qquad k_{3'3}^* \approx k_{3'3}$$

When the pH is decreased, hydrolysis loses even more in importance but protolysis gains so that $k_{33'} + k_{313'} \gg k_{323'}$ and $k_{3'3} + k_{3'13} \gg k_{3'23}$, and hence

$$k_{33'}^* = k_{33'} + k_{313'}, \qquad k_{3'3}^* = k_{3'3} + k_{3'13}$$

Conversely, as the pH moves from 7 to 10 it is the terms $k_{323'}$ and $k_{3'23}$ which may no longer be negligible.

From these considerations it becomes obvious that it is relatively simple to determine the rate coefficients $k_{33'}$ and $k_{3'3}$ for direct proton transfer between acid–base pairs with pK values not too far from 7 since at pH near 7 it is the only process contributing significantly to the single measurable relaxation time of the system. When pK_{AH} and pK_{BH} are both known, the concentration of all species appearing in Eq. 4.78 can be calculated at any pH value. In this case there remains only one unknown in Eq. 4.78 because of the relation

$$k_{3'3} = k_{33'}/K_{33'} = k_{33'}(K_{BH}/K_{AH}) \tag{4.92}$$

In principle one measurement at a single pH value would suffice.

If, on the other hand, only pK_{AH} but not pK_{BH} is known accurately, one may conduct the experiment with $[AH]_0 \gg [BH]_0$ so that Eq. 4.78 simplifies to

$$1/\tau = 1/\tau_{33'} \approx k_{33'}\bar{c}_{AH} + k_{3'3}\bar{c}_A$$

Here τ^{-1} must be determined at least at two different pH values because $K_{33'}$ is unknown and hence $k_{3'3}$ cannot be found via Eq. 4.92.

(2) pH $\lesssim$ 4. Here the concentration of H^+ is usually comparable or even larger than the concentration of one or several of the species AH, A, BH, and B. As a consequence state ① in scheme 4.75 can no longer be treated as a steady state.

On the other hand, state ② can be neglected altogether because of the extremely low c_{HO}. Thus the system can be described by the upper half of scheme 4.75. It is characterized by two relaxation times, both of which are observable with the appropriate experimental techniques.

The general mathematical treatment of this system presents no more difficulties than that of system 4.64 and is not presented here; however, it is discussed in the literature (*9, 10*). Two special cases, both commonly encountered in practical work, are discussed here instead. The first is the case where one of the acid–base pairs acts as an indicator to monitor spectrophotometrically the (relatively slower) proton transfers to and from the other acid–base pair. The second refers to proton transfers to and from AH and A in the presence of BH/B acting as a buffer system.

A. Proton Transfer Monitored by Indicator

Although indicators can serve to monitor proton transfer reactions as rapid as those of the indicator itself (*11*) they are more commonly used to monitor relatively slow proton transfer reactions such as deprotonation of C—H acids [for an example see Stuehr (*12*)] or acids from which proton abstraction is slowed down because of an intramolecular hydrogen bond [for an example see Eigen and Kruse (*13*)].

Let us rewrite the upper half of scheme 4.75 as

$$\begin{array}{ccc} & \underset{}{\overset{①}{A + H^+ + In}} & \\ {}^{k_{31}}\nearrow\swarrow_{k_{13}} & & {}_{k_{3'1}}\nwarrow\searrow^{k_{13'}} \\ \underset{③}{AH + In} & \underset{k_{3'3}}{\overset{k_{33'}}{\rightleftharpoons}} & \underset{③'}{A + InH} \end{array} \tag{4.93}$$

where In and InH refer to an indicator. We define the part relaxation times

$$1/\tau_{13} = k_{13}(\bar{c}_H + \bar{c}_A) + k_{31}$$
$$1/\tau_{13'} = k_{13'}(\bar{c}_H + \bar{c}_{In}) + k_{3'1}$$
$$1/\tau_{33'} = k_{33'}(\bar{c}_{AH} + \bar{c}_{In}) + k_{3'3}(\bar{c}_A + \bar{c}_{InH})$$

If the proton transfers involving the indicator are much more rapid than those for AH/A, we have

$$1/\tau_{13'} \gg 1/\tau_{13}, 1/\tau_{33'}$$

Note that $\tau_{33'}^{-1}$ is usually relatively small because the rate coefficients of direct proton transfer between a general acid and a general base ($k_{33'}$, $k_{3'3}$) are usually one to two orders of magnitude lower than the rate coefficients of protonation by H^+ (*9, 14*).

Whenever one of the part relaxation times of a complex reaction system is much shorter than all the others, the shortest relaxation time of the entire system coincides with this short part relaxation time. Thus

$$1/\tau_1 = 1/\tau_{13'} = k_{13'}(\bar{c}_H + \bar{c}_{In}) + k_{3'1}$$

For the second relaxation time, which is usually the one of greater interest, the derivation is quite similar to that for scheme 3.65; the main difference is that the linearized rate equation now contains four additional terms arising from the direct transfer ③ ⇌ ③′. Thus

$$\frac{d\,\Delta c_{AH}}{dt} = -k_{31}\,\Delta c_{AH} + k_{13}\bar{c}_{H}\,\Delta c_{A} + k_{13}\bar{c}_{A}\,\Delta c_{H}$$

$$- k_{33'}\bar{c}_{AH}\,\Delta c_{In} - k_{33'}\bar{c}_{In}\,\Delta c_{AH} + k_{3'3}\bar{c}_{A}\,\Delta c_{InH} + k_{3'3}\bar{c}_{InH}\,\Delta c_{A} \quad (4.94)$$

The stoichiometric relations are the same as for scheme 3.65, i.e.,

$$\Delta c_{In} = -\Delta c_{InH} \quad (4.95)$$

$$\Delta c_{A} = -\Delta c_{AH} \quad (4.96)$$

$$\Delta c_{H} + \Delta c_{InH} + \Delta c_{AH} = 0 \quad (4.97)$$

whereas Eq. 3.71 in terms of the present symbolism becomes

$$\Delta c_{InH} = \frac{\bar{c}_{In}}{K_{3'1} + \bar{c}_{H}}\,\Delta c_{H} = \alpha\,\Delta c_{H} \quad (4.98)$$

Inserting Eq. 4.98 into Eq. 4.97 affords

$$\Delta c_{H} = -\Delta c_{AH}/(1 + \alpha) \quad (4.99)$$

After the necessary substitutions, Eq. 4.94 becomes

$$\frac{d\,\Delta c_{AH}}{dt} = -\frac{1}{\tau_2}\,\Delta c_{AH} \quad (4.100)$$

with

$$\frac{1}{\tau_2} = k_{31} + k_{13}\left(\bar{c}_{H} + \bar{c}_{A}\frac{1}{1+\alpha}\right) + k_{33'}\left(\bar{c}_{In} + \bar{c}_{AH}\frac{\alpha}{1+\alpha}\right)$$

$$+ k_{3'3}\left(\bar{c}_{InH} + \bar{c}_{A}\frac{\alpha}{1+\alpha}\right) \quad (4.101)$$

Owing to their high extinction coefficients, the indicators can usually be kept at very low concentrations; i.e., $\bar{c}_{In}, \bar{c}_{InH} \ll \bar{c}_{AH}, \bar{c}_{A}$. As a consequence one often has $\alpha \ll 1$, as can be shown as follows: $\bar{c}_{In}$ expressed in terms of the stoichiometric concentration $[In]_0$ is given by

$$\bar{c}_{In} = \frac{K_{3'1}}{K_{3'1} + \bar{c}_{H}}\,[In]_0 \quad (4.102)$$

Note that $K_{3'1}$ is the acid dissociation constant of InH. From Eqs. 4.98 and 4.102 we obtain

$$\alpha = \frac{\bar{c}_{In}}{K_{3'1} + \bar{c}_{H}} = \frac{K_{3'1}}{(K_{3'1} + \bar{c}_{H})^2}\,[In]_0$$

We see that for pH $\leq$ 4 and pK_{InH} $(= -\log K_{3'1}) \leq 4$, α indeed becomes quite small. For example, pH = pK_{InH} = 4, $[In]_0 = 10^{-5}\,M$, $\alpha = 0.025$; or pH = pK_{InH} = 3, $[In]_0 = 10^{-5}$, $\alpha = 0.0025$.

Thus all concentration terms associated with $k_{33'}$ and $k_{3'3}$ are very small. Further, we have typically $k_{13} \gg k_{33'}, k_{3'3}$ so that Eq. 4.101 simplifies to

$$1/\tau_2 = k_{31} + k_{13}(\bar{c}_H + \bar{c}_A) \tag{4.103}$$

which is the same as Eq. 3.75 derived by neglecting the reaction ③ ⇌ ③′ from the start.

In the alkaline region one comes to an analogous conclusion for $pK_{InH} \geq 10$, pH ≥ 10 $[\tau_2^{-1} = k_{3'2} + k_{23'}(\bar{c}_{OH} + \bar{c}_{AH})]$.

B. Slow Proton Transfer in the Presence of a Buffer

The mathematical procedure for deriving τ_2^{-1} is exactly the same as that for scheme 4.93. Hence, after substituting the symbols B and BH (buffer) for In and InH, respectively, we obtain, in analogy to Eq. 4.101,

$$\frac{1}{\tau_2} = k_{31} + k_{13}\left(\bar{c}_H + \bar{c}_A \frac{1}{1+\alpha}\right) + k_{33'}\left(\bar{c}_B + \bar{c}_{AH}\frac{\alpha}{1+\alpha}\right) + k_{3'3}\left(\bar{c}_{BH} + \bar{c}_A \frac{\alpha}{1+\alpha}\right) \tag{4.104}$$

with

$$\alpha = \frac{\bar{c}_B}{K_{BH} + \bar{c}_H} = \frac{K_{BH}}{(K_{BH} + \bar{c}_H)^2}[B]_0 \tag{4.105}$$

However, in this situation we have $\bar{c}_B, \bar{c}_{BH} \gg \bar{c}_A, \bar{c}_{AH}, \bar{c}_H$, and $\alpha \gg 1$ so that Eq. 4.104 simplifies to

$$\tau_2^{-1} = k_{31} + k_{13}\bar{c}_H + k_{33'}\bar{c}_B + k_{3'3}\bar{c}_{BH} \tag{4.106}$$

In practical work one typically performs experiments at constant pH and varying buffer concentration. A plot of τ_2^{-1} versus $\bar{c}_B$ or $\bar{c}_{BH}$ yields a straight line with a slope $k_{33'} + k_{3'3}\bar{c}_H/K_{BH}$, or $k_{33'}K_{BH}/\bar{c}_H + k_{3'3}$, respectively, and an intercept $k_{31} + k_{13}\bar{c}_H$. By plotting intercepts of such plots at different pH values versus $\bar{c}_H$, k_{31} and k_{13} can be obtained. Finally, when the slopes of the buffer plots are plotted versus $\bar{c}_H$ or $1/\bar{c}_H$, respectively, $k_{33'}$ and $k_{3'3}$ can also be calculated provided K_{BH} is known. [For an application see Bernasconi (*15*).]

4.5 Miscellaneous Multistep Schemes. Two Calculated Examples from Organic Chemistry

It would be a formidable task to try to treat all possible or even only "chemically reasonable" multistep schemes. Since the principles involved are always the same the student who has mastered the contents of the preceding chapters should be able to tackle almost any scheme, provided it is "solvable", i.e., enough steps are either very fast or very slow so that there are at most only *pairs* of strongly coupled differential equations with which to deal. There are also numerous articles in the literature which deal in considerable detail with the mathematical analysis of complex relaxation spectra. Some of these are listed as references (*1–3, 5–8, 16–21*) at the end of this chapter; among these Czerlinski's book (*19*) is the most extensive collection of possible reaction schemes.

For the practicing student who may want to work through a few additional non-trivial problems there follow two instructive examples from our own laboratory. The discussion includes an analysis of how the experimental data were treated in order to obtain the desired rate constants.

4.5.1 Electron Transfer Reactions. One Measurable Relaxation Time

Scheme 4.107 is typical for the comproportionation–disproportionation of

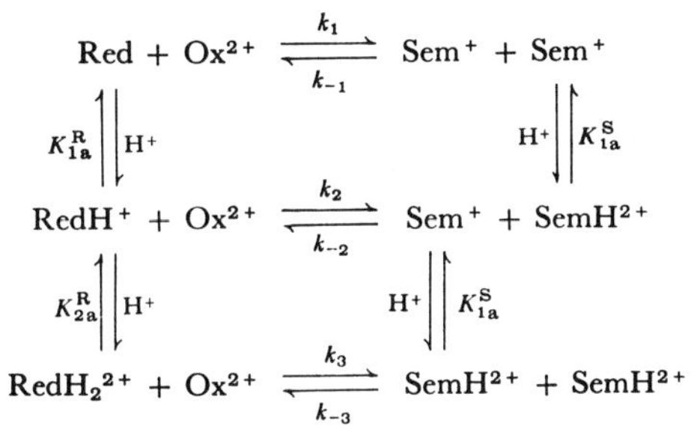

$$
\begin{array}{ccccc}
\mathrm{Red} + \mathrm{Ox}^{2+} & \underset{k_{-1}}{\overset{k_1}{\rightleftharpoons}} & \mathrm{Sem}^{+} + \mathrm{Sem}^{+} \\
K_{1a}^{R} \Big\updownarrow \mathrm{H}^{+} & & \mathrm{H}^{+} \Big\updownarrow K_{1a}^{S} \\
\mathrm{RedH}^{+} + \mathrm{Ox}^{2+} & \underset{k_{-2}}{\overset{k_2}{\rightleftharpoons}} & \mathrm{Sem}^{+} + \mathrm{SemH}^{2+} \\
K_{2a}^{R} \Big\updownarrow \mathrm{H}^{+} & & \mathrm{H}^{+} \Big\updownarrow K_{1a}^{S} \\
\mathrm{RedH_2}^{2+} + \mathrm{Ox}^{2+} & \underset{k_{-3}}{\overset{k_3}{\rightleftharpoons}} & \mathrm{SemH}^{2+} + \mathrm{SemH}^{2+}
\end{array}
\qquad (4.107)
$$

azaviolenes, a class of compounds represented by the general structures (*22*)

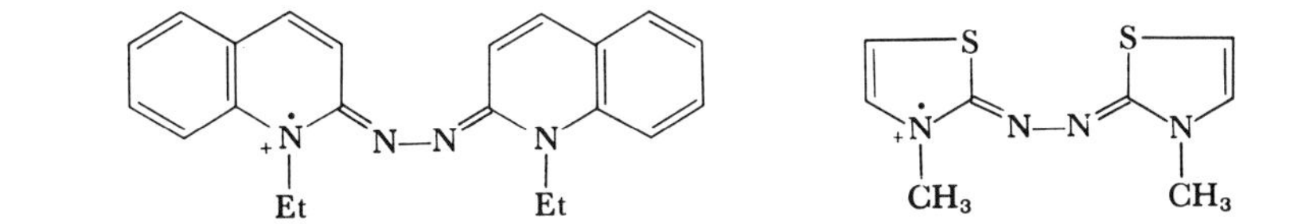

$$
\underset{\mathrm{Ox}^{2+}}{\overset{+}{\mathrm{N}}(\mathrm{R})\!-\!\mathrm{N{=}N}\!-\!\overset{+}{\mathrm{N}}(\mathrm{R})}
\ \underset{-\bar{e}}{\overset{+\bar{e}}{\rightleftharpoons}}\
\underset{\mathrm{Sem}^{+}}{\overset{+}{\mathrm{N}}(\mathrm{R})\!=\!\mathrm{N{-}N}\!=\!\mathrm{N}(\mathrm{R})}
\ \underset{-\bar{e}}{\overset{+\bar{e}}{\rightleftharpoons}}\
\underset{\mathrm{Red}}{\mathrm{N}(\mathrm{R})\!=\!\mathrm{N{-}N}\!=\!\mathrm{N}(\mathrm{R})}
$$

where Ox^{2+} stands for oxidized, Red for reduced, and Sem^{+} for semireduced. The structures of two representative azaviolenes are shown in their Sem^{+} forms

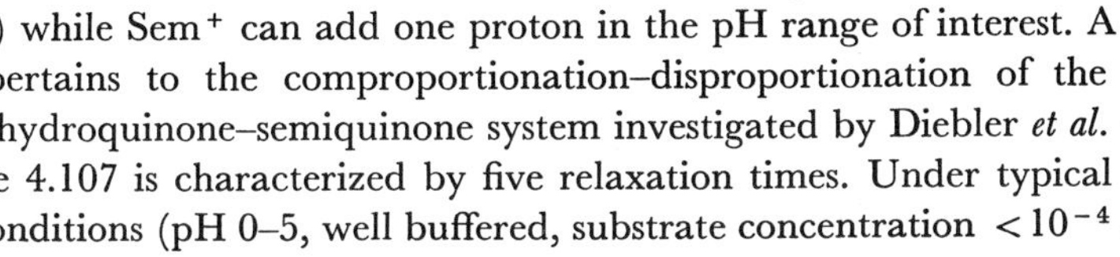

Structure I: quinoline-type rings, $\overset{\bullet}{\mathrm{N}}^{+}$(Et)—N—N—N(Et)

Structure II: thiazoline-type rings (S), $\overset{\bullet}{\mathrm{N}}^{+}$($CH_3$)—N—N—N($CH_3$)

I II

As shown in scheme 4.107 Red can add one or two protons (to the azo bridge nitrogens) while Sem^{+} can add one proton in the pH range of interest. A similar scheme pertains to the comproportionation–disproportionation of the benzoquinone–hydroquinone–semiquinone system investigated by Diebler *et al.* (*23*).

Scheme 4.107 is characterized by five relaxation times. Under typical experimental conditions (pH 0–5, well buffered, substrate concentration $< 10^{-4}\ M$) the proton transfer equilibria are all very rapidly established and the four relaxation times associated with them are too rapid to be measured by the experimental methods at hand (stopped flow and temperature jump) (*24*). The fifth relaxation time is measurable; it is associated with the three horizontal electron transfer pathways. From a mere inspection of scheme 4.107 one expects that at high pH the first pathway (k_1, k_{-1}) will contribute mostly or exclusively to τ^{-1}, at very low pH τ^{-1} should be determined mainly by the third pathway, while at intermediate pH values all pathways may contribute. Thus the study of the pH dependence of τ^{-1} should allow a determination of some or hopefully all six rate coefficients for electron transfer. These qualitative expectations are borne out by a quantitative treatment which now follows.

In deriving τ^{-1} we choose to express the rate in terms of Ox^{2+}:

$$\begin{aligned} dc_{Ox}/dt = & -k_1 c_R c_{Ox} - k_2 c_{RH} c_{Ox} - k_3 c_{RH_2} c_{Ox} \\ & + k_{-1}(c_S)^2 + k_{-2} c_S c_{SH} + k_{-3}(c_{SH})^2 \end{aligned} \tag{4.108}$$

where Ox = Ox^{2+}, R = Red, S = Sem^+, etc. Note that

$$dc_{Ox}/dt = d(c_R + c_{RH} + c_{RH_2})/dt = -\tfrac{1}{2}\, d(c_S + c_{SH})/dt.$$

For small perturbations we obtain the linearized equation

$$\begin{aligned} d\,\Delta c_{Ox}/dt = & -k_1 \bar{c}_R\,\Delta c_{Ox} - k_1 \bar{c}_{Ox}\,\Delta c_R - k_2 \bar{c}_{RH}\,\Delta c_{Ox} - k_2 \bar{c}_{Ox}\,\Delta c_{RH} \\ & - k_3 \bar{c}_{RH_2}\,\Delta c_{Ox} - k_3 \bar{c}_{Ox}\,\Delta c_{RH_2} + 2k_{-1}\bar{c}_S\,\Delta c_S \\ & + k_{-2}\bar{c}_S\,\Delta c_{SH} + k_{-2}\bar{c}_{SH}\,\Delta c_S + 2k_{-3}\bar{c}_{SH}\,\Delta c_{SH} \end{aligned} \tag{4.109}$$

In buffered solutions the following equilibrium relationships hold true:

$$\bar{c}_R = \frac{1}{D_R}(\bar{c}_R)_{tot} \tag{4.110}$$

$$\bar{c}_{RH} = \frac{\bar{c}_H/K_{1a}^R}{D_R}(\bar{c}_R)_{tot} \tag{4.111}$$

$$\bar{c}_{RH_2} = \frac{(\bar{c}_H)^2/K_{1a}^R K_{2a}^R}{D_R}(\bar{c}_R)_{tot} \tag{4.112}$$

$$\bar{c}_S = \frac{1}{D_S}(\bar{c}_S)_{tot} \tag{4.113}$$

$$\bar{c}_{SH} = \frac{\bar{c}_H/K_{1a}^S}{D_S}(\bar{c}_S)_{tot} \tag{4.114}$$

with

$$(\bar{c}_R)_{tot} = \bar{c}_R + \bar{c}_{RH} + \bar{c}_{RH_2}, \qquad (\bar{c}_S)_{tot} = \bar{c}_S + \bar{c}_{SH}$$

$$D_R = 1 + \frac{\bar{c}_H}{K_{1a}^R} + \frac{(\bar{c}_H)^2}{K_{1a}^R K_{2a}^R}, \qquad D_S = 1 + \frac{\bar{c}_H}{K_{1a}^S}$$

Since the reaction solutions were prepared by dissolving Sem^+ into the medium,

Ox^{2+} and Red in its various protonated forms were only produced by disproportionation. Consequently, we have

$$(\bar{c}_{\mathrm{R}})_{\mathrm{tot}} = \bar{c}_{\mathrm{Ox}} \tag{4.115}$$

From stoichiometry or mass balance we have

$$\Delta c_{\mathrm{Ox}} + \Delta c_{\mathrm{R}} + \Delta c_{\mathrm{RH}} + \Delta c_{\mathrm{RH_2}} + \Delta c_{\mathrm{S}} + \Delta c_{\mathrm{SH}} = 0 \tag{4.116}$$

$$\Delta c_{\mathrm{Ox}} = \Delta c_{\mathrm{R}} + \Delta c_{\mathrm{RH}} + \Delta c_{\mathrm{RH_2}} = -\tfrac{1}{2}(\Delta c_{\mathrm{S}} + \Delta c_{\mathrm{SH}}) \tag{4.117}$$

Equilibrium considerations similar to the ones leading from Eq. 3.10 to 3.11 provide

$$\Delta c_{\mathrm{R}} = \frac{1}{D_{\mathrm{R}}} \Delta c_{\mathrm{Ox}} \tag{4.118}$$

$$\Delta c_{\mathrm{RH}} = \frac{\bar{c}_{\mathrm{H}}/K_{1a}^{\mathrm{R}}}{D_{\mathrm{R}}} \Delta c_{\mathrm{Ox}} \tag{4.119}$$

$$\Delta c_{\mathrm{RH_2}} = \frac{(\bar{c}_{\mathrm{H}})^2/K_{1a}^{\mathrm{R}} K_{2a}^{\mathrm{R}}}{D_{\mathrm{R}}} \Delta c_{\mathrm{Ox}} \tag{4.120}$$

$$\Delta c_{\mathrm{S}} = -\frac{2}{D_{\mathrm{S}}} \Delta c_{\mathrm{Ox}} \tag{4.121}$$

$$\Delta c_{\mathrm{SH}} = -\frac{2\bar{c}_{\mathrm{H}}/K_{1a}^{\mathrm{S}}}{D_{\mathrm{S}}} \Delta c_{\mathrm{Ox}} \tag{4.122}$$

With the necessary substitutions from Eqs. 4.110–4.122, Eq. 4.109 can be expressed as

$$\frac{d\,\Delta c_{\mathrm{Ox}}}{dt} = -\tau^{-1} \Delta c_{\mathrm{Ox}}$$

with

$$\tau^{-1} = \frac{2}{D_{\mathrm{R}}}\left(k_1 + k_2 \frac{\bar{c}_{\mathrm{H}}}{K_{1a}^{\mathrm{R}}} + k_3 \frac{(\bar{c}_{\mathrm{H}})^2}{K_{1a}^{\mathrm{R}} K_{2a}^{\mathrm{R}}}\right)\bar{c}_{\mathrm{Ox}} + \frac{4}{(D_{\mathrm{S}})^2}\left(k_{-1} + k_{-2}\frac{\bar{c}_{\mathrm{H}}}{K_{1a}^{\mathrm{S}}} + k_{-3}\frac{(\bar{c}_{\mathrm{H}})^2}{(K_{1a}^{\mathrm{S}})^2}\right)(\bar{c}_{\mathrm{S}})_{\mathrm{tot}} \tag{4.123}$$

Although K_{1a}^{R} and K_{2a}^{R} could be measured independently, thus also providing D_{R}, Eq. 4.123 still contains too many unknowns for a direct evaluation of the various rate coefficients from a concentration and pH dependence of τ^{-1}. But since $K_1 = k_1/k_{-1}$ could also be measured independently, the number of unknowns could be reduced to a manageable level as follows. Defining $k_2/k_{-2} = K_2$ and $K_3 = k_3/k_{-3}$, substituting k_1/K_1 for k_{-1}, k_2/K_2 for k_{-2}, k_3/K_3 for k_{-3}, and making use of the relationships $K_2 = K^{\mathrm{R}}_{1\mathrm{a}}K_1/K^{\mathrm{S}}_{1\mathrm{a}}$, $K_3 = K^{\mathrm{R}}_{2\mathrm{a}}K_2/K^{\mathrm{S}}_{1\mathrm{a}} = K^{\mathrm{R}}_{1\mathrm{a}}K^{\mathrm{R}}_{2\mathrm{a}}K_1/(K^{\mathrm{S}}_{1\mathrm{a}})^2$ one may write Eq. 4.123 as

$$\tau^{-1} = \left(k_1 + k_2\frac{\bar{c}_{\mathrm{H}}}{K^{\mathrm{R}}_{1\mathrm{a}}} + k_3\frac{(\bar{c}_{\mathrm{H}})^2}{K^{\mathrm{R}}_{1\mathrm{a}}K^{\mathrm{R}}_{2\mathrm{a}}}\right)\left(\frac{2}{D_{\mathrm{R}}}\bar{c}_{\mathrm{Ox}} + \frac{4}{K_1(D_{\mathrm{S}})^2}(\bar{c}_{\mathrm{S}})_{\mathrm{tot}}\right) \quad (4.124)$$

Finally, taking advantage of the relations

$$\bar{c}_{\mathrm{S}} = (K_1\bar{c}_{\mathrm{R}}\bar{c}_{\mathrm{Ox}})^{1/2} = \left(\frac{K_1}{D_{\mathrm{R}}}\right)^{1/2}\bar{c}_{\mathrm{Ox}} \quad (4.125)$$

$$(\bar{c}_{\mathrm{S}})_{\mathrm{tot}} + \bar{c}_{\mathrm{Ox}} + (\bar{c}_{\mathrm{R}})_{\mathrm{tot}} = (\bar{c}_{\mathrm{S}})_{\mathrm{tot}} + 2\bar{c}_{\mathrm{Ox}} = [\mathrm{Sem}^+]_0 \quad (4.126)$$

(where $[\mathrm{Sem}^+]_0$ is the stoichiometric concentration of Sem^+) and of Eq. 4.113, $\bar{c}_{\mathrm{Ox}}$ and $(\bar{c}_{\mathrm{S}})_{\mathrm{tot}}$ can be expressed in terms of $[\mathrm{Sem}^+]_0$ as

$$\bar{c}_{\mathrm{Ox}} = \frac{1}{2 + D_{\mathrm{S}}(K_1/D_{\mathrm{R}})^{1/2}}[\mathrm{Sem}^+]_0 \quad (4.127)$$

$$(\bar{c}_{\mathrm{S}})_{\mathrm{tot}} = \frac{D_{\mathrm{S}}(K_1/D_{\mathrm{R}})^{1/2}}{2 + D_{\mathrm{S}}(K_1/D_{\mathrm{R}})^{1/2}}[\mathrm{Sem}^+]_0 \quad (4.128)$$

Substituting Eqs. 4.127 and 4.128 into Eq. 4.124 affords

$$\tau^{-1} = \left(k_1 + k_2\frac{\bar{c}_{\mathrm{H}}}{K^{\mathrm{R}}_{1\mathrm{a}}} + k_3\frac{(\bar{c}_{\mathrm{H}})^2}{K^{\mathrm{R}}_{1\mathrm{a}}K^{\mathrm{R}}_{2\mathrm{a}}}\right)$$
$$\times\left(\frac{2}{D_{\mathrm{R}}[2 + D_{\mathrm{S}}(K_1/D_{\mathrm{R}})^{1/2}]} + \frac{4(K_1/D_{\mathrm{R}})^{1/2}}{K_1D_{\mathrm{S}}[2 + D_{\mathrm{S}}(K_1/D_{\mathrm{R}})^{1/2}]}\right)[\mathrm{Sem}^+]_0 \quad (4.129)$$

Equation 4.129 predicts that a plot of τ^{-1} versus $[\mathrm{Sem}^+]_0$ should be linear and pass through the origin, with a slope given by the expression enclosed in the brackets. Figure 4.1 shows such plots at various pH values for the specific case of

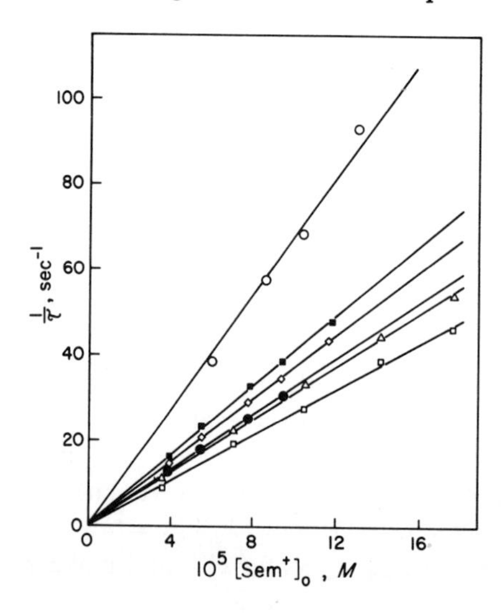

FIGURE 4.1. Representative plots of $1/\tau$ versus $[\mathrm{Sem}^+]_0$ in 50% 2-methoxyethanol–50% water for azaviolene derived from 1-ethyl-2-quinolonazine: ○, pH 4.79; ■, pH 1.11; ◊, pH 1.43; ●, pH 3.86; △, pH 2.37; □, pH 3.19. [From Bernasconi *et al.* (*24*), by permission of the American Chemical Society.]

the azaviolene (**I**) measured in 50% 2-methoxyethanol–50 water% (*24*). The various rate coefficients were evaluated as follows.

At pH > 3 the third pathway (k_3, k_{-3}) makes a negligible contribution to τ^{-1}. Also, since Sem^+ is not significantly protonated at these pH values, $D_S \approx 1$. Thus Eq. 4.129 simplifies to

$$\tau^{-1} = \left(k_1 + k_2 \frac{\bar{c}_H}{K_{1a}^R}\right)\left(\frac{2}{D_R[1 + (K_1/D_R)^{1/2}]} + \frac{4(K_1/D_R)^{1/2}}{K_1[2 + (K_1/D_R)^{1/2}]}\right)[\text{Sem}^+]_0 \tag{4.130}$$

with k_1 and k_2 the only remaining unknowns. From five slopes at pH > 3, k_1 and k_2 were evaluated by a computer fitting procedure.

When data at pH < 3 were included, a satisfactory fit with Eq. 4.130 could no longer be obtained, indicating that $D_S > 1$ below pH 3. Thus D_S was now allowed to vary; i.e., the data were fitted to Eq. 4.129 minus the k_3 term. There resulted an excellent fit down to pH ~ 1 without changing k_1 and k_2; this provided a value for K_{1a}^S and with it K_2 since $K_2 = K_1 K_{1a}^R / K_{1a}^S$.

Below pH 1 the fit again became unsatisfactory, indicating that the third pathway becomes important. When k_1 and k_2 were kept constant and K_{1a}^S and k_3 were allowed to vary, an excellent fit to Eq. 4.129 with the data at all 15 pH values was obtained without significantly changing K_{1a}^S from its value found at pH > 1. Values for K_3 and k_{-3} were finally calculated by the relationships $K_3 = K_2 K_{2a}^R / K_{1a}^S$ and $k_{-3} = k_3/K_3$.

With $pK_{1a}^R = 6.67$, $pK_{2a}^R = 2.47$, and $K_1 = 2.42 \times 10^5$ determined by a spectrophotometric method, the best fit with Eq. 4.129 was found with the following rate constants:

$$\begin{aligned} k_1 &= 1.56 \times 10^9\ M^{-1}\,\text{sec}^{-1}, & k_{-1} &= 6.45 \times 10^3\ M^{-1}\,\text{sec}^{-1} \\ k_2 &= 1.14 \times 10^6\ M^{-1}\,\text{sec}^{-1}, & k_{-2} &= 1.8 \times 10^7\ M^{-1}\,\text{sec}^{-1} \\ k_3 &= 2.7 \times 10^3\ M^{-1}\,\text{sec}^{-1}, & k_{-3} &= 1.1 \times 10^7\ M^{-1}\,\text{sec}^{-1} \end{aligned}$$

4.5.2 Anionic σ Complexes in Mixed Hydroxylic Solvents. Four Measurable Relaxation Times

In basic alcohol–water mixtures 1,3,5-trinitrobenzene (TNB) forms several 1:1 and 1:2 σ complexes according to the reactions summarized as follows:

$$\begin{array}{ccccc} \text{T} & \underset{k_{-1}}{\overset{k_1\bar{c}_{\text{RO}^-}}{\rightleftharpoons}} & \text{R} & \underset{k_{-3}}{\overset{k_3\bar{c}_{\text{RO}^-}}{\rightleftharpoons}} & \text{RR} \\ {\scriptstyle k_{-2}}\Big\uparrow\Big\downarrow{\scriptstyle k_2\bar{c}_{\text{HO}^-}} & & {\scriptstyle k_{-4'}}\Big\uparrow\Big\downarrow{\scriptstyle k_{4'}\bar{c}_{\text{HO}^-}} & & \\ \text{H} & \underset{k_{-3'}}{\overset{k_{3'}\bar{c}_{\text{RO}^-}}{\rightleftharpoons}} & \text{RH} & & \\ {\scriptstyle k_{-4}}\Big\uparrow\Big\downarrow{\scriptstyle k_4\bar{c}_{\text{HO}^-}} & & & & \\ \text{HH} & & & & \end{array} \tag{4.131}$$

T stands for TNB and the other symbols refer to the following complexes:

R H RR

HH RH

The reactions of scheme 4.131 are all slow enough to be measurable by the temperature-jump or stopped-flow methods (*25*, *26*). Of the five expected relaxation times four could actually be detected and evaluated. All experiments were

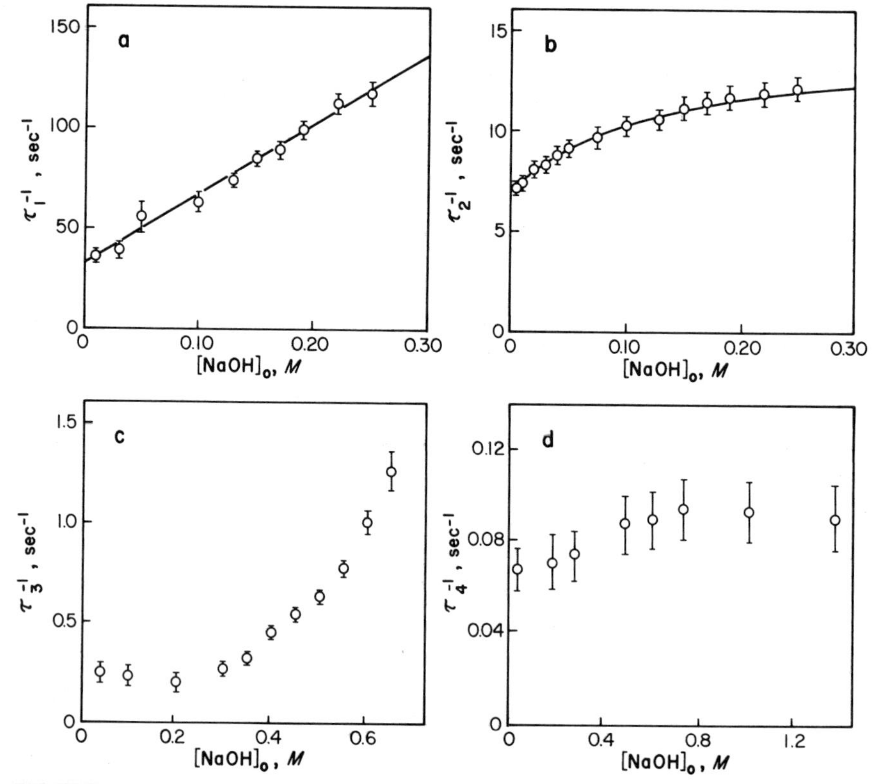

FIGURE 4.2. Relaxation times for anionic σ complexes of 1,3,5-trinitrobenzene in 19% ethanol–81% water. [From Bernasconi and Bergstrom (*25*, *26*), by permission of the American Chemical Society.]

performed with the base (RO^- and HO^-) in large excess over TNB. Respective plots of τ_1^{-1}, τ_2^{-1}, τ_3^{-1}, and τ_4^{-1} versus $[NaOH]_0$ for a 19% EtOH–81% H_2O mixture are shown in Fig. 4.2.

From data obtained in pure water and pure alcohol it was possible to conclude that $k_1\bar{c}_{RO} + k_{-1} \gg k_2\bar{c}_{HO} + k_{-2} \gg k_3\bar{c}_{RO} + k_{-3} \gg k_4\bar{c}_{HO} + k_{-4}$; further, it was assumed on chemical grounds that $k_{3'} \approx k_3$; $k_{-3'} \approx k_{-3}$; $k_{4'} \approx k_4$; $k_{-4'} \approx k_{-4}$.

With these assumptions the derivation of the five relaxation times associated with scheme 4.131 is now straightforward; it will become apparent which one of the five is undetectable.

The shortest relaxation time, τ_1, refers simply to the equilibration $T + RO^- \rightleftharpoons R$ and is given by

$$\tau_1^{-1} = k_1\bar{c}_{RO} + k_{-1} \tag{4.132}$$

Taking into consideration the rapid equilibrium reaction

$$ROH + HO^- \overset{K'}{\rightleftharpoons} RO^- + H_2O \tag{4.133}$$

in the mixed solvent and expressing

$$\bar{c}_{RO}/\bar{c}_{HO} = K'\chi_{ROH}/\chi_{H_2O} = K$$

where χ_{ROH} and χ_{H_2O} are mole fractions, we find

$$\bar{c}_{RO} = [NaOH]_0K/(1 + K), \qquad \bar{c}_{HO} = [NaOH]_0/(1 + K)$$

Thus Eq. 4.132 becomes

$$\tau_1^{-1} = \frac{k_1K}{1 + K}[NaOH]_0 + k_{-1} \tag{4.134}$$

The plot of τ_1^{-1} versus $[NaOH]_0$ in Fig. 4.2a is linear as called for by Eq. 4.134, with a slope $k_1K/(1 + K)$ and an intercept k_{-1}. With $K = 0.047$ we obtained $k_1 = 7700\ M^{-1}\ sec^{-1}$ and $k_{-1} = 32\ sec^{-1}$ (*25*).

The second relaxation time refers to the reaction $T + HO^- \rightleftharpoons H$, with the reaction $T + RO^- \rightleftharpoons R$ at equilibrium. Thus we now must consider the scheme

$$R \underset{k_{1R}}{\overset{k_{-1}}{\rightleftharpoons}} T \underset{k_{-2}}{\overset{k_{2H}}{\rightleftharpoons}} H$$

with the pseudounimolecular rate constants $k_{1R} = k_1\bar{c}_{RO}$ and $k_{2H} = k_2\bar{c}_{HO}$. This is analogous to case 1 in Table 3.1 and thus

$$\tau_2^{-1} = \frac{k_2\bar{c}_{HO}}{1 + K_1\bar{c}_{RO}} + k_{-2} \tag{4.135}$$

with $K_1 = k_1/k_{-1}$. In terms of the stoichiometric NaOH concentration Eq. 4.135 becomes

$$\tau_2^{-1} = \frac{k_2[NaOH]_0}{1 + K + KK_1[NaOH]_0} + k_{-2} \tag{4.136}$$

which is consistent with the plot shown in Fig. 4.2b.

Evaluation of k_{-2} is performed by using the intercept of Fig. 4.2b ($k_{-2} = 6.8\ \text{sec}^{-1}$) while $k_2 = 70.2\ M^{-1}\ \text{sec}^{-1}$ is found from the slope of an inversion plot according to

$$\frac{1}{\tau_2^{-1} - k_{-2}} = \frac{1 + K}{k_2[\text{NaOH}]_0} + \frac{KK_1}{k_2} \tag{4.137}$$

With k_2 determined, the intercept of the inversion plot, KK_1/k_2, allows the calculation of $K_1 = 248\ M^{-1}$, which compares very well with $K = k_1/k_{-1} = 241\ M^{-1}$ obtained from τ_1^{-1}.

The third and fourth relaxation times, τ_3 and $\tau_{3'}$, refer to the reactions R + RO$^-$ $\rightleftharpoons$ RR and H + RO$^-$ $\rightleftharpoons$ RH. Since the rates of these processes are very similar ($k_{3'} \approx k_3$; $k_{-3'} \approx k_{-3}$) the two reactions arc strongly coupled. Consequently, in deriving τ_3^{-1} and $\tau_{3'}^{-1}$ we must use the determinant procedure outlined in Section 3.2.2.

The two coupled linearized rate equations are

$$\frac{d\,\Delta c_{\text{RR}}}{dt} = k_{3\text{R}}\,\Delta c_{\text{R}} - k_{-3}\,\Delta c_{\text{RR}} \tag{4.138}$$

$$\frac{d\,\Delta c_{\text{RH}}}{dt} = k_{3'\text{R}}\,\Delta c_{\text{H}} - k_{-3'}\,\Delta c_{\text{RH}} + k_{4'\text{H}}\,\Delta c_{\text{R}} - k_{-4'}\,\Delta c_{\text{RH}} \tag{4.139}$$

with $k_{3\text{R}} = k_3\bar{c}_{\text{RO}}$, $k_{3'\text{R}} = k_{3'}\bar{c}_{\text{RO}}$, $k_{4'\text{H}} = k_{4'}\bar{c}_{\text{HO}}$. Since $k_{4'\text{H}}$ and $k_{-4'}$ are much smaller than the other rate constants, the last two terms in Eq. 4.139 can be neglected. From stoichiometric and equilibrium considerations similar to those leading to Eq. 3.11, we have

$$\Delta c_{\text{T}} + \Delta c_{\text{R}} + \Delta c_{\text{H}} + \Delta c_{\text{RR}} + \Delta c_{\text{RH}} = 0 \tag{4.140}$$

$$\Delta c_{\text{R}} = K_1\bar{c}_{\text{RO}}\,\Delta c_{\text{T}} = K_{1\text{R}}\,\Delta c_{\text{T}} \tag{4.141}$$

$$\Delta c_{\text{H}} = K_2\bar{c}_{\text{HO}}\,\Delta c_{\text{T}} = K_{2\text{H}}\,\Delta c_{\text{T}} \tag{4.142}$$

Combining Eqs. 4.140–4.142 affords

$$\Delta c_{\text{T}} = -D_1(\Delta c_{\text{RR}} + \Delta c_{\text{RH}}) \tag{4.143}$$

with

$$D_1 = (1 + K_{1\text{R}} + K_{2\text{H}})^{-1}$$

Substituting Eqs. 4.141–4.143 into Eqs. 4.138 and 4.139 leads to

$$d\,\Delta c_{\text{RR}}/dt + a_{11}\,\Delta c_{\text{RR}} + a_{12}\,\Delta c_{\text{RH}} = 0 \tag{4.144}$$

$$d\,\Delta c_{\text{RH}}/dt + a_{21}\,\Delta c_{\text{RR}} + a_{22}\,\Delta c_{\text{RH}} = 0 \tag{4.145}$$

with

$$a_{11} = k_{3\text{R}}K_{1\text{R}}D_1 + k_{-3}, \qquad a_{12} = k_{3\text{R}}K_{1\text{R}}D_1$$

$$a_{21} = k_{3'\text{R}}K_{2\text{H}}D_1, \qquad a_{22} = k_{3'\text{R}}K_{2\text{H}}D_1 + k_{-3'}$$

Equations 4.144 and 4.145 are of the form of Eqs. 3.27 and 3.28. Thus the two relaxation times are given by Eqs. 3.47 and 3.48 (τ_1^{-1} replaced by τ_3^{-1}, τ_2^{-1} replaced by $\tau_{3'}^{-1}$).

If we now assume $k_{-3'}$ and k_{-3} not only to be similar but equal in magnitude, the analytical expressions for τ_3^{-1} and $\tau_{3'}^{-1}$ become very simple and are given by

$$1/\tau_3 = (k_{3R}K_{1R} + k_{3'R}K_{2H})D_1 + k_{-3} \tag{4.146}$$

$$1/\tau_{3'} = k_{-3} \tag{4.147}$$

The similarity of the present situation with the simpler two-step scheme represented by case 1a in Table 3.2 is to be noted.

For an evaluation of the rate constants the additional assumption $k_{3'} = k_3$ is now made, which simplifies Eq. 4.146 to

$$1/\tau_3 = k_{3R}(K_{1R} + K_{2H})D_1 + k_{-3} \tag{4.148}$$

In terms of $[NaOH]_0$ this becomes

$$\tau_3^{-1} = \frac{k_3K(KK_1 + K_2)[NaOH]_0^2}{(1 + K)\{1 + K + (KK_1 + K_2)[NaOH]_0\}} + k_{-3} \tag{4.149}$$

Of the two relaxation times only τ_3 could be observed (Fig. 4.2c); $\tau_{3'}$ remained undetectable because of a vanishing relaxation amplitude (*26*) (see Chapter 7).

The concentration dependence of τ_3^{-1} is consistent with Eq. 4.149. From the intercept of the plot we obtain $k_{-3} = 0.20\ \text{sec}^{-1}$. k_3 was evaluated at relatively high $[NaOH]_0$ where $(KK_1 + K_2)\,[NaOH]_0 \gg 1 + K$ so that Eq. 4.149 reduces to the linear relationship 4.150. From the slope $k_3 = 45\ M^{-1}\ \text{sec}^{-1}$

$$\frac{1}{\tau_3} = \frac{k_3K}{1 + K}[NaOH]_0 + k_{-3} \tag{4.150}$$

was obtained.

The fifth relaxation time, τ_4, is easily derived by assuming that the species T, R, RR, H, and RH are in equilibrium with each other with respect to the time scale of τ_4. The linearized rate equation is given by

$$d\,\Delta c_{HH}/dt = k_{4H}\,\Delta c_H - k_{-4}\,\Delta c_{HH} \tag{4.151}$$

The stoichiometric relationship is now given by

$$\Delta c_T + \Delta c_R + \Delta c_H + \Delta c_{RR} + \Delta c_{RH} + \Delta c_{HH} = 0 \tag{4.152}$$

Furthermore, besides Eqs. 4.141 and 4.142 there are now also

$$\Delta c_{RR} = K_{3R}\,\Delta c_R = K_{1R}K_{3R}\,\Delta c_T \tag{4.153}$$

$$\Delta c_{RH} = K_{3'R}\,\Delta c_H = K_{2H}K_{3'R}\,\Delta c_T \tag{4.154}$$

Combining Eqs. 4.141, 4.142, 4.153, and 4.154 with Eq. 4.152 affords

$$\Delta c_T = -D_2\,\Delta c_{HH} \tag{4.155}$$

with

$$D_2 = (1 + K_{1R} + K_{2H} + K_{1R}K_{3R} + K_{2H}K_{3'R})^{-1}$$

Combining Eq. 4.142 with 4.155 leads to

$$\Delta c_H = -K_{2H}D_2\, \Delta c_{HH} \tag{4.156}$$

Thus Eq. 4.151 becomes

$$d\, \Delta c_{HH}/dt = -\tau_4^{-1}\, \Delta c_{HH}$$

with

$$1/\tau_4 = k_{4H}K_{2H}D_2 + k_{-4}$$

or

$$\frac{1}{\tau_4} = \frac{k_4K_2[\text{NaOH}]_0^2}{(1 + K)^2 + (1 + K)(KK_1 + K_2)[\text{NaOH}]_0 + K(KK_1K_3 + K_2K_{3'R})[\text{NaOH}]_0^2} + k_{-4}$$

The experimental values of τ_4^{-1} varied only little with $[\text{NaOH}]_0$ (Fig. 4.2d), indicating that $\tau_4^{-1} \approx k_{-4} = 0.07\ \text{sec}^{-1}$.

Problems

1. A temperature-jump study of the pH indicator alizarin yellow R in aqueous solution gave the following results:

pH	$10^{-4} \times \tau^{-1}$ (sec^{-1})
10.30	4.37
10.49	4.62
10.60	5.41
10.70	5.81

The stoichiometric indicator concentration was $9.8 \times 10^{-5}\ M$, its spectrophotometrically determined $pK_a = 10.75$, and the solution contained only NaOH as base (no buffer added). What reaction is responsible for the relaxation effect and what are its rate constants? Compare your answer with that of Rose and Stuehr (*27*).

2. The formation of a hemimercaptal by the addition of 2-mercaptoethanol to the aldehydic group of pyridoxal-5′-phosphate has been shown (*28*) to proceed by the following mechanism in aqueous solution:

$$RSH + OH^- \rightleftharpoons RS^- + H_2O \qquad \text{(fast)}$$

$$R'CH{=}O + RS^- \rightleftharpoons R'CH\begin{matrix} O^- \\ SR \end{matrix} \qquad \text{(slow)}$$

$$R'CH\begin{matrix} O^- \\ SR \end{matrix} + H_2O \rightleftharpoons R'CH\begin{matrix} OH \\ SR \end{matrix} + OH^- \qquad \text{(fast)}$$

Derive the longest relaxation time in a buffered solution.

3. Derive the longest relaxation time for the system

$$\begin{array}{ccccccc}
E + S & \rightleftharpoons & ES & \rightleftharpoons & EP & \rightleftharpoons & P + E \\
\updownarrow H^+ & & \updownarrow H^+ & & \updownarrow H^+ & & \updownarrow H^+ \\
EH^+ + S & \rightleftharpoons & EHS^+ & \rightleftharpoons & EHP^+ & \rightleftharpoons & P + EH^+ \\
\updownarrow H^+ & & \updownarrow H^+ & & \updownarrow H^+ & & \updownarrow H^+ \\
EH_2^{2+} + S & \rightleftharpoons & EH_2S^{2+} & \rightleftharpoons & EH_2P^{2+} & \rightleftharpoons & P + EH_2^{2+}
\end{array}$$

Assume that all equilibria except $ES \rightleftharpoons EP$, $EHS^+ \rightleftharpoons EHP^+$, and $EH_2S^{2+} \rightleftharpoons EH_2P^{2+}$ are rapidly established, that the enzyme is present in a large excess over S and P, and the solution is buffered. Compare your result with that of Hammes and Schimmel (*8*).

4. Strehlow (*29*) has investigated the hydration of pyruvic acid by the pressure-jump method. The reaction system can be described by the following scheme:

$$\begin{array}{ccc}
CH_3C(=O)COOH & \overset{H_2O}{\rightleftharpoons} & CH_3C(OH)_2COOH \\
\updownarrow H^+ & & \updownarrow H^+ \\
CH_3C(=O)COO^- & \overset{H_2O}{\rightleftharpoons} & CH_3C(OH)_2COO^-
\end{array}$$

Derive all the relaxation times by assuming that the proton transfer equilibria are much more rapidly established than the hydration equilibria.

5. The kinetics of the proton transfer reactions of polyacrylic acid (PAH) have been reported by Weiss *et al.* (*30*). They found one relaxation time in the range of 20–250 μsec; it was measured by the temperature-jump method in the presence of phenol red ($pK_a = 7.7$) whose concentration was always small compared to the concentrations of PAH and PA^-. Their results are summarized in the following table:

pH	$10^3 \times [PAH]_0$ (*M*)	$10^{-3} \times \tau^{-1}$ (sec^{-1})	pH	$10^3 \times [PAH]_0$ (*M*)	$10^{-3} \times \tau^{-1}$ (sec^{-1})
6.20	0.2	24	7.10	0.5	7.8
6.20	0.5	31	7.10	1.0	9.7
6.20	1.0	36	7.10	2.0	13.2
6.70	0.2	13.3	7.10	5.0	18.5
6.70	0.5	14.9	7.90	0.5	4.5
6.70	1.0	17.9	7.90	1.0	5.3
6.70	2.0	22.0	7.90	2.0	6.8
			7.90	5.0	11.4

Evaluate $k_{33'}$ and $k_{3'3}$ (see scheme 4.75) for the proton transfer between PAH and the indicator.

6. The kinetics of the keto–enol tautomerism of barbituric acid in aqueous solution has been investigated by Eigen *et al.* (*31*). The authors concluded that the mechanism is as follows:

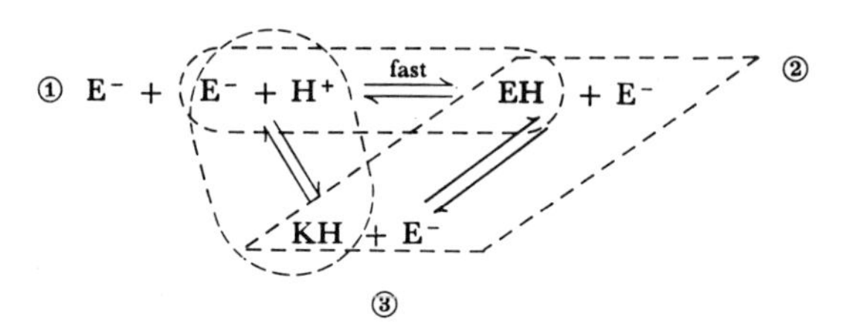

where EH is the enol, E^- the enolate ion, and KH the keto form; in the reaction ② ⇌ ③ E^- acts as a general base catalyst.

(a) Derive the long relaxation time under the assumption that equilibrium ① ⇌ ② is rapid.

(b) Derive the long relaxation time under assumption (a) and in the presence of a pH indicator.

7. Assume you were able to measure all three relaxation times as a function of $\bar{c}_A + \bar{c}_B$ for the reaction

$$A + B \rightleftharpoons C \rightleftharpoons D \rightleftharpoons E$$

Show how in case the τ values are not very well separated you could determine the six rate coefficients by plotting A_1, A_2, and A_3 (Eqs. 4.45–4.47) versus $\bar{c}_A + \bar{c}_B$.

References

1. G. W. Castellan, *Ber. Bunsenges. Phys. Chem.* **67**, 898 (1963).
2. K. Kustin, D. Shear, and K. Kleitman, *J. Theoret. Biol.* **9**, 186 (1965).
3a. G. G. Hammes and P. R. Schimmel, *J. Phys. Chem.* **70**, 2319 (1966).
3b. G. G. Hammes and P. R. Schimmel, *J. Phys. Chem.* **71**, 917 (1967).
4. M. Eigen and K. Tamm, *Z. Elektrochem.* **66**, 93, 107 (1962).
5. J. L. Haslam, *J. Phys. Chem.* **76**, 366 (1972).
6. G. G. Hammes and J. L. Haslam, *Biochemistry* **8**, 1591 (1969).
7. M. Eigen and L. DeMaeyer, *in* "Techniques of Chemistry" (G. G. Hammes, ed.), Vol. VI, part 2, p. 90. Wiley (Interscience), New York, 1973.
8. G. G. Hammes and P. R. Schimmel, *in* "The Enzymes" (P. D. Boyer, ed.), Vol. 2, p. 67. Academic Press, New York, 1970.
9. M. Eigen, *Angew. Chem. Int. Ed.* **3**, 1 (1964).
10. A. F. Yapel, Jr. and R. Lumry, *Methods Biochem. Anal.* **20**, 169 (1971).
11. J. J. Auborn, P. Warrick, Jr., and E. M. Eyring, *J. Phys. Chem.* **75**, 2488 (1971).
12. J. Stuehr, *J. Amer. Chem. Soc.* **89**, 2826 (1967).
13. M. Eigen and W. Kruse, *Z. Naturforsch.* **18B**, 857 (1963).
14. M.-L. Ahrens and G. Maass, *Angew. Chem. Int. Ed.* **7**, 818 (1968).
15. C. F. Bernasconi, *J. Phys. Chem.* **75**, 3636 (1971).

16. M. Eigen and L. DeMaeyer, *in* "Technique of Organic Chemistry" (S. L. Friess, E. S. Lewis, and A. Weissberger, eds.), Vol. VIII, part 2, p. 895. Wiley (Interscience), New York, 1963.
17. K. Kustin, D. Shear, and D. Kleitman, *J. Theoret. Biol.* **9**, 186 (1965).
18. R. A. Alberty, G. Yagil, W. F. Diven, and M. Takahashi, *Acta Chem. Scand.* **17**, 34 (1963).
19. G. H. Czerlinski, "Chemical Relaxation," Dekker, New York, 1966.
20. G. G. Hammes, *Advan. Protein Chem.* **23**, 1 (1968).
21. M. Eigen, *Quart. Rev. Biophys.* **1**, 3 (1968).
22. S. Hünig, *Pure Appl. Chem.* **15**, 109 (1967).
23. H. Diebler, M. Eigen, and P. Matthies, *Z. Naturforsch.* **16B**, 629 (1961).
24. C. F. Bernasconi, R. G. Bergstrom, and W. J. Boyle, Jr., *J. Amer. Chem. Soc.* **96**, 4643 (1974).
25. C. F. Bernasconi and R. G. Bergstrom, *J. Org. Chem.* **36**, 1325 (1971).
26. C. F. Bernasconi and R. G. Bergstrom, *J. Amer. Chem. Soc.* **96**, 2397 (1974).
27. M. C. Rose and J. Stuehr, *J. Amer. Chem. Soc.* **90**, 7205 (1968).
28. P. Schuster and H. Winkler, *Tetrahedron* 2249 (1970).
29. H. Strehlow, *Z. Elektrochem.* **66**, 392 (1962).
30. S. Weiss, H. Diebler, and I. Michaeli, *J. Phys. Chem.* **75**, 267 (1971).
31. M. Eigen, G. Ilgenfritz, and W. Kruse, *Chem. Ber.* **98**, 1623 (1965).

Chapter 5 | What Is a Small Perturbation?

In Section 1.1 it was pointed out that linearization of a rate equation is possible for "small" perturbations. In most treatments of the subject a small perturbation is defined by assuming that $|\Delta c_j| \ll \bar{c}_j$ for all species partaking in the reaction; $|\Delta c_j| \leq 0.05\bar{c}_j$ is usually considered "safe." We shall see in this chapter that this condition is often too restrictive and that, depending on the specific situation, linearization of the rate equation is still permissible even for relatively large perturbations. On the other hand, in systems such as $n\text{A} \rightleftharpoons \text{A}_n$ where n is a large number, $|\Delta c_j| \leq 0.05\bar{c}_j$ may be too large for linearization to be permissible.

We first discuss the $\text{A} + \text{B} \rightleftharpoons \text{C}$ system and later extend our conclusions to some other common reaction schemes.

5.1 Conditions for Linearization in the A + B ⇌ C System

5.1.1 Error Introduced by the Square Term

From

$$\frac{d\,\Delta c_{\text{A}}}{dt} = -\frac{1}{\tau}\Delta c_{\text{A}} - k_1(\Delta c_{\text{A}})^2 \tag{5.1}$$

which is essentially the same as Eq. 1.16 with τ^{-1} given by Eq. 1.19 it is clear that the condition for linearization is that

$$k_1(\Delta c_{\text{A}})^2 \ll (1/\tau)|\Delta c_{\text{A}}|$$

For "small" perturbations the $k_1(\Delta c_{\text{A}})^2$ term is in fact usually negligible.

It is the size of the error we are willing to tolerate in the evaluation of τ^{-1} which determines what we want to call "negligibly small." The problem can be treated on different levels of sophistication; the main difficulty in a rigorous treatment lies in the fact that Δc_{A} is not constant but a function of time. That is, in the early parts of the relaxation curve the term $k_1(\Delta c_{\text{A}})^2$ can be relatively large but will eventually become negligible regardless of the size of the initial perturbation, $\Delta c_{\text{A}}{}^0$. Consequently, the error will greatly depend on what part of the relaxation curve and on how wide a range of this curve is used for the determination of τ^{-1}. For example, if $k_1(\Delta c_A{}^0)^2$ is quite large, the error would be substantial if the portion between, say, $t = 0$ and $t = 2\tau$ were used for evaluation; it would become much smaller if the portion between $t = \tau$ and $t = 3\tau$ were used instead.

The crudest approach is to equate the expression

$$\frac{k_1(\Delta c_A{}^0)^2}{\tau^{-1}\,\Delta c_A{}^0} = k_1\tau\,\Delta c_A{}^0 \tag{5.2}$$

with the relative error. Since the gradual decrease of $k_1\tau\,\Delta c_A$ along the relaxation curve is not taken into account, this crude approach greatly overestimates the error.

A more realistic approach is based on the following. Integration of Eq. 5.1 yields

$$\frac{\Delta c_A}{1 + k_1\tau\,\Delta c_A} = \frac{\Delta c_A{}^0}{1 + k_1\tau\,\Delta c_A{}^0}\exp\left(\frac{-t}{\tau}\right) \tag{5.3}$$

After rearranging we obtain

$$\Delta c_A = \Delta c_A{}^0\,\frac{\exp(-t/\tau)}{1 + q\,\Delta c_A{}^0[1 - \exp(-t/\tau)]} \tag{5.4}$$

with

$$q = k_1\tau \tag{5.5}$$

Note that for Δc_C we obtain

$$\Delta c_C = \Delta c_C{}^0\,\frac{\exp(-t/\tau)}{1 - q\,\Delta c_C{}^0[1 - \exp(-t/\tau)]} \tag{5.6}$$

which differs from Eq. 5.4 by the minus sign in the denominator.

In a typical relaxation experiment τ^{-1} is evaluated from a semilog plot of the relaxation curve which is assumed to be exponential (Eq. 1.33). For not too large values of $q\,\Delta c_A{}^0$ ($q\,\Delta c_C{}^0$) the function of Eq. 5.4 (5.6) is in fact difficult to distinguish from a pure exponential, particularly when there is some experimental scatter. That is, a plot of log Δc_A (or the logarithm of a physical property proportional to Δc_A) versus t is almost linear and may induce the experimenter to believe he deals with a genuine first-order process. This is best illustrated by some numerical examples. Figure 5.1 shows plots of $\log(\Delta c_A/\Delta c_A{}^0)$ generated according to Eq. 5.4 for $q\,\Delta c_A{}^0 = \pm 0.1$ and 0 (pure exponential), respectively. It is apparent that for $q\,\Delta c_A{}^0 = \pm 0.1$ the plots are practically linear after $t = 0.5\tau$.

Let us estimate the error introduced into the value of $1/\tau$ when $q\,\Delta c_A{}^0 = \pm 0.1$. In a typical experimental situation one would draw the best straight line through the points of the semilog plot made from the observed relaxation curve in the range $t = 0$ to $t = 2\tau$ and would equate the slope with $-1/2.303\tau$. The error that would ensue from this procedure is about $+4.2\%$ for $q\,\Delta c_A{}^0 = +0.1$, and about -4.5% for $q\,\Delta c_A{}^0 = -0.1$. If one were to draw the best line in the range

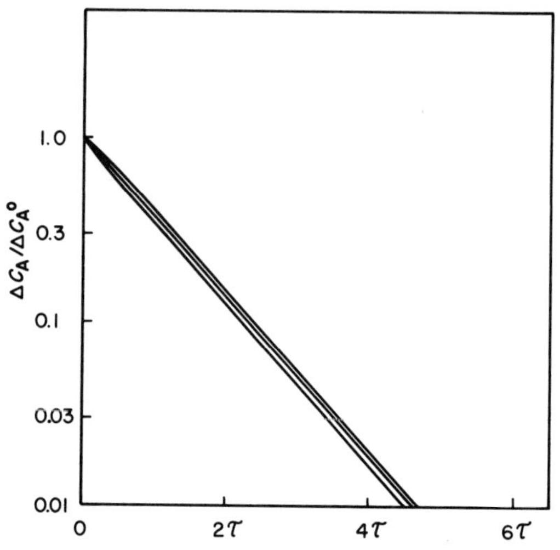

FIGURE 5.1. Plot of $\log(\Delta c_A/\Delta c_A{}^0)$ versus time according to Eq. 5.4 with $q\,\Delta c_A{}^0 = \pm 0.1$ and 0. Upper curve: $[\exp(-t/\tau)]/\{1 - 0.1[1 - \exp(-t/\tau)]\}$; middle curve: $\exp(-t/\tau)$; lower curve: $[\exp(-t/\tau)]/\{1 + 0.1[1 - \exp(-t/\tau)]\}$.

between 0.5τ and 2.5τ instead, the error drops to +2.4% for $q\,\Delta c_A{}^0 = +0.1$, and −2.8% for $q\,\Delta c_A{}^0 = -0.1$.

Note that equating $q\,\Delta c_A{}^0$ (simplest approach) with the error in $1/\tau$ would estimate it to be ±10% for $q\,\Delta c_A{}^0 = \pm 0.1$, which is a two- to fourfold overestimate.

Table 5.1 summarizes the result of similar error estimates as a function of $q\,\Delta c_A{}^0$ and of the portion of the relaxation curve used for the evaluation.

Table 5.1. Approximate Percent Errors in $1/\tau$ as Function of $q\,\Delta c_A{}^0$ and Time Range[a]

$q\,\Delta c_A{}^0$	0–2τ	0.5τ–2.5τ	τ–3τ	1.5τ–3.5τ	2τ–4τ	2.5τ–4.5τ	3τ–5τ
+0.1	+4.2	+2.4	+1.4				
−0.1	−4.5	−2.8	−1.7				
+0.2	+8.0	+4.6	+2.8				
−0.2	−9.5	−6.1	−3.8				
+0.3	+11.5	+6.5	+3.8	+2.3			
−0.3	−15.0	−9.8	−6.3	−4.1			
+0.4		+8.3	+4.8	+2.7	+1.9		
−0.4		−14.3	−9.3	−5.9	−3.6		
+0.5			+5.6	+3.4	+2.3	+1.4	
−0.5			−13.3	−8.6	−6.3	−3.5	
+1.0			+8.9	+4.0	+3.0	+1.9	
+2.0			+12.4	+7.0	+4.1	+2.6	+1.4
+3.0			+14.2	+8.0	+4.7	+2.8	+1.7
+8.0			+17.5	+9.6	+5.6	+3.6	+2.0
$+10^2$			+20.1	+10.9	+6.2	+3.7	+2.2
$+10^3$			+20.3	+11.1	+6.3	+3.8	+2.2
$+10^4$			+20.4	+11.1	+6.3	+3.8	+2.2

[a] When equating the slope of the "best straight line" of a plot of $\log \Delta c_A$ versus time.

Needless to say, since these are based on perfect "data" the true error will usually be somewhat larger due to imperfect experimental data, particularly if the late portions of the relaxation curve are used where Δc_A (or any measured physical property related to it) is becoming very small and difficult to determine with good precision.

Several features of this table call for comment.

1. When $\Delta c_A{}^0$ and thus $q\ \Delta c_A{}^0$ is positive, τ^{-1} is always overestimated; when $\Delta c_A{}^0$ is negative, τ^{-1} is always underestimated. This makes qualitative sense since $\Delta c_A{}^0 = \bar{c}_A{}^i - \bar{c}_A{}^f > 0$ means that there is an excess of A (and of B as well since $\Delta c_A{}^0 = \Delta c_B{}^0 = \bar{c}_B{}^i - c_B{}^f$) at the beginning and relaxation induces the equilibrium position to be shifted toward C. The initially too high reactant (A and B) concentrations lead to a too high rate of the bimolecular process [by $k_1(\Delta c_A)^2$] which speeds up the relaxation process. When $\Delta c_A{}^0 = \bar{c}_A{}^i - \bar{c}_A{}^f < 0$ there is a shortage of A and B and relaxation pushes the equilibrium toward A and B. Here the contribution of the bimolecular step to τ^{-1} is too small by $k_1(\Delta c_A)^2$.

2. For a given time range a positive $\Delta c_A{}^0$ leads to a smaller error than an equally large negative $\Delta c_A{}^0$. This is reasonable since we just stated that for $\Delta c_A{}^0 > 0$ relaxation is speeded up but slowed down for $\Delta c_A{}^0 < 0$. This means that the $k_1(\Delta c_A)^2$ term becomes negligibly small earlier when $\Delta c_A{}^0 > 0$ but later when $\Delta c_A{}^0 < 0$.

3. The most negative value for $q\ \Delta c_A{}^0$ in the table is -0.5; this is the theoretical limit and corresponds to a situation where the reaction is started with species C alone (e.g., by dissolving C in the solvent), i.e., $\bar{c}_A{}^i = \bar{c}_B{}^i = 0$. In this case $\Delta c_A{}^0 = -\bar{c}_A = -\bar{c}_B$. Thus

$$q\ \Delta c_A{}^0 = -k_1 \tau \bar{c}_A = -\frac{k_1 \bar{c}_A}{2k_1 \bar{c}_A + k_{-1}}$$

and $q\ \Delta c_A{}^0$ reaches its minimum of -0.5 when $2k_1\bar{c}_A \gg k_{-1}$.

No such limit on $q\ \Delta c_A{}^0$ exists for positive $\Delta c_A{}^0$ and in fact very high values are conceivable. For example, if the reaction is started by mixing A and B and the equilibrium favors C very strongly so that A is practically completely converted to C, we have $\Delta c_A{}^0 \approx c_A{}^0$, but $\bar{c}_A \ll c_A{}^0$, and thus $q\ \Delta c_A{}^0 \approx 0.5 c_A{}^0/\bar{c}_A$ (assuming $\bar{c}_A = \bar{c}_B$, and $k_{-1} \ll 2k\bar{c}_A$).

4. Even for very large positive perturbations, the function 5.4 becomes approximately exponential after a relatively short period of time and the errors approach a limiting value for any given segment of the relaxation curve. This again is a consequence of the rapid drop of the $k_1(\Delta c_A)^2$ term because of the greatly accelerated bimolecular step. Note, however, that since Δc_A drops so fast in the initial phases, by the time the $k_1(\Delta c_A)^2$ term has become negligible Δc_A may at the same time become undetectably small. Whether this occurs depends mainly on the relative magnitude of the $k_1(\bar{c}_A + \bar{c}_B)$ and the k_{-1} terms, respectively, a subject discussed in the next section.

5.1.2 Permissible Perturbation as a Function of the Equilibrium Position

In the preceding section we have seen that if we want to evaluate the relaxation curve in the time range $t = 0$ to $t = 2\tau$ and are willing to tolerate an error in τ^{-1} no greater than 4–5%, the following relation must be fulfilled:

$$|q\,\Delta c_A{}^0| = |k_1\tau\,\Delta c_A{}^0| \leq 0.1 \tag{5.7}$$

To what amount of equilibrium perturbation as measured by $\Delta c_A{}^0$ or $\Delta c_A{}^0/\bar{c}_A$ does this correspond? There is no simple answer to this question since it will depend on which side of the equilibrium is favored in the final equilibrium state.

Equation 5.7 can be rewritten as

$$|\Delta c_A{}^0| \leq 0.1(\bar{c}_A + \bar{c}_B + K_1^{-1}) \tag{5.8}$$

with $K_1 = k_1/k_{-1}$. For simplicity let us assume $\bar{c}_A = \bar{c}_B$. If $\bar{c}_A + \bar{c}_B \gg K_1^{-1}$, which means that the equilibrium position favors C over A and B, Eq. 5.8 becomes

$$|\Delta c_A{}^0| \leq 0.2\bar{c}_A \tag{5.9}$$

That is, $\Delta c_A{}^0$ is permitted to be as large as 20% of $\bar{c}_A$. Note that this is the "worst" case since the relaxation of the system is entirely dominated by the bimolecular step $(1/\tau \approx 2k_1\bar{c}_A)$, which is the one responsible for the $q\,\Delta c_A{}^0$ term.

At the other extreme, when $\bar{c}_A + \bar{c}_B \ll K_1^{-1}$, the equilibrium position favors the reactants, relaxation is governed mainly by the unimolecular step $(1/\tau \approx k_{-1})$, and Eq. 5.8 simplifies to

$$|\Delta c_A{}^0| \leq 0.1K_1^{-1} \tag{5.10}$$

Here $|\Delta c_A{}^0|$ may be very large (e.g., if $K_1 = 10^{-2}$, $|\Delta c_A{}^0| \leq 10\,M$), in fact many-fold larger than the equilibrium concentration of any of the species partaking in the system, just as in a purely first-order (irreversible) reaction.

5.2 Other Systems

In systems such as

$$n\text{A} \underset{k_{-1}}{\overset{k_1}{\rightleftharpoons}} \text{B} \tag{5.11}$$

where the rate law equation

$$-\frac{1}{n}\frac{d\,\Delta c_A}{dt} = k_1(\bar{c}_A + \Delta c_A)^n - k_{-1}(\bar{c}_B + \Delta c_B) \tag{5.12}$$

precludes pseudo-first-order conditions the considerations just discussed are of even more practical import. System 5.11 also allows us to illustrate the effect of a reaction with a molecularity $n > 2$ on the linearization requirements which become stricter the higher n is.

Substituting $\Delta c_B = -\Delta c_A/n$ and expanding Eq. 5.12 affords

$$\frac{d\,\Delta c_A}{dt} = -\frac{1}{\tau}\,\Delta c_A - \frac{n^2(n-1)}{2}\,k_1(\bar{c}_A)^{n-2}(\Delta c_A)^2 + \cdots \tag{5.13}$$

with

$$\tau^{-1} = n^2 k_1(\bar{c}_A)^{n-1} + k_{-1} \tag{5.14}$$

If we define

$$q = \frac{n^2(n-1)}{2}\,k_1(\bar{c}_A)^{n-2}\tau \tag{5.15}$$

we can apply Eq. 5.4 and in principle use Table 5.1 for estimating the error in $1/\tau$ to be expected for a given $q\,\Delta c_A{}^0$. Note, however, that since the higher order terms in Eq. 5.13 can only be neglected for $\Delta c_A{}^0 < c_A{}^0$—the next higher term in Eq. 5.13 reads $[n^2(n-1)(n-2)/3!]k_1(\bar{c}_A)^{n-3}(\Delta c_A)^3$—or even $\Delta c_A{}^0 \ll c_A{}^0$ for large n there are obvious limitations to the range of validity of Eq. 5.4 and thus to the use of Table 5.1.

It is now interesting to consider how great a perturbation is allowed in order to still assure

$$|q\,\Delta c_A{}^0| \leq 0.1 \tag{5.16}$$

In the unfavorable case where the forward rate dominates the relaxation $[1/\tau \approx n^2 k_1(\bar{c}_A)^{n-1}]$ this relation becomes

$$|\Delta c_A{}^0| \leq \frac{0.2}{n-1}\,\bar{c}_A \tag{5.17}$$

which shows that when n is very large only a tiny perturbation is permitted. Failure to appreciate this fact can lead to very gross errors. An example of a reaction with a rate law which is formally the same as Eq. 5.12 is the mechanism of micelle formation proposed by Kresheck *et al.* (*1*); here $n = 87$. Thus if the forward step were to dominate, a perturbation of only $|\Delta\bar{c}_A{}^0|/\bar{c}_A \leq 0.0023$, i.e., $\leq 0.23\%$ change in $\bar{c}_A$, would be permissible. For more details see Chapter 10.

Another interesting situation arises in the common reaction scheme

$$A + B \underset{k_{-1}}{\overset{k_1}{\rightleftharpoons}} 2C \tag{5.18}$$

The rate equation is given by

$$\frac{d\,\Delta c_A}{dt} = -\frac{1}{\tau}\,\Delta c_A - (k_1 - 4k_{-1})(\Delta c_A)^2 \tag{5.19}$$

with

$$1/\tau = k_1(\bar{c}_A + \bar{c}_B) + 4k_{-1}\bar{c}_C \tag{5.20}$$

Here we have

$$q = (k_1 - 4k_{-1})\tau \tag{5.21}$$

It is noteworthy that the term $q\,\Delta c_A{}^0$ becomes small not only for a small $\Delta c_A{}^0$ but for any $\Delta c_A{}^0$ if $k_1 \approx 4k_{-1}$. This is typical for all systems where the reactions are bimolecular in both directions. For example, for the system

$$\mathrm{A + B} \underset{k_{-1}}{\overset{k_1}{\rightleftharpoons}} \mathrm{C + D} \tag{5.22}$$

$q = 0$ when $k_1 = k_{-1}$.

Thus if in reaction 5.18 the value of the equilibrium constant is not too far from 4 (or not too far from 1 for Eq. 5.22), q will be relatively small, thereby allowing a relatively large $\Delta c_A{}^0$ without introducing a large error.

A situation of special interest prevails when the equilibrium position can be shifted back and forth because one of the species is part of a rapid acid–base equilibrium, as in the scheme

$$\begin{array}{c} \mathrm{A + B} \underset{k_{-1}}{\overset{k_1}{\rightleftharpoons}} \mathrm{2C} \\ K_{BH} \Big\Updownarrow \mathrm{H^+} \\ \mathrm{BH^+} \end{array} \tag{5.23}$$

In our research we encountered several such cases (*2*) where relaxation of the system was induced by a pH jump in the stopped-flow apparatus. The question here is how large a pH jump can be applied so as to still be able to linearize the rate equation. Let us assume we are willing to tolerate $|q\,\Delta c_A{}^0| \leq 0.2$, which according to Table 5.1 would introduce a maximum error of less than 10% in τ^{-1}.

The rate equation for scheme 5.23, under the assumption of pH buffering and assuming that $\bar{c}_A = \bar{c}_B + \bar{c}_{BH}$, is given by

$$\frac{d\,\Delta c_A}{dt} = -\frac{1}{\tau}\,\Delta c_A - \left(\frac{K_{BH}k_1}{K_{BH} + \bar{c}_H} - 4k_{-1}\right)(\Delta c_A)^2 \tag{5.24}$$

with

$$\frac{1}{\tau} = \frac{2K_{BH}k_1}{K_{BH} + \bar{c}_H}\,\bar{c}_A + 4k_{-1}\bar{c}_C \tag{5.25}$$

K_{BH} is defined as the acid dissociation constant of BH. We can again define a q as

$$q = \left(\frac{K_{BH}k_1}{K_{BH} + \bar{c}_H} - 4k_{-1}\right)\tau \tag{5.26}$$

which will allow us to calculate $q\,\Delta c_A{}^0$ for a variety of possible pH jumps.

In a typical pH jump experiment one starts with a solution at $\overline{\mathrm{pH}}^{\mathrm{i}}$, containing the various species at their initial concentrations $\bar{c}_A{}^{\mathrm{i}}$, $\bar{c}_B{}^{\mathrm{i}}$, $\bar{c}_{BH}{}^{\mathrm{i}}$, $\bar{c}_C{}^{\mathrm{i}}$. The pH jump is produced by rapidly adding some excess acid or base which leads to the $\overline{\mathrm{pH}}^{\mathrm{f}}$ corresponding to final equilibrium. Chemical relaxation then sets in to finally produce the equilibrium concentrations $\bar{c}_A{}^{\mathrm{f}}$, $\bar{c}_B{}^{\mathrm{f}}$, $\bar{c}_{BH}^{\mathrm{f}}$, and $\bar{c}_C{}^{\mathrm{f}}$. $\Delta c_A{}^0$ is calculated as

$$\Delta c_A{}^0 = \tfrac{1}{2}\bar{c}_A{}^{\mathrm{i}} - \bar{c}_A{}^{\mathrm{f}} \tag{5.27}$$

The factor $\frac{1}{2}$ comes in since in a typical stopped-flow experiment the excess acid or base is added to an equal volume of the solution at $p\bar{H}^{i}$, which leads to a two-fold dilution.

The variation of $q\ \Delta c_A{}^0$ as a function of the pH jump is best illustrated with a specific example, as given in Table 5.2 for the case where $K_1 = k_1/k_{-1} = 10^4$,

Table 5.2. qc_0, $\Delta c_A{}^0/c_0$, and $q\ \Delta c_A{}^0$ as Functions of pH Jump[a]

$p\bar{H}^{i}$	$p\bar{H}^{f}$	qc_0	$\Delta c_A{}^0/c_0$	$q\ \Delta c_A{}^0$
9	8	35.3	−0.00348	−0.123
7	8	35.3	+0.0173	−0.612
8	7	15.0	−0.0173	−0.260
6	7	15.0	+0.526	+0.789
7	6	4.77	−0.0526	−0.250
5	6	4.77	+0.110	+0.523
6	5	0.95	−0.110	−0.104
4	5	0.95	+0.140	+0.133
5.8	4.8	0.462	−0.121	−0.056
3.8	4.8	0.462	+0.136	+0.063
Any	4.602	0	Any	0
5.4	4.4	−0.469	−0.136	+0.064
3.4	4.4	−0.469	+0.121	−0.057
5.2	4.2	−0.957	−0.140	+0.133
3.2	4.2	−0.957	+0.110	−0.105
5	4	−1.50	−0.140	+0.210
3	4	−1.50	+0.0983	−0.147
4	3	−6.17	−0.0983	+0.607
2	3	−6.17	+0.0445	−0.275
3	2	−19.9	−0.0445	+0.888
1	2	−19.9	+0.0160	−0.319

[a] With $K_1 = 10^4$, $K_{BH} = 10^{-8}$.

$K_{BH} = 10^{-8}$. It is also instructive to look at q and $\Delta c_A{}^0$ separately and see how they vary with the pH jump so as to better appreciate whether for a given pH jump $q\ \Delta c_A{}^0$ is small because of a small $\Delta c_A{}^0$ or because of a small q. Hence these quantities, normalized for a given total substrate concentration $c_0 = \bar{c}_A + \bar{c}_B + \bar{c}_{BH} + \bar{c}_C$ (i.e., q replaced by qc_0, $\Delta c_A{}^0$ replaced by $\Delta c_A{}^0/c_0$), are also included in the table.

We note the following features.

1. At very high $p\bar{H}^{f}$ ($\bar{c}_H \ll K_{BH}$) $q\ \Delta c_A{}^0$ is quite small because $\Delta c_A{}^0$ is small. This is understandable since at high pH the acid–base equilibrium strongly favors B over BH^+, essentially converting system 5.23 into 5.18 and thus making it virtually pH independent. Hence experiments in this region are not indicated, not because of a prohibitively high $q\ \Delta c_A{}^0$ but because $\Delta c_A{}^0/c_0$ is too small to measure.

2. At $\overline{\mathrm{pH}}^{\mathrm{f}} = 4.602$, $qc_0 = 0$ and thus $q\,\Delta c_{\mathrm{A}}{}^0 = 0$ for a pH jump of any size, assuring a perfectly exponential relaxation curve. pH jumps (ΔpH) of one unit toward $\overline{\mathrm{pH}}^{\mathrm{f}} = 4.6 \pm 0.5$ are all associated with rather small qc_0 values and thus lead to quite acceptable $q\,\Delta c_{\mathrm{A}}{}^0$. This is therefore the optimal range for doing experiments. Jumps toward $\overline{\mathrm{pH}}^{\mathrm{f}} < 4$, though still mostly associated with easily measurable concentration changes down to $\overline{\mathrm{pH}}^{\mathrm{f}} \geq 2$, all have $|q\,\Delta c_{\mathrm{A}}{}^0| > 0.2$.

3. It makes a big difference whether one jumps "up" or "down" toward a given $\overline{\mathrm{pH}}^{\mathrm{f}}$ because $\Delta c_{\mathrm{A}}{}^0/c_0$ depends on both $\overline{\mathrm{pH}}^{\mathrm{i}}$ and $\overline{\mathrm{pH}}^{\mathrm{f}}$. For $\overline{\mathrm{pH}}^{\mathrm{f}} > 4.602$ jumping down is accompanied by a smaller $|\Delta c_{\mathrm{A}}{}^0|/c_0$ and thus a smaller $|q\,\Delta c_{\mathrm{A}}{}^0|$ than jumping up. Thus the linearization condition is better fulfilled in the former case but of course at the expense of a more difficult detectability. For $\overline{\mathrm{pH}}^{\mathrm{f}} < 4.602$ the opposite is true.

Problems

1. Chemical relaxation in aqueous amylamine solutions has been interpreted as being partly due to self-association of the amine molecules to form pentamers

$$5\mathrm{A} \underset{k_{-1}}{\overset{k_1}{\rightleftharpoons}} \mathrm{A}_5$$

with $k_1 = 1.2 \times 10^7\ M^{-4}\ \mathrm{sec}^{-1}$ and $k_{-1} = 5 \times 10^7\ \mathrm{sec}^{-1}$ at 25° (*3*). What is the largest permissible perturbation $(\Delta c_{\mathrm{A}}{}^0/\bar{c}_{\mathrm{A}})$ if you are willing to tolerate an error of 2% in the relaxation time for (a) $[\mathrm{A}]_0 = 0.01\ M$, (b) $[\mathrm{A}]_0 = 3\ M$?

2. The dimerization of the tetrasodium salt of Co(II)-4,4′,4″,4‴-tetrasulfophthalocyanine (M = monomer, D = dimer)

$$2\mathrm{M} \underset{k_{-1}}{\overset{k_1}{\rightleftharpoons}} \mathrm{D}$$

has been investigated by the solvent-jump (dilution by adding solvent) technique in the stopped-flow apparatus (*4*). The spectrophotometrically determined equilibrium constant $K_1 = k_1/k_{-1} = 2.05 \times 10^{-5}\ M^{-1}$. Suggest experimental conditions (such as total substrate concentration and size of solvent jumps) that would be appropriate for the determination of both k_1 and k_{-1}.

References

1. G. C. Kresheck, E. Hamori, G. Davenport, and H. A. Scheraga, *J. Amer. Chem. Soc.* **88**, 246 (1966).
2. C. F. Bernasconi, R. G. Bergstrom, and W. J. Boyle, Jr., *J. Amer. Chem. Soc.* **96**, 4643 (1974).
3. S. Nishikawa, T. Yasunaga, and K. Takahashi, *Bull. Chem. Soc. Japan* **46**, 2992 (1973).
4. Z. A. Schelly, R. D. Farina, and E. M. Eyring, *J. Phys. Chem.* **74**, 617 (1970).

Chapter 6 | Relaxation Amplitudes in Single-Step Systems

6.1 The Extent of Equilibrium Displacement

The techniques most commonly referred to as "relaxation techniques" are based on the perturbation of an already existing equilibrium state by means of an external "forcing parameter." Thc commonly used forcing parameters are changes of temperature, pressure, or electric field strength. They are either applied once (jump or transient techniques) or in an oscillatory fashion (stationary techniques).

In the temperature-jump method the temperature dependence of the equilibrium constant according to the van't Hoff equation

$$\frac{\partial \ln K}{\partial T} = \frac{\Delta H}{RT^2} \tag{6.1}$$

is exploited; ΔH is the molar enthalpy change of the reaction and R is the gas constant.

Rearrangement of Eq. 6.1 affords

$$\partial \ln K = \frac{\partial K}{K} = \frac{\Delta H}{RT^2}\,\partial T \tag{6.2}$$

or, for finite but small changes in K ($\Delta K \ll K$), we can write

$$\frac{\Delta K}{K} = \frac{\Delta H}{RT^2}\,\Delta T \tag{6.3}$$

It is evident that the relative change in K for a given temperature jump ΔT is proportional to ΔH.

Similarly, if the external forcing parameter is a change in pressure or a change in the electric field intensity, corresponding thermodynamic relationships exist to describe the extent by which the equilibrium constant changes for a given change in the forcing parameter (e.g., $\partial \ln K/\partial p = -\Delta V/RT$), as is discussed in Chapters 12 and 13.

Since the relations to be derived in the following pertain equally to any type of forcing parameter, we will restrict ourselves to the most popular temperature-jump method for illustration. We shall mention a few applications which illustrate how the information obtained from temperature-jump relaxation amplitudes has been used. Additional examples, relating also to other relaxation methods, are presented in the discussions of the applications of the various experimental techniques.

It is important to appreciate that even a large ΔH is no guarantee that a temperature jump will displace the equilibrium strongly enough to produce a measurable relaxation effect. The magnitude of the change in the concentrations very much depends on the position of the equilibrium which is a function of the equilibrium constant, and often also of the total concentration of the reactants. This is in contrast to relaxation experiments based on concentration jumps (i.e., initiated by mixing) where much larger equilibrium displacements can usually be achieved if necessary.

Related to this is the question of which species participating in the equilibrium is the most suitable for monitoring the relaxation, obviously a question of great practical importance. Let us illustrate these problems with two examples.

6.1.1 The A $\rightleftharpoons$ B System

Though this is the simplest possible system most of the conclusions reached here are qualitatively applicable to any other single-step equilibrium.

Let us write the mass law referring to the equilibrium before the jump,

$$\bar{c}_B{}^i = K^i \bar{c}_A{}^i \tag{6.4}$$

and after relaxation is complete

$$\bar{c}_B{}^f = K^f \bar{c}_A{}^f \tag{6.5}$$

Expressing $\bar{c}_A{}^i = \bar{c}_A{}^f + \Delta c_A{}^0$, $\bar{c}_B{}^i = \bar{c}_B{}^f + \Delta c_B{}^0$, and $K^i = K^f + \Delta K$ allows us to write Eq. 6.4 as

$$\bar{c}_B{}^f + \Delta c_B{}^0 = \bar{c}_B{}^f - \Delta c_A{}^0 = (K^f + \Delta K)(\bar{c}_A{}^f + \Delta c_A{}^0) \tag{6.6}$$

Expanding leads to

$$\Delta c_B{}^0 = -\Delta c_A{}^0 = K^f\,\Delta c_A{}^0 + \Delta K \bar{c}_A{}^f + \Delta K\,\Delta c_A{}^0 \tag{6.7}$$

which for small perturbations can be linearized ($\Delta K\,\Delta c_A{}^0$ negligible) and after rearrangement affords

$$-\Delta c_A{}^0 = \frac{\Delta K}{1+K}\bar{c}_A = \frac{K}{1+K}\bar{c}_A\frac{\Delta K}{K} \tag{6.8}$$

Note that for simplicity we henceforth again omit the superscripts i and f. Note also that Eq. 6.8 could have been obtained more directly by differentiation of Eq. 6.5, i.e.,

$$\Delta c_B{}^0 = \left(\frac{\partial \bar{c}_B{}^f}{\partial \bar{c}_A{}^f}\right)_{K^f} \Delta c_A{}^0 + \left(\frac{\partial \bar{c}_B{}^f}{\partial K^f}\right)_{\bar{c}_A{}^f} \Delta K = K^f \, \Delta c_A{}^0 + \Delta K \bar{c}_A{}^f$$

Equation 6.8 is of the general form

$$-\Delta c_A{}^0 = \Gamma \, \Delta K/K \tag{6.9}$$

It can easily be shown (and will be demonstrated with another example) that one always obtains an equation of this form regardless of the type of equilibrium considered; the function Γ is characteristic of the type of equilibrium and includes a concentration term. In the present case we have

$$\Gamma = \frac{K}{1 + K} \bar{c}_A \tag{6.10}$$

If $\bar{c}_A$ is expressed in terms of the total concentration, $c_0 = \bar{c}_A + \bar{c}_B$, we have

$$\bar{c}_A = c_0/(1 + K) \tag{6.11}$$

and thus

$$\Gamma = \frac{K}{(1 + K)^2} c_0 \tag{6.12}$$

In explicit terms $\Delta c_A{}^0$ then becomes, after combining Eqs. 6.3, 6.9, and 6.12,

$$-\Delta c_A{}^0 = \frac{K}{(1 + K)^2} c_0 \frac{\Delta H}{RT^2} \Delta T \tag{6.13}$$

(Note that in order to be consistent with the definition of $\Delta K = K^i - K^f$, the definition of ΔT in Eq. 6.13 and in subsequent equations must be $\Delta T = T_i - T_f$, and not $\Delta T = T_f - T_i$.) Thus, apart from being proportional to c_0, ΔH, and ΔT, the displacement $\Delta c_A{}^0$ depends greatly on the magnitude of K. Γ is small for $K \ll 1$ and $K \gg 1$, but goes through a maximum, $\Gamma = 0.25 c_0$, when $K = 1$ (from $d\Gamma/dK = 0$).

An alternative way of expressing Γ is in terms of $\alpha = \bar{c}_A/c_0$, i.e., the fraction of c_0 that is in the form of A at equilibrium. From Eq. 6.11 it follows

$$\alpha = 1/(1 + K); \qquad K = (1 - \alpha)/\alpha \tag{6.14}$$

and thus

$$\Gamma = \alpha(1 - \alpha) c_0 \tag{6.15}$$

Figure 6.1 shows a plot of Γ/c_0 as a function of α.

It is apparent that the largest concentration changes are achieved when the equilibrium is balanced, i.e., $\alpha \approx 0.5$ or $K \approx 1$. When the equilibrium position is very much one-sided $\Delta c_A{}^0$ is greatly reduced. (When talking about the size of $\Delta c_A{}^0$, $\Delta c_B{}^0$, etc., or related quantities discussed below, we mean *absolute* values, irrespective of the sign.) For example, when $\alpha = 0.1$ or 0.9 ($K = 9$ or 0.11, respectively) $\Gamma/c_0 = 0.09$ which is not yet too bad compared to the maximum of 0.25; however, for $\alpha = 0.01$ or 0.99 ($K = 99$ or 0.01, respectively) Γ/c_0 drops to ≈ 0.01, i.e., only 4% of the maximum. This illustrates one of the limitations of

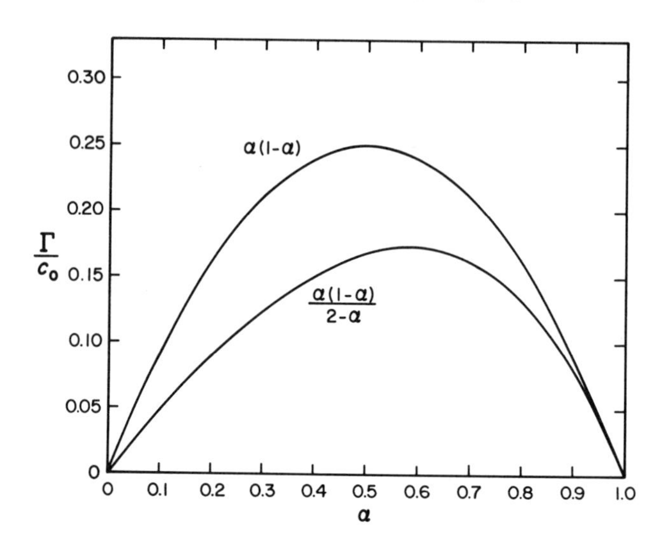

FIGURE 6.1. Γ/c_0 as a function of α for the $A \rightleftharpoons B$ and the $A + B \rightleftharpoons C$ ($\bar{c}_A = \bar{c}_B$) systems.

relaxation techniques, namely that in order to obtain appreciable equilibrium displacements, the equilibirum position should not be too greatly one-sided.

In the transient or jump techniques relaxation is usually measured by monitoring the change in the physical property related to the concentration of one species only. In such a case the *relative* change of concentration, say $\Delta c_A{}^0/\bar{c}_A$, is a more relevant criterion of whether a given experiment may prove successful. From Eq. 6.8 we obtain

$$-\frac{\Delta c_A{}^0}{\bar{c}_A} = \frac{K}{1+K}\frac{\Delta K}{K} = (1-\alpha)\frac{\Delta K}{K} \tag{6.16}$$

whereas for the relative change in c_B we have

$$\frac{\Delta c_B{}^0}{\bar{c}_B} = \frac{1}{1+K}\frac{\Delta K}{K} = \alpha\frac{\Delta K}{K} \tag{6.17}$$

From Eq. 6.16 we see that when $K \gg 1$ ($\alpha \ll 1$)

$$-\frac{\Delta c_A{}^0}{\bar{c}_A} = \frac{\Delta K}{K} \tag{6.18}$$

which is the maximum possible value, whereas for $K \ll 1$ ($\alpha \approx 1$) the relative change becomes very small; for $\Delta c_B{}^0/\bar{c}_B$ the opposite is true.

When $K = 1$ ($\alpha = 0.5$) we have

$$-\frac{\Delta c_A{}^0}{\bar{c}_A} = \frac{\Delta c_B{}^0}{\bar{c}_B} = 0.5\frac{\Delta K}{K} \tag{6.19}$$

It is evident *that the species disfavored by the equilibrium is always subject to a larger relative change than the favored species,* which is intuitively obvious and immediately suggests which species one should try to monitor in a relaxation experiment. This is essentially true for any single-step equilibrium no matter how many species are involved. Note, however, that if the equilibrium is very greatly one-sided one may have to choose a rather high c_0 in order to assure that the concentration of the minor component is measurably large. This often sets a practical limit to how one-sided an equilibrium is allowed to be.

In practice it is unfortunately not always possible to monitor the minor species. Assume the concentration changes are detected spectrophotometrically. Only if there is a wavelength where the minor component absorbs significantly without serious interference by the absorption of the major species is it possible to get a satisfactory relaxation amplitude, i.e., a large relative change in OD. The problem can be formulated as follows. The difference in the OD between the initial and final state is given by ($\Delta OD^0 = OD^i - OD^f$)

$$\Delta OD^0 = l\epsilon_A \, \Delta c_A{}^0 + l\epsilon_B \, \Delta c_B{}^0 = l(\epsilon_A - \epsilon_B) \, \Delta c_A{}^0 \tag{6.20}$$

whereas the equilibrium OD is

$$\overline{OD} = l\epsilon_A \bar{c}_A + l\epsilon_B \bar{c}_B = l(\epsilon_A + K\epsilon_B)\bar{c}_A \tag{6.21}$$

where l is the path length, and ϵ_A and ϵ_B are the molar extinction coefficients. Thus the relative change in OD is

$$\frac{\Delta OD^0}{\overline{OD}} = \frac{\epsilon_A - \epsilon_B}{\epsilon_A + K\epsilon_B} \frac{\Delta c_A{}^0}{\bar{c}_A} = -\frac{\epsilon_A - \epsilon_B}{\epsilon_A + K\epsilon_B} \frac{K}{1 + K} \frac{\Delta K}{K} \tag{6.22}$$

where $\Delta c_A{}^0/\bar{c}_A$ is taken from Eq. 6.16.

If A is the minor component ($K \gg 1$), Eq. 6.22 simplifies to

$$\frac{\Delta OD^0}{\overline{OD}} = -\frac{\epsilon_A - \epsilon_B}{\epsilon_A + K\epsilon_B} \frac{\Delta K}{K} \tag{6.23}$$

and the best wavelength is obviously where $\epsilon_B = 0$. If no such wavelength can be found, it might in some cases be the lesser evil to select a wavelength where $\epsilon_B > \epsilon_A$ or even $\epsilon_B \gg \epsilon_A$. For example, if $K = 10$ and the spectra of A and B are such that at no wavelength the ratio ϵ_A/ϵ_B is greater than 1.7, we calculate $\Delta OD^0/\overline{OD} = -0.06\Delta K/K$ at the "optimal" wavelength where $\epsilon_A/\epsilon_B = 1.7$. It turns out it would be better if one chooses a wavelength where A does not absorb at all since then $\Delta OD^0/\overline{OD}$ would still be $0.1\Delta K/K$.

In practical work one may not know K or ϵ_A and ϵ_B beforehand. The optimal wavelength is then determined empirically.

6.1.2 The A + B $\rightleftharpoons$ C System

Qualitatively we will reach the same conclusions as for the A $\rightleftharpoons$ B system. However, because the system is not symmetrical, the equations relating to the quantitative aspects will reflect this asymmetry.

The equation corresponding to 6.6 is now

$$\bar{c}_C + \Delta c_C{}^0 = (K + \Delta K)(\bar{c}_A + \Delta c_A{}^0)(\bar{c}_B + \Delta c_B{}^0) \tag{6.24}$$

which after linearization affords

$$-\Delta c_A{}^0 = -\Delta c_B{}^0 = \Delta c_C{}^0 = \Gamma(\Delta K/K) \tag{6.25}$$

with

$$\Gamma = \frac{\bar{c}_C}{1 + K(\bar{c}_A + \bar{c}_B)} = \frac{K\bar{c}_A\bar{c}_B}{1 + K(\bar{c}_A + \bar{c}_B)} \tag{6.26}$$

The two special cases where (a) $\bar{c}_B \gg \bar{c}_A$ and (b) $\bar{c}_A = \bar{c}_B$ are of particular interest and are now discussed.

(a) $\bar{c}_B \gg \bar{c}_A$. Equation 6.26 reduces to

$$\Gamma = \frac{K'}{1 + K'}\bar{c}_A \tag{6.27}$$

with $K' = K\bar{c}_B$. This is identical to Eq. 6.10 except that K is replaced by K', and thus all considerations applying to the A $\rightleftharpoons$ B system are equally valid here. The only difference is that K' is adjustable by varying $\bar{c}_B$. This can be a considerable advantage since one may be able to choose whether A or C should be the minor component, depending on the spectral characteristics of A and C. For example, let us assume C has a strong absorption in the visible region and a weak one in the near ultraviolet, whereas A absorbs strongly in the near uv but not at all in the visible region; B is assumed to absorb only in the far uv. For $K = 1000\ M^{-1}$ and $\bar{c}_B \geq 10^{-2}\ M$ the equilibrium would strongly favor C over A so that monitoring A seems indicated. But since there is so much more C than A the absorption of C in the near uv could be quite significant despite $\epsilon_A \gg \epsilon_C$, which could lead to small relaxation amplitudes. Thus choosing concentrations $\bar{c}_B \leq 10^{-3}\ M$ and monitoring the relaxation in the visible would be more satisfactory. Again, in the absence of detailed knowledge of K, ϵ_A, and ϵ_B the best wavelength and the best concentration of the reactants must be found empirically.

(b) $\bar{c}_A = \bar{c}_B$. Equation 6.26 becomes

$$\Gamma = K\bar{c}_A{}^2/(1 + 2K\bar{c}_A) \tag{6.28}$$

Introducing $\alpha = \bar{c}_A/c_0$ with $c_0 = 0.5\bar{c}_A + 0.5\bar{c}_B + \bar{c}_C = \bar{c}_A + \bar{c}_C$ allows K to be expressed as

$$K = (1 - \alpha)/\alpha^2 c_0 \tag{6.29}$$

and thus

$$\Gamma = \frac{\alpha(1 - \alpha)}{2 - \alpha}c_0 \tag{6.30}$$

The function $\Gamma/c_0 = \alpha(1 - \alpha)/(2 - \alpha)$ versus α is shown in Fig. 6.1. It is similar to $\Gamma/c_0 = \alpha(1 - \alpha)$ for the A $\rightleftharpoons$ B system but somewhat asymmetrical. From $d\Gamma/d\alpha$ we obtain the maximum $\Gamma/c_0 = 0.173$ at $\alpha = 0.586$ which corresponds to $Kc_0 = 1.21$.

For the relative changes we have

$$-\frac{\Delta c_A{}^0}{\bar{c}_A} = -\frac{\Delta c_B{}^0}{\bar{c}_B} = \frac{K\bar{c}_A}{1 + 2K\bar{c}_A}\frac{\Delta K}{K} = \frac{1-\alpha}{2-\alpha}\frac{\Delta K}{K} \tag{6.31}$$

$$\frac{\Delta c_C{}^0}{\bar{c}_C} = \frac{1}{1 + 2K\bar{c}_A}\frac{\Delta K}{K} = \frac{\alpha}{2-\alpha}\frac{\Delta K}{K} \tag{6.32}$$

Again we see that the favored species show small relative changes, the disfavored species large relative changes. Note, however, that here the maximum relative change in the concentration of C (for $\alpha \approx 1$ or $2K\bar{c}_A \ll 1$) is

$$\frac{\Delta c_C{}^0}{\bar{c}_C} = \frac{\Delta K}{K} \tag{6.33}$$

whereas for A ($\alpha \approx 0$ or $2K\bar{c}_A \gg 1$) it is only one-half that (absolute) value:

$$-\frac{\Delta c_A{}^0}{\bar{c}_A} = 0.5\frac{\Delta K}{K} \tag{6.34}$$

6.1.3 Generalization to Other Single-Step Systems

The derivation of the Γ factor as described in detail for the $A \rightleftharpoons B$ system illustrates well the physical principles involved. However, Eigen and DeMaeyer (*1*) give a useful general expression which allows the rapid derivation of Γ for any single-step equilibrium of the form

$$a\text{A} + b\text{B} + c\text{C} + \cdots \rightleftharpoons u\text{U} + v\text{V} + w\text{W} + \cdots \tag{6.35}$$

without having to go through the physical reasoning every time. It is given by

$$\Gamma = \left(\sum_j \nu_j{}^2/\bar{c}_j\right)^{-1} \tag{6.36}$$

whereas the general expression for $\Delta c_j{}^0$ is

$$\Delta c_j{}^0 = \nu_j \Gamma(\Delta K/K) \tag{6.37}$$

where the ν_j are the stoichiometric coefficients. For example, in the $A + B \rightleftharpoons C$ system formula 6.36 takes the form

$$\Gamma = \left(\frac{1}{\bar{c}_A} + \frac{1}{\bar{c}_B} + \frac{1}{\bar{c}_C}\right)^{-1} = \left(\frac{1}{\bar{c}_A} + \frac{1}{\bar{c}_B} + \frac{1}{K\bar{c}_A\bar{c}_B}\right)^{-1}$$

$$= \frac{K\bar{c}_A\bar{c}_B}{1 + K(\bar{c}_A + \bar{c}_B)}$$

which is the same as Eq. 6.26.

6.2 Determination of ΔH from Relaxation Amplitudes

The direct proportionality between the relaxation amplitudes in a temperature-jump experiment and the reaction enthalpy suggests a method for determining ΔH directly from such amplitudes. But since there are usually other ways to determine ΔH (e.g., the temperature dependence of K, or from the differences in the activation enthalpies, $\Delta H = \Delta H_1^\ddagger - \Delta H_{-1}^\ddagger$, of the two rate constants) which are often simpler and more accurate, not much use has been made of this possibility. However, a check by relaxation amplitude measurements of ΔH determined by one of the more common methods may help to support or reject an assumed mechanism, for example in a case where the experimenter believes he is dealing with a one-step process whereas in fact there are two or more steps. Let us therefore consider briefly how ΔH is determined from amplitudes.

If K is known and the relative change in concentration of species j, $\Delta c_j^0/\bar{c}_j$, can be measured directly, for example spectrophotometrically, at a wavelength where j is the only absorbing species ($\Delta \text{OD}^0/\overline{\text{OD}} = \Delta c_j^0/\bar{c}_j$), one just solves equations such as 6.16, 6.17, 6.31, or 6.32 for $\Delta K/K$, which is related to ΔH through Eq. 6.3. Note that here and in the following we are not only interested in the absolute magnitude of Δc_j^0 but in its sign as well. In fact even if one is not interested in determining the value of ΔH, the sign of the amplitude tells one whether the reaction is exothermic or endothermic.

In case one does not know K it is still possible to find $\Delta K/K$ by choosing conditions strongly favoring one side of the equilibrium and monitoring a minor component so that equations such as 6.18, 6.33, and 6.34 are valid; here $\Delta c_j^0/\bar{c}_j$ is independent of K. For example, in the system $\text{A} + \text{B} \rightleftharpoons \text{C}$ where $\bar{c}_\text{B} \gg \bar{c}_\text{A}$ one could monitor C and determine empirically the concentration of B where the relaxation amplitude reaches its maximum ($K' = K\bar{c}_\text{B} \ll 1$).

If more than one species absorbs at the wavelength in question, knowledge of the respective extinction coefficients and of K nevertheless allows equations such as 6.22 to be solved for $\Delta K/K$.

It should be pointed out that in a temperature-jump experiment there are contributions to ΔOD^0 from a variety of sources such as volume expansion effects which have nothing to do with the actual reaction. Special care must therefore be applied to assure that one measures only the relevant part of ΔOD^0. This is also the reason why the seemingly much simpler method of determining ΔOD^0 by just measuring the OD in an ordinary spectrophotometer at two different temperatures would lead to completely erroneous conclusions; in some cases the contribution to ΔOD^0 from volume expansion effects may be larger than that from chemical relaxation! Since this expansion is very rapid and thus usually more rapid than the relaxation effect under study, the two effects can easily be separated in the temperature-jump spectrophotometer (for more details see Section 11.2.1).

Applications. Examples illustrating the use of relaxation amplitudes from temperature-jump experiments to determine ΔH include the complex formation between Mg^{2+} and murexide (*2, 3*), and the complex formation between proflavin and trypsin (*4a, b*). Using a somewhat different approach (*5*), Czerlinski and Malkewitz (*6*) determined ΔH of the reaction between glutamic aspartic aminotransferase and DL-erythro-β-hydroxyaspartic acid. Chock (*7*) has concluded from a *non*linear dependence of the relaxation amplitude versus $1/T^2$ (Eq. 6.3) that the mechanism of complex formation between monovalent cations and cyclic polyethers must involve at least two steps.

6.3 Determination of *K* from Relaxation Amplitudes

Though less obvious than for ΔH, relaxation amplitudes can in fact be used to calculate the equilibrium constant in some cases.

For example, let us assume that in the $A \rightleftharpoons B$ system $\Delta OD^0/\overline{OD}$ (Eq. 6.22) can be measured at two different wavelengths. At the first $\epsilon_A \gg \epsilon_B$, $K\epsilon_B$ so that $(\Delta OD^0/\overline{OD})_1 = -\Delta K/(1 + K)$; at the second $\epsilon_A \ll \epsilon_B$, $K\epsilon_B$ so that $(\Delta OD^0/\overline{OD})_2 = \Delta K/(1 + K)K$. The ratio $(\Delta OD^0/\overline{OD})_1/(\Delta OD^0/\overline{OD})_2$ is then simply $-K$.

Another possibility is to exploit the concentration dependence of Γ in the $A + B \rightleftharpoons C$ system. In the case where $\bar{c}_B \gg \bar{c}_A$ the function Γ goes through a maximum when $K' = 1$ (Eq. 6.27) and thus $K = 1/\bar{c}_B$ at the maximum. Since $\Delta OD^0 = (\epsilon_A + \epsilon_B - \epsilon_C)l\,\Delta c_A{}^0$, and $\Delta c_A{}^0$ is proportional to Γ (Eq. 6.25) Γ goes through its maximum at the same time as ΔOD^0. Similarly, in the case where $\bar{c}_A = \bar{c}_B$ one obtains $K = 1.21/c_0$, where $c_0 = \bar{c}_A + \bar{c}_C$ at the maximum amplitude.

It should be pointed out that just as for ΔH, the determination of equilibrium constants from relaxation amplitudes is not very common since there are usually other more precise methods available for achieving the same goal. The amplitude method has perhaps its greatest potential when K in an equilibrium such as $A + B \rightleftharpoons C$ is very large (i.e., favoring C). In such situations the more common methods (titration, Benesi–Hildebrand, etc.) become increasingly more inaccurate as K gets larger, whereas the precision with which K can be obtained from amplitude measurements increases, at least up to a certain point which depends on the sensitivity of the detection system at hand.

When K is very large, the method of determining K from the concentration ($\bar{c}_B$ or c_0) at which Γ is at its maximum is not practical because the maximum is only reached at such low concentrations that the relaxation effect becomes too small to measure; e.g., if $K = 10^6$, $c_0 \sim 10^{-6}\,M$ at the maximum when $\bar{c}_A = \bar{c}_B$. Hence a different approach is needed. We discuss it for the $A + B \rightleftharpoons C$ system and follow essentially the treatment of Winkler (*2*).

We assume that the reaction solutions are prepared by mixing A and B so that the relations

$$\bar{c}_A + \bar{c}_C = [A]_0 \tag{6.38}$$

$$\bar{c}_B + \bar{c}_C = [B]_0 \tag{6.39}$$

hold where $[A]_0$ and $[B]_0$ are the analytical concentrations of A and B, respectively. The mass law can then be written as

$$K = \frac{\bar{c}_C}{([A]_0 - \bar{c}_C)([B]_0 - \bar{c}_C)} \tag{6.40}$$

Solving for $\bar{c}_C$ affords

$$\bar{c}_C = \tfrac{1}{2}([A]_0 + [B]_0 + K^{-1} - [([A]_0 + [B]_0 + K^{-1})^2 - 4[A]_0[B]_0)]^{1/2}) \tag{6.41}$$

or after rearrangement

$$[A]_0 + [B]_0 + K^{-1} - 2\bar{c}_C = [([A]_0 - [B]_0)^2 + 2K^{-1}([A]_0 + [B]_0) + K^{-2}]^{1/2} \quad (6.42)$$

After adding K^{-1} to both sides of Eq. 6.38 and then adding the thus modified Eq. 6.38 to 6.39 yields

$$\bar{c}_A + \bar{c}_B + K^{-1} = [A]_0 + [B]_0 + K^{-1} - 2\bar{c}_C \quad (6.43)$$

Next we write Eq. 6.26 in the form of

$$\Gamma = \bar{c}_C K^{-1}/(\bar{c}_A + \bar{c}_B + K^{-1}) \quad (6.44)$$

and combine Eqs. 6.42–6.44 which leads to

$$\Gamma = \frac{1}{2K}\left(\frac{[A]_0 + [B]_0 + K^{-1}}{[([A]_0 - [B]_0)^2 + 2K^{-1}([A]_0 + [B]_0) + K^{-2}]^{1/2}} - 1\right) \quad (6.45)$$

If we concentrate on cases where K is very large, i.e., $K \gg [A]_0 + [B]_0$, which is equivalent to $K^{-1} \ll [A]_0 + [B]_0$, there are two limiting situations that are of special interest.

1. $[A]_0$ and $[B]_0$ are sufficiently different from each other so that $([A]_0 - [B]_0)^2 \gg 2K^{-1}([A]_0 + [B]_0)$. In this case Eq. 6.45 simplifies to

$$\Gamma = \frac{1}{2K}\left(\frac{[A]_0 + [B]_0}{|[A]_0 - [B]_0|} - 1\right) \quad (6.46)$$

For $[A]_0 \gg [B]_0$ this leads to

$$\Gamma = \frac{1}{K}\frac{[B]_0}{[A]_0} \quad \text{or} \quad \frac{\Gamma}{[A]_0} = \frac{1}{K[A]_0}\frac{[B]_0}{[A]_0} \quad (6.47)$$

whereas for $[B]_0 \gg [A]_0$

$$\Gamma = \frac{1}{K}\frac{[A]_0}{[B]_0} \quad \text{or} \quad \frac{\Gamma}{[A]_0} = \frac{1}{K[A]_0}\frac{[A]_0}{[B]_0} \quad (6.48)$$

2. When $[A]_0 \approx [B]_0$ so that $([A]_0 - [B]_0)^2 \ll 2K^{-1}([A]_0 + [B]_0)$ Eq. 6.45 takes the form

$$\Gamma \approx \frac{1}{2K}\left(\frac{[A]_0}{(K^{-1}[A]_0)^{1/2}} - 1\right) \approx \frac{1}{2}\left(\frac{[A]_0}{K}\right)^{1/2} \quad (6.49)$$

or

$$\frac{\Gamma}{[A]_0} \approx \frac{1}{2(K[A]_0)^{1/2}} \quad (6.50)$$

Since $K^{-1} \ll [A]_0 + [B]_0$ it is apparent that Γ for $[A]_0 \approx [B]_0$ is much larger than when $[B]_0 \gg [A]_0$ or $[A]_0 \gg [B]_0$; that is, for a given $[A]_0$ the function Γ versus the ratio $[B]_0/[A]_0$ goes through a maximum at $[B]_0/[A]_0 = 1$.

Figure 6.2 shows the function $\Gamma/[A]_0$ plotted versus $[B]_0/[A]_0$ for a number of

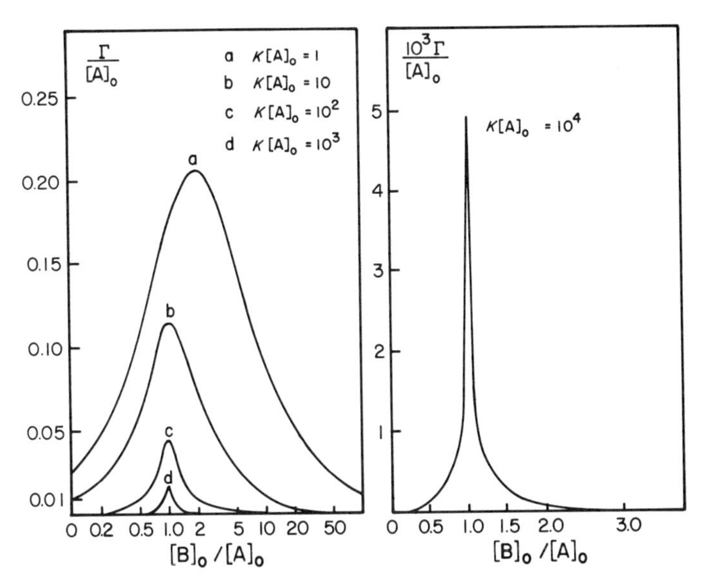

FIGURE 6.2. $\Gamma/[A]_0$ as a function of $[B]_0/[A]_0$ for the $A + B \rightleftharpoons C$ system. [From Winkler (*2*), by permission of the author.]

different $K[A]_0$. Note that although $\Gamma/[A]_0$ becomes smaller for increasing $K[A]_0$, which is in agreement with the principle that $\Delta c_j{}^0$ decreases as the equilibrium becomes more one-sided, at the same time the maximum becomes sharper. It is this characteristic shape of the $\Gamma/[A]_0$ function for high $K[A]_0$ values which allows K to be determined from the relaxation amplitude as follows.

The change in OD is given by

$$\Delta \text{OD}^0 = l(\epsilon_A + \epsilon_B - \epsilon_C)\Gamma(\Delta K/K) \tag{6.51}$$

Although the proportionality factor $l(\epsilon_A + \epsilon_B - \epsilon_C)\,\Delta K/K$ will in general not be known, the *shape* of the curve of ΔOD^0 versus $[B]_0/[A]_0$ is the same as that of Γ. Thus the K for which a theoretical plot of Γ (or $\Gamma/[A]_0$) versus $[B]_0/[A]_0$ has the same shape as the ΔOD^0 plot is chosen as the true K.

Let us illustrate the procedure with a numerical example. Assume $K = 10^6\ \text{M}^{-1}$, $\Delta K/K = 0.2$, $l\epsilon_A = 3 \times 10^4\ \text{M}^{-1}$, $l\epsilon_C = 10^4\ M^{-1}$, and thus $l(\epsilon_A - \epsilon_C) = 2 \times 10^4\ M^{-1}$, $\epsilon_B = 0$. $[A]_0$ is kept constant at $4 \times 10^{-4}\ M$, and thus $K[A]_0 = 400$, whereas $[B]_0$ is varied from $4 \times 10^{-5}\ M$ ($[B]_0/[A]_0 = 0.1$) to $4 \times 10^{-3}\ M$ ($[B]_0/[A]_0 = 10$). At the maximum ($[B]_0/[A]_0 = 1$) we calculate for the theoretical curve $\Gamma = 10^{-5}\ M^{-1}$ or $\Gamma/[A]_0 = 0.025$ from Eq. 6.49 or 6.50, respectively, whereas for $[B]_0 = 4 \times 10^{-5}$ and $4 \times 10^{-3}\ M$, $\Gamma = 10^{-7}\ M^{-1}$ or $\Gamma/[A]_0 = 2.5 \times 10^{-4}$ from Eqs. 6.47 and 6.48, respectively.

From Eq. 6.51 we find $\Delta\text{OD}^0 = 0.04$ at the maximum, $\Delta\text{OD}^0 = 4 \times 10^{-4}$ at $[B]_0 = 4 \times 10^{-5}\ M$, and $[B]_0 = 4 \times 10^{-3}\ M$. We can bring the curve of ΔOD^0 and that of $\Gamma/[A]_0$ to the same scale by multiplying each $\Gamma/[A]_0$ value by the scale factor 0.04/0.025. The two curves thus become superimposable.

In practice it is unlikely that one would immediately guess the right K on inspection of the ΔOD^0 curve. Hence one would calculate a theoretical $\Gamma/[A]_0$ curve for an estimated K value, determine the scale factor from a comparison of the maxima in the ΔOD^0 and the $\Gamma/[A]_0$ curves, and try to superimpose them. If K has been guessed too high, the $\Gamma/[A]_0$ curve near the maximum is steeper than the ΔOD^0 curve; if K has been underestimated, the $\Gamma/[A]_0$ curve is less steep.

It should be noted that except near the maximum ΔOD^0 values are typically quite small, 4×10^{-4} at the two extremes in our example. The photometric accuracy of the average spectrophotometer is around 10^{-3} OD units and would not be sufficient to measure the very small ΔOD^0. However, the best temperature-jump spectrophotometers available today allow $\Delta OD^0 = 10^{-4}$ to be measured, mainly due to the use of very high intensity lamps which increases the signal-to-noise ratio (see Section 11.2.1).

Comparison with Other Methods and Applications. As seen in Fig. 6.2 with increasing $K[A]_0$ (e.g., with increasing K for a given $[A]_0$) the function $\Gamma/[A]_0$ becomes sharper and thus more characteristic of the underlying value of K. This means that the precision with which K can be determined increases with increasing K, at least up to a certain point which depends on the photometric accuracy of the temperature-jump spectrophotometer. This is in contrast to the more common methods of evaluating K which become *less* accurate the higher K becomes.

Let us illustrate this with one of the spectrophotometric methods, using the same numerical example as before. Assuming $[B]_0 \gg [A]_0$ throughout we can write

$$\frac{\overline{OD} - OD_A}{OD_C - \overline{OD}} = K[B]_0 \tag{6.52}$$

with $OD_A = l\epsilon_A[A]_0$, $OD_C = l\epsilon_C[A]_0$. If ϵ_A and ϵ_C are known, plotting the left-hand side of Eq. 6.52 versus $[B]_0$ yields a straight line with a slope of K. Since $l\epsilon_A = 3 \times 10^4$ and $l\epsilon_B = 10^4$ we have $OD_A = 3 \times 10^4[A]_0$ and $OD_C = 10^4[A]_0$, respectively. $\overline{OD}$ is calculated as

$$\overline{OD} = l\epsilon_A\bar{c}_A + l\epsilon_C\bar{c}_C \tag{6.53}$$

and varies with $[B]_0$ [$\bar{c}_A = [A]_0/(1 + K[B]_0)$, $\bar{c}_C = K[A]_0[B]_0/(1 + K[B]_0)$]. For $[B]_0 = 5 \times 10^{-5}\,M$ we obtain $\overline{OD} = 1.04 \times 10^4[A]_0$ and thus $OD_C - \overline{OD} = (1.00 - 1.04) \times 10^4[A]_0$; for $[B]_0 = 10^{-4}\,M$, $\overline{OD} = 1.02 \times 10^4[A]_0$ and thus $OD_C - \overline{OD} = (1.00 - 1.02) \times 10^4[A]_0$; for $[B]_0 = 3 \times 10^{-4}\,M$, $\overline{OD} = 1.007 \times 10^4[A]_0$ and thus $OD_C - \overline{OD} = (1.000 - 1.007) \times 10^4[A]_0$; and so on. It is obvious that the denominator $(OD_C - \overline{OD})$ in Eq. 6.52 is a very small difference between two large numbers and therefore very uncertain

The amplitude method outlined above has been applied to determine the stability constants of complexes between alkaline earth metals and murexide (*2, 3*), with association constants on the order of 10^5–$10^6\,M^{-1}$.

Problems

1. Derive Γ as function of K and c_0 for the reactions

$$\text{(a)}\quad A + B \rightleftharpoons C + D \qquad (c_A = c_B;\ c_C = c_D)$$

$$\text{(b)}\quad 2A \rightleftharpoons A_2$$

$$\text{(c)}\quad 2A \rightleftharpoons B + C \qquad (c_B = c_C)$$

Introduce α if necessary.

2. Assume $K = 10$ for reactions (a) and (c), $K = 10^3\ M^{-1}$ for reaction (b) in Problem 1. Determine for each case the range of c_0 in which the relative change in the concentration of the *reactant(s)*, and the range in which the relative change in the *product(s)*, is larger in a relaxation experiment.

3. Is there a way to determine K and/or ΔH from the relaxation amplitude in any of the reaction schemes of Problem 1? If yes, suggest how you would proceed.

References

1. M. Eigen and L. DeMaeyer, *in* "Technique of Organic Chemistry" (S. L. Friess, E. S. Lewis, and A. Weissberger, eds.), Vol. VIII, part 2, pp. 895. Wiley (Interscience), New York, 1963.
2. R. Winkler, Ph.D. Thesis, Max Planck Inst., Göttingen and Tech. Univ. of Vienna (1969).
3. M. Eigen and R. Winkler, *Neurosci. Res. Program Bull.* **9**, 330 (1971).

4a. F. Guillain and D. Thusius, *J. Amer. Chem. Soc.* **92**, 5534 (1970).

4b. D. Thusius, *Biochimie* **55**, 277 (1973).

5. G. Czerlinski, "Chemical Relaxation," Dekker, New York, 1966.
6. G. H. Czerlinski and J. Malkewitz, *Biochemistry* **4**, 1127 (1965).
7. P. B. Chock, *Proc. Nat. Acad. Sci. U.S.* **69**, 1939 (1972).

Chapter 7 | Relaxation Amplitudes in Multistep Systems

As we have seen in the preceding chapter relaxation amplitudes can be used to obtain reaction enthalpies and even equilibrium constants, but perhaps the most useful feature is that a qualitative understanding of the factors determining amplitudes helps in planning intelligent experiments. This is even more true for multistep equilibria. It is usually not worth the great effort necessary to obtain thermodynamic parameters or equilibrium constants from relaxation amplitudes, except where they cannot be determined by simpler methods. On the other hand, since there is a greater number of factors influencing the amplitudes in multistep equilibria compared to single-step systems, the experimenter has a greater choice in varying the experimental conditions so as to optimize them in order to obtain the highest possible amplitude for the process under study and the smallest possible amplitude for the processes possibly interfering with his measurements.

Furthermore, there are certain systems where an understanding of the factors governing amplitudes is crucial for the interpretation of experimental results, for example in situations where a relaxation effect expected on the basis of an assumed mechanism is missing because of a zero or very small amplitude. Since relaxation amplitudes to a considerable extent depend on the method of equilibrium perturbation, such an understanding can lead the experimenter to choose a more appropriate method which can make the missing relaxation effect visible. Some examples illustrating these points are discussed in Section 7.3.

Most of our discussion in this chapter focuses on two-step equilibria but some of the qualitative insights can easily be applied to larger systems. First, we shall deal with the relatively simple situation where one step equilibrates much faster than the other. We then will treat the general case and introduce the concept of normal modes of reactions as applied to various limiting situations derived from the general case. The physical interpretation of these normal modes and their analogy to the normal modes of virbrating molecules will be stressed.

7.1 Two-Step Systems with Rapid Equilibration of One Step

7.1.1 The A $\rightleftharpoons$ M $\rightleftharpoons$ B System

Most of the principles can be illustrated with the simple system

$$\begin{array}{c} \mathrm{A} \\ k_1 \nearrow\!\!\swarrow k_{-1} \\ \mathrm{M} \\ k_{-2} \nwarrow\!\!\searrow k_2 \\ \mathrm{B} \end{array} \tag{7.1}$$

The way the rate coefficients are defined in Eq. 7.1 leads to more symmetrical mathematical expressions than a definition such as

$$\mathrm{A} \underset{k_{-1}}{\overset{k_1}{\rightleftharpoons}} \mathrm{M} \underset{k_{-2}}{\overset{k_2}{\rightleftharpoons}} \mathrm{B}$$

and is therefore preferred here.

Just as in single-step systems where the principles developed for the simple A $\rightleftharpoons$ B reaction are applicable to reactions of higher molecularity, the principles developed for system 7.1 can be applied to almost any two-step reaction. They are, for example, directly applicable to systems 2, 3, and 4 in Table 3.1 except that in dealing with systems 2 and 3 we substitute $k_1' = k_1(\bar{c}_A + \bar{c}_B)$ for k_{-1}, in system 4 we substitute $k_2' = k_2(\bar{c}_B + \bar{c}_C)$ for k_2, and so on. The same holds true for system 5 of Table 3.1 if D is kept quasi constant (large excess or buffering) so that $k_2' = k_2\bar{c}_D$, or for system 6 if both B and D are quasi constant.

Let us first derive the general expressions for $\Delta c_A{}^0$, $\Delta c_B{}^0$, and $\Delta c_M{}^0$ which prevail regardless of the relative values of the four rate constants. Applying the same arguments that led to Eq. 6.7 we obtain the two (already linearized for small ΔK_1, ΔK_2) simultaneous equations

$$\Delta c_A{}^0 = K_1\,\Delta c_M{}^0 + \Delta K_1\bar{c}_M \tag{7.2}$$

$$\Delta c_B{}^0 = K_2\,\Delta c_M{}^0 + \Delta K_2\bar{c}_M \tag{7.3}$$

Note that if, for example, the k_1 and k_2 steps were bimolecular (M + N $\rightleftharpoons$ A, M + N $\rightleftharpoons$ B), Eqs. 7.2 and 7.3 would read

$$\Delta c_A{}^0 = K_1'\,\Delta c_M{}^0 + \Delta K_1\bar{c}_M\bar{c}_N \tag{7.4a}$$

$$\Delta c_B{}^0 = K_2'\,\Delta c_M{}^0 + \Delta K_2\bar{c}_M\bar{c}_N \tag{7.4b}$$

with $K_1' = K_1(\bar{c}_M + \bar{c}_N)$, $K_2' = K_2(\bar{c}_M + \bar{c}_N)$.

Taking into account the mass balance equations

$$\Delta c_M{}^0 = -\Delta c_A{}^0 - \Delta c_B{}^0 \tag{7.5}$$

$$c^0 = \bar{c}_A + \bar{c}_B + \bar{c}_M \tag{7.6}$$

and solving the two simultaneous equations 7.2 and 7.3 leads to

$$\Delta c_A{}^0 = \bar{c}_A \frac{(1 + K_2)\,\Delta K_1/K_1 - K_2\,\Delta K_2/K_2}{1 + K_1 + K_2}$$

$$= c_0 K_1 \frac{(1 + K_2)\,\Delta K_1/K_1 - K_2\,\Delta K_2/K_2}{(1 + K_1 + K_2)^2} \tag{7.7}$$

$$\Delta c_B{}^0 = \bar{c}_B \frac{(1 + K_1)\,\Delta K_2/K_2 - K_1\,\Delta K_1/K_1}{1 + K_1 + K_2}$$

$$= c_0 K_2 \frac{(1 + K_1)\,\Delta K_2/K_2 - K_1\,\Delta K_1/K_1}{(1 + K_1 + K_2)^2} \tag{7.8}$$

$$\Delta c_M{}^0 = -\bar{c}_M \frac{K_1\,\Delta K_1/K_1 + K_2\,\Delta K_2/K_2}{1 + K_1 + K_2}$$

$$= -c_0 \frac{K_1\,\Delta K_1/K_1 + K_2\,\Delta K_2/K_2}{(1 + K_1 + K_2)^2} \tag{7.9}$$

As previously (Eq. 6.3), in the case of a temperature-jump experiment the relative changes in the equilibrium constants are

$$\frac{\Delta K_1}{K_1} = \frac{\Delta H_1}{RT^2}\,\Delta T \tag{7.10}$$

$$\frac{\Delta K_2}{K_2} = \frac{\Delta H_2}{RT^2}\,\Delta T \tag{7.11}$$

It is easy to modify Eqs. 7.7–7.9 for the case where the k_1 and k_2 steps are bimolecular and the experiment is conducted under pseudo-first-order conditions ($\bar{c}_N \gg \bar{c}_M$ in Eqs. 7.4).

It is important to appreciate that the expressions 7.7–7.9 refer to the *total* displacement, that is, the sum of the amplitudes of the individual relaxation processes, for example

$$\Delta c_A{}^0 = \Delta c_A^{01} + \Delta c_A^{02} = \bar{c}_A{}^i - \bar{c}_A{}^f \tag{7.12}$$

where Δc_A^{01} refers to the faster, Δc_A^{02} to the slower process. Expressions such as 7.7–7.9 therefore give no insight into the individual amplitudes; additional information relating to the relative rates of the various reactions is needed for obtaining Δc_A^{01} and Δc_A^{02} individually.

We assume now that the $M \rightleftharpoons A$ step equilibrates much more rapidly than the $M \rightleftharpoons B$ step, i.e., $k_1 + k_{-1} \gg k_2, k_{-2}$.

In deriving the relaxation *time* expressions for such a situation we had argued (Section 3.2.1) that since the new equilibrium position of the first step is reached before equilibration of the slow step has made any significant progress, the rapid step can be treated as if we were dealing with a single-step equilibrium. This is equally true with respect to the relaxation *amplitude* associated with τ_1, $\Delta c_A^{01} = -\Delta c_M^{01}$ ($\Delta c_B^{01} = 0$), which is thus simply given by

$$\Delta c_A^{01} = \bar{c}_A{}^1 \frac{1}{1 + K_1} \frac{\Delta K_1}{K_1} = c_0 \frac{K_1}{(1 + K_1)(1 + K_1 + K_2)} \frac{\Delta K_1}{K_1} \tag{7.13}$$

or

$$\Delta c_M^{01} = -\bar{c}_M{}^1 \frac{K_1}{1 + K_1} \frac{\Delta K_1}{K_1} = -c_0 \frac{K_1}{(1 + K_1)(1 + K_1 + K_2)} \frac{\Delta K_1}{K_1} \tag{7.14}$$

in analogy to Eqs. 6.17 and 6.16, respectively. Note that as in Section 3.2.1 $\bar{c}_A{}^1$ and $\bar{c}_M{}^1$ refer to the equilibrium concentration reached when $\tau_2 \gg t \gg \tau_1$ and not to the final equilibrium state; usually there is no great difference between $\bar{c}_j{}^1$ and $\bar{c}_j{}^f$ and the distinction between the two is academic unless one deals with large perturbations which also render Eqs. 7.2 and 7.3 invalid because of non-negligible $\Delta K\, \Delta c$ terms (see Eq. 6.7).

The amplitude associated with τ_2 for the three concentration variables are then simply given by

$$\Delta c_j^{02} = \Delta c_j{}^0 - \Delta c_j^{01} \tag{7.15}$$

For $\bar{c}_A{}^1 \approx \bar{c}_A{}^f = \bar{c}_A$ and $\bar{c}_M{}^1 \approx \bar{c}_M{}^f = \bar{c}_M$ Eq. 7.15 becomes

$$\Delta c_A^{02} = \bar{c}_A \frac{K_2(K_1\, \Delta K_1/K_1 - \Delta K_2/K_2 - K_1\, \Delta K_2/K_2)}{(1 + K_1)(1 + K_1 + K_2)}$$

$$= c_0 \frac{K_1 K_2(K_1\, \Delta K_1/K_1 - \Delta K_2/K_2 - K_1\, \Delta K_2/K_2)}{(1 + K_1)(1 + K_1 + K_2)^2} \tag{7.16}$$

$$\Delta c_B^{02} = \Delta c_B{}^0 \qquad \text{(see Eq. 7.8)} \tag{7.17}$$

$$\Delta c_M^{02} = \bar{c}_M \frac{K_2(K_1\, \Delta K_1/K_1 - \Delta K_2/K_2 - K_1\, \Delta K_2/K_2)}{(1 + K_1)(1 + K_1 + K_2)}$$

$$= c_0 \frac{K_2(K_1\, \Delta K_1/K_1 - \Delta K_2/K_2 - K_1\, \Delta K_2/K_2)}{(1 + K_1)(1 + K_1 + K_2)^2} \tag{7.18}$$

Note that $\Delta c_A^{02} = K_1\, \Delta c_M^{02}$, as it should.

There are several points of interest with respect to Eqs. 7.16–7.18.

1. Again in reference to the temperature-jump method we see that it is not necessary that the second step has a finite enthalpy of reaction (ΔH_2 or $\Delta K_2/K_2 \neq 0$) for the second relaxation process to be observable since Δc_A^{02}, Δc_B^{02}, Δc_M^{02} do not vanish for $\Delta K_2/K_2 = 0$ as long as $\Delta K_1/K_1 \neq 0$. This is of considerable practical importance; in fact, sometimes an intrinsically small relaxation amplitude can be enhanced by coupling the reaction under study to a fast equilibrium with a large enthalpy of reaction. This common situation is treated in Section 7.1.2.

2. It is instructive to consider a few limiting situations with respect to Eqs. 7.16, 7.17 (7.8, respectively), and 7.18. When $K_1 \ll 1$, Eq. 7.8 simplifies to

$$\Delta c_B{}^0 = \Delta c_B^{02} = \bar{c}_B \frac{\Delta K_2/K_2 - K_1\, \Delta K_1/K_1}{1 + K_2} \tag{7.19}$$

Unless $\Delta K_1/K_1$ is very much larger than $\Delta K_2/K_2$ so that the term $K_1\ \Delta K_1/K_1$ is not negligible compared to $\Delta K_2/K_2$, Eq. 7.19 can be further simplified to

$$\Delta c_B^{02} = \bar{c}_B \frac{1}{1 + K_2} \frac{\Delta K_2}{K_2} \tag{7.20}$$

which of course is the equation for a single-step equilibrium (see Eq. 6.17) as one would expect since $K_1 \ll 1$ means that A is a very minor component of the system.

When $K_1 \gg 1$ the opposite is the case; Eq. 7.8 becomes

$$\Delta c_B{}^0 = \Delta c_B^{02} = \bar{c}_B \frac{K_1(\Delta K_2/K_2 - \Delta K_1/K_1)}{K_1 + K_2} \tag{7.21}$$

and ΔH_1 is just as important as ΔH_2 in determining Δc_B^{02}. In the event that $\Delta H_1 \approx \Delta H_2$ we have $\Delta K_2/K_2 \approx \Delta K_1/K_1$ and the amplitude becomes vanishingly small.

7.1.2 Amplitude Enhancement through Coupling to a Buffer Equilibrium

Amplitude enhancement of a weak relaxation effect is most often feasible in acid–base reactions when studied in the presence of large concentrations of a buffer that has a strong temperature dependence. The effect can be understood in the following qualitative terms. The temperature jump causes the buffer to shift its equilibrium position which leads to a rapid pH change as the buffer relaxes toward its new equilibrium. This pH change now acts as a pH jump on the reaction under study; it may represent a much larger perturbation than the one caused directly by the temperature jump. It can enhance or oppose the effect of the temperature jump, depending on the signs of the reaction enthalpies of the buffer and the reaction under study, and on which species is coupled to the buffer equilibrium.

Let us develop the mathematical formalism for the following simple case

$$BH \rightleftharpoons B^- + H^+ \tag{7.22a}$$

$$H^+ + A^- \rightleftharpoons AH \tag{7.22b}$$

where B^- and BH are the buffer species and reaction 7.22a equilibrates much faster than reaction 7.22b. We are interested in finding an expression for $\Delta c_A^{02} = \Delta c_A{}^0 = -\Delta c_{AH}^{02} = -\Delta c_{AH}^0$. From

$$(\bar{c}_{BH} + \Delta c_{BH}^0)(K_B + \Delta K_B) = (\bar{c}_B + \Delta c_B{}^0)(\bar{c}_H + \Delta c_H{}^0) \tag{7.23}$$

$$(\bar{c}_{AH} + \Delta c_{AH}^0)(K_A + \Delta K_A) = (\bar{c}_A + \Delta c_A{}^0)(\bar{c}_H + \Delta c_H{}^0) \tag{7.24}$$

where K_A and K_B are defined as acid dissociation constants, we find for small perturbations

$$\Delta c_B{}^0(\bar{c}_H + K_B) = \Delta K_B \bar{c}_{BH} - \bar{c}_B\ \Delta c_H \tag{7.25}$$

$$\Delta c_A{}^0(\bar{c}_H + K_A) = \Delta K_A \bar{c}_{AH} - \bar{c}_A\ \Delta c_H \tag{7.26}$$

Stoichiometric considerations provide

$$\Delta c_{\mathrm{B}}{}^0 = -\Delta c_{\mathrm{BH}}^0 \tag{7.27}$$

$$\Delta c_{\mathrm{A}}{}^0 = -\Delta c_{\mathrm{AH}}^0 \tag{7.28}$$

$$\Delta c_{\mathrm{H}}{}^0 = -\Delta c_{\mathrm{AH}}^0 - \Delta c_{\mathrm{BH}}^0 = \Delta c_{\mathrm{A}}{}^0 + \Delta c_{\mathrm{B}}{}^0 \tag{7.29}$$

Substituting $\Delta c_{\mathrm{H}}{}^0$ from Eq. 7.29 into Eqs. 7.25 and 7.26 affords

$$\Delta c_{\mathrm{B}}{}^0(\bar{c}_{\mathrm{H}} + \bar{c}_{\mathrm{B}} + K_{\mathrm{B}}) = \Delta K_{\mathrm{B}}\bar{c}_{\mathrm{BH}} - \bar{c}_{\mathrm{B}}\,\Delta c_{\mathrm{A}}{}^0 \tag{7.30}$$

$$\Delta c_{\mathrm{A}}{}^0(\bar{c}_{\mathrm{H}} + \bar{c}_{\mathrm{A}} + K_{\mathrm{A}}) = \Delta K_{\mathrm{A}}\bar{c}_{\mathrm{AH}} - \bar{c}_{\mathrm{A}}\,\Delta c_{\mathrm{B}}{}^0 \tag{7.31}$$

Combining Eqs. 7.30 and 7.31 leads to

$$\Delta c_{\mathrm{A}}{}^0 = \frac{\Delta K_{\mathrm{A}}\bar{c}_{\mathrm{AH}}(\bar{c}_{\mathrm{H}} + \bar{c}_{\mathrm{B}} + K_{\mathrm{B}}) - \Delta K_{\mathrm{B}}\bar{c}_{\mathrm{A}}\bar{c}_{\mathrm{BH}}}{(\bar{c}_{\mathrm{H}} + \bar{c}_{\mathrm{A}} + K_{\mathrm{A}})(\bar{c}_{\mathrm{H}} + \bar{c}_{\mathrm{B}} + K_{\mathrm{B}}) - \bar{c}_{\mathrm{A}}\bar{c}_{\mathrm{B}}} \tag{7.32}$$

The following two limiting cases are of interest.

1. In the presence of very little buffer ($\bar{c}_{\mathrm{B}}, \bar{c}_{\mathrm{BH}} \ll \bar{c}_{\mathrm{A}}, \bar{c}_{\mathrm{AH}}, \bar{c}_{\mathrm{H}}$) Eq. 7.32 simplifies to

$$\Delta c_{\mathrm{A}}{}^0 = \frac{\Delta K_{\mathrm{A}}\bar{c}_{\mathrm{AH}}}{\bar{c}_{\mathrm{H}} + \bar{c}_{\mathrm{A}} + K_{\mathrm{A}}} = \frac{\bar{c}_{\mathrm{A}}\bar{c}_{\mathrm{H}}}{\bar{c}_{\mathrm{H}} + \bar{c}_{\mathrm{A}} + K_{\mathrm{A}}}\,\frac{\Delta K_{\mathrm{A}}}{K_{\mathrm{A}}} \tag{7.33}$$

which is essentially the same as Eq. 6.25 except that K_{A} is defined as a dissociation constant, K in 6.25 (6.26) as an association constant. We see that the buffer has no influence on the amplitude of the relaxation of process 7.22b as expected.

2. At high buffer concentration ($\bar{c}_{\mathrm{B}}, \bar{c}_{\mathrm{BH}} \gg \bar{c}_{\mathrm{A}}, \bar{c}_{\mathrm{AH}}, \bar{c}_{\mathrm{H}}$) Eq. 7.32 becomes

$$\Delta c_{\mathrm{A}}{}^0 = \frac{\Delta K_{\mathrm{A}}\bar{c}_{\mathrm{AH}}\bar{c}_{\mathrm{B}} - \Delta K_{\mathrm{B}}\bar{c}_{\mathrm{A}}\bar{c}_{\mathrm{BH}}}{\bar{c}_{\mathrm{B}}(\bar{c}_{\mathrm{H}} + K_{\mathrm{A}})} = \frac{\bar{c}_{\mathrm{A}}\bar{c}_{\mathrm{H}}}{\bar{c}_{\mathrm{H}} + K_{\mathrm{A}}}\left(\frac{\Delta K_{\mathrm{A}}}{K_{\mathrm{A}}} - \frac{\Delta K_{\mathrm{B}}}{K_{\mathrm{B}}}\right) \tag{7.34}$$

It is obvious that if $\Delta K_{\mathrm{A}}/K_{\mathrm{A}}$ and $\Delta K_{\mathrm{B}}/K_{\mathrm{B}}$ have different signs, the amplitude is enhanced; if they have the same sign, the amplitude is reduced by the presence of the buffer. Since the enthalpy ΔH_{A} is often not known, the effect of a specific buffer must usually be determined empirically.

7.1.3 Applications (Two- or Multistep Reactions)

Just as in single-step systems, amplitudes in temperature-jump experiments can be exploited for a calculation of reaction enthalpies. For ΔH of the fast step the principles are identical with those of single-step reactions; for ΔH of the slow step equations such as 7.16–7.21 can be used provided the equilibrium constants and $\Delta K_1/K_1$ are known. Examples illustrating the procedure are the competitive binding of proflavin and benzamidine to trypsin (*1*, *2*), or the binding of alkali ions to macrocyclic compounds such as monactin in the presence of the indicator murexide (*3*, *4*).

Winkler (*3*) has also developed a procedure to obtain equilibrium constants from amplitude measurements; it is an extension of the curve-shape-adjusting method developed for single-step systems which was discussed in Section 6.3.

Based on the same principles reaction volumes can be determined from pressure-jump amplitudes ($\partial \ln K/\partial p = -\Delta V/RT$), for example in the association of

Be^{2+} and SO_4^{2-} (*5*). Conversely, pressure-jump amplitudes along with an estimate of reaction volumes have been used to estimate equilibrium constants of the complex formation between Al^{3+} or Ga^{3+} and weak ligands (*6*).

Relaxation amplitudes of ultrasonic experiments have also been exploited; as will be shown in Chapter 15 they can provide ΔH and/or ΔV though the precision is frequently not very satisfactory.

The concentration dependence of relaxation amplitudes is sometimes invoked to confirm or discard an assumed mechanism (without evaluating the thermodynamic parameters). Examples are the study of the interaction of base with 2,4,6-trinitrotoluene in methanol (*7*), and the binding of proflavin to DNA (*8*).

Robinson *et al.* (*9*) have developed the formalism which relates the relaxation amplitudes to the thermodynamic parameters in aggregating systems such as the self-association of acridine orange to dimers and trimers. Assuming different mechanistic models different amplitude expressions were derived. Comparison of the theoretical predictions from the different models with the experimental amplitudes allowed the choice of the most likely mechanism, and also the calculation of the thermodynamic parameters.

A further example that is of considerable current interest refers to the relaxation effects observed in solutions of polypeptides or proteins. There is no general agreement whether relaxation, as detected by ultrasonic methods, is a consequence of a conformational transition or of proton transfers (*10*). It has been shown that a careful analysis of the pH dependence of the relaxation amplitude may provide a method for distinguishing between the possible mechanisms (*11*). For more on conformational transitions of polypeptides see Chapters 10, 13, 15, and 16. Other applications of relaxation amplitudes are mentioned in our discussion of the experimental techniques.

7.2 Normal Modes of Reactions

Let us return to reaction scheme 7.1. When the two steps equilibrate at similar rates the problem of finding the various Δc_j^{01} and Δc_j^{02} values is considerably more complicated because of the strong coupling between the two equilibria. This coupling, we recall, is described mathematically by the two simultaneous differential equations

$$dx_1/dt + a_{11}x_1 + a_{12}x_2 = 0 \tag{7.35}$$

$$dx_2/dt + a_{21}x_1 + a_{22}x_2 = 0 \tag{7.36}$$

(which are the same as Eqs. 3.27 and 3.28, respectively) where the a_{ij} terms are functions of the various rate coefficients and sometimes of equilibrium concentrations as well. For the system 7.1 we have

$$x_1 = \Delta c_{\mathrm{A}} \tag{7.37}$$

$$x_2 = \Delta c_{\mathrm{B}} \tag{7.38}$$

$$a_{11} = k_1 + k_{-1} \tag{7.39}$$

$$a_{12} = k_1 \tag{7.40}$$
$$a_{21} = k_2 \tag{7.41}$$
$$a_{22} = k_2 + k_{-2} \tag{7.42}$$

We further recall that the general solution of this system of differential equations is of the form

$$x_1 = x_1^{01} \exp(-t/\tau_1) + x_1^{02} \exp(-t/\tau_2) \tag{7.43}$$
$$x_2 = x_2^{02} \exp(-t/\tau_1) + x_2^{02} \exp(-t/\tau_2) \tag{7.44}$$

It is x_1^{01}, x_1^{02}, x_2^{01}, and x_2^{02} which represent the respective amplitudes associated with a given concentration variable and a given relaxation time, for example $x_1^{01} = \Delta c_A^{01}$ or $x_2^{02} = \Delta c_B^{02}$.

The problem of expressing the individual relaxation amplitudes in strongly coupled multistep systems has been treated by several authors with various degrees of mathematical sophistication and explicitness, and using a variety of approaches (*12–18*). We shall use the approach that needs the least mathematical sophistication on the part of the reader.

The starting point is the realization that one can treat coupled chemical reactions in very much the same way as one treats the coupled oscillators in a vibrating molecule. Just as it is useful to deal with normal modes of vibration and normal mode frequencies rather than with the vibration of isolated bonds one may introduce the concept of "normal mode of reactions" or "normal reactions."

By their very definition normal modes are not coupled (orthogonality). Mathematically this is expressed by equations such as

$$\frac{dy_1}{dt} = -\frac{1}{\tau_1} y_1; \qquad y_1 = y_1{}^0 \exp \frac{-t}{\tau_1} \tag{7.45}$$

$$\frac{dy_2}{dt} = -\frac{1}{\tau_2} y_2; \qquad y_2 = y_2{}^0 \exp \frac{-t}{\tau_2} \tag{7.46}$$

which are of the form of rate equations for single-step equilibria; y_1 and y_2 are called the normal concentration variables or eigenconcentrations and correspond to the normal coordinates in the vibrating molecule

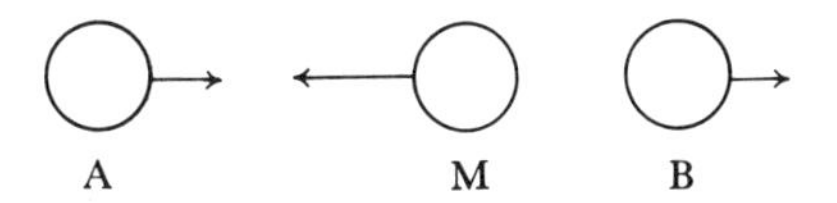

First normal mode of vibration with normal coordinate y_1.

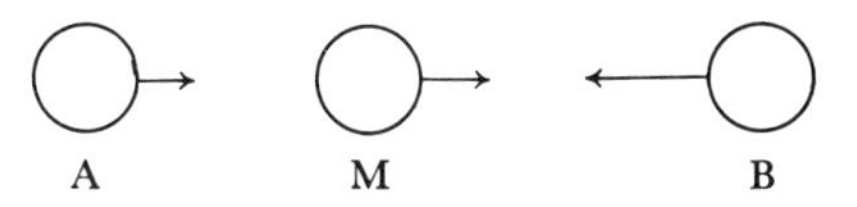

Second normal mode of vibration with normal coordinate y_2.

Thus the y_j values can be considered the "principal axes" of the system, obtained by an "affine coordinate transformation" of the x_j values whereas the reciprocal relaxation times correspond to the time constants of the equilibration (frequency of vibration in the vibrating molecule) along these principal axes.

The normal concentration variables are linear combinations of the actual concentration variables

$$y_1 = m_{11}x_1 + m_{12}x_2 \tag{7.47}$$

$$y_2 = m_{21}x_1 + m_{22}x_2 \tag{7.48}$$

Vice versa the x_j terms are also linear combinations of the y_j terms

$$x_1 = m_{-11}y_1 + m_{-12}y_2 \tag{7.49}$$

$$x_2 = m_{-21}y_1 + m_{-22}y_2 \tag{7.50}$$

If we substitute the respective exponential expressions of Eqs. 7.45 and 7.46 for y_1 and y_2, we obtain

$$x_1 = m_{-11}y_1{}^0 \exp(-t/\tau_1) + m_{-12}y_2{}^0 \exp(-t/\tau_2) \tag{7.51}$$

$$x_2 = m_{-21}y_1{}^0 \exp(-t/\tau_1) + m_{-22}y_2{}^0 \exp(-t/\tau_2) \tag{7.52}$$

Comparison with Eqs. 7.43 and 7.44, respectively, affords

$$x_1^{01} = m_{-11}y_1{}^0, \qquad x_1^{02} = m_{-12}y_2{}^0, \qquad x_2^{01} = m_{-21}y_1{}^0, \qquad x_2^{02} = m_{-22}y_2{}^0$$

Hence in order to determine the individual relaxation amplitudes we must first transform our set of x_j values into a set of y_j values (i.e., find the m_{ij}), and then transform the y_j back to the x_j (i.e., find the m_{-ij}).

The mathematical operations for these transformations involve some matrix algebra, which scares a good number of organic chemists. Since the derivations to follow are not essential for an understanding of the application to reaction systems, the casual reader may want to proceed directly to Eqs. 7.95–7.96 and 7.105–7.106 (7.104).

Our derivation is essentially an elaboration and slight modification of that given by Eigen and Ilgenfritz (*18*). The problem of finding the coefficients m_{ij} can be formulated by the simple equation

$$\mathbf{y} = \mathbf{M}\mathbf{x} \tag{7.53}$$

That is, we regard x_1 and x_2 as the components of a column vector $\mathbf{x}$,

$$\mathbf{x} = \begin{pmatrix} x_1 \\ x_2 \end{pmatrix} \tag{7.54}$$

which upon multiplication with a (square) transformation matrix $\mathbf{M}$,

$$\mathbf{M} = \begin{pmatrix} m_{11} & m_{12} \\ m_{21} & m_{22} \end{pmatrix} \tag{7.55}$$

yields a new column vector $\mathbf{y}$ with the components y_1 and y_2:

$$\mathbf{y} = \begin{pmatrix} y_1 \\ y_2 \end{pmatrix} \tag{7.56}$$

thus Eq. 7.53 is just a short form of writing the two equations 7.47 and 7.48.

For an affine transformation the relationship

$$\mathbf{x} = \mathbf{M}^{-1}\mathbf{y} \tag{7.57}$$

must also be valid with

$$\mathbf{M}^{-1} = \begin{pmatrix} m_{-11} & m_{-12} \\ m_{-21} & m_{-22} \end{pmatrix} \tag{7.58}$$

Equation 7.57 is the short form for the two equations 7.49 and 7.50. $\mathbf{M}^{-1}$ is the inverse matrix of $\mathbf{M}$; since by definition

$$\mathbf{M}\mathbf{M}^{-1} = \mathbf{I} \tag{7.59}$$

where $\mathbf{I}$ is the diagonal matrix of ones

$$\mathbf{I} = \begin{pmatrix} 1 & 0 \\ 0 & 1 \end{pmatrix} \tag{7.60}$$

$\mathbf{M}^{-1}$ can be expressed in terms of the elements of $\mathbf{M}$ (the m_{ij} terms) by

$$\mathbf{M}^{-1} = \begin{pmatrix} m_{22} & -m_{12} \\ -m_{21} & m_{11} \end{pmatrix} \Big/ \det \mathbf{M} \tag{7.61}$$

with

$$\det \mathbf{M} = m_{11}m_{22} - m_{12}m_{21} \tag{7.62}$$

Thus $m_{-11} = m_{22}/\det \mathbf{M}$, $m_{-12} = -m_{12}/\det \mathbf{M}$, $m_{-21} = -m_{21}/\det \mathbf{M}$, and $m_{-22} = m_{11}/\det \mathbf{M}$.

In a similar way we can regard dx_1/dt and dx_2/dt as the components of a vector $\dot{\mathbf{x}}$; and dy_1/dt, dy_2/dt as the components of a vector $\dot{\mathbf{y}}$. We can therefore formulate Eqs. 7.35–7.36 and Eqs. 7.45–7.46, respectively, as

$$\dot{\mathbf{x}} = -\mathbf{A}\mathbf{x} \tag{7.63}$$

with

$$\mathbf{A} = \begin{pmatrix} a_{11} & a_{12} \\ a_{21} & a_{22} \end{pmatrix} \tag{7.64}$$

and

$$\dot{\mathbf{y}} = -\mathbf{B}\mathbf{y} \tag{7.65}$$

with

$$\mathbf{B} = \begin{pmatrix} 1/\tau_1 & 0 \\ 0 & 1/\tau_2 \end{pmatrix} \tag{7.66}$$

Combining Eq. 7.63 with Eq. 7.57 affords

$$\dot{\mathbf{x}} = -\mathbf{A}\mathbf{x} = -\mathbf{A}\mathbf{M}^{-1}\mathbf{y} \tag{7.67}$$

Since Eq. 7.53 also requires

$$\dot{\mathbf{y}} = \mathbf{M}\dot{\mathbf{x}} \tag{7.68}$$

to be true, it follows with Eq. 7.67 that

$$\dot{\mathbf{y}} = -\mathbf{M}\mathbf{A}\mathbf{M}^{-1}\mathbf{y} \tag{7.69}$$

is valid. Comparison of Eq. 7.69 with Eq. 7.65 leads to

$$\mathbf{MAM}^{-1} = \mathbf{B} \tag{7.70}$$

which, after multiplication of each side with $\mathbf{M}$, can also be written as

$$\mathbf{MA} = \mathbf{BM} \tag{7.71}$$

or alternatively, after transposing, as

$$\mathbf{A}'\mathbf{M}' = \mathbf{M}'\mathbf{B}' \tag{7.72}$$

where the transposed matrices are

$$\mathbf{A}' = \begin{pmatrix} a_{11} & a_{21} \\ a_{12} & a_{22} \end{pmatrix} \tag{7.73}$$

$$\mathbf{M}' = \begin{pmatrix} m_{11} & m_{21} \\ m_{12} & m_{22} \end{pmatrix} \tag{7.74}$$

$$\mathbf{B}' = \begin{pmatrix} 1/\tau_1 & 0 \\ 0 & 1/\tau_2 \end{pmatrix} = \mathbf{B} \tag{7.75}$$

Before we can proceed any further we have to introducc the concept of *eigenvectors*. An eigenvector is a vector which, when multiplied by a matrix, yields the same vector except for being multiplied by a simple number called the *eigenvalue* of the matrix. For example, in the case of $\mathbf{A}$, the matrix of the coefficients a_{ij} (Eq. 7.64), we can write

$$\mathbf{Am}_1 = \frac{1}{\tau_1}\mathbf{m}_1 \quad \text{or} \quad \left(\mathbf{A} - \frac{1}{\tau_1}\mathbf{I}\right) = \mathbf{0} \tag{7.76}$$

$$\mathbf{Am}_2 = \frac{1}{\tau_2}\mathbf{m}_2 \quad \text{or} \quad \left(\mathbf{A} - \frac{1}{\tau_2}\mathbf{I}\right) = \mathbf{0} \tag{7.77}$$

where $\mathbf{m}_1$ and $\mathbf{m}_2$ are the eigenvectors, and τ_1^{-1} and τ_2^{-1} the eigenvalues of $\mathbf{A}$; the latter are found, we recall, by solving the determinantal equation 3.45 which in matrix notation reads

$$\left|\mathbf{A} - \frac{1}{\tau}\mathbf{I}\right| = 0 \tag{7.78}$$

Breaking down the two eigenvectors into their elements

$$\mathbf{m}_1 = \begin{pmatrix} m_{11} \\ m_{21} \end{pmatrix}, \quad \mathbf{m}_2 = \begin{pmatrix} m_{12} \\ m_{22} \end{pmatrix}$$

allows the formulation of a matrix comprising all the elements of the eigenvectors, that is, $\mathbf{M}$ according to Eq. 7.55. Equations 7.76–7.77 can then be written as

$$\mathbf{AM} = \left(\frac{1}{\tau_1}\mathbf{m}_1, \frac{1}{\tau_2}\mathbf{m}_2\right) \tag{7.79}$$

Furthermore, since

$$\left(\frac{1}{\tau_1}\mathbf{m}_1, \frac{1}{\tau_2}\mathbf{m}_2\right) = (\mathbf{m}_1, \mathbf{m}_2)\begin{pmatrix} 1/\tau_1 & 0 \\ 0 & 1/\tau_2 \end{pmatrix} = \mathbf{MB} \tag{7.80}$$

where $\mathbf{B}$ is the diagonal matrix of Eq. 7.66, we find

$$\mathbf{AM} = \mathbf{MB} \tag{7.81}$$

This is identical to relation 7.72 except for the fact that 7.72 refers to the transposed matrices.

From this we conclude that in order to obtain the transformation matrix $\mathbf{M}$ we have to proceed in two steps.

1. Find the components of the eigenvectors associated with $\mathbf{A}'$; they provide the elements of $\mathbf{M}'$.
2. Transpose $\mathbf{M}'$ in order to obtain $\mathbf{M}$.

The components of the eigenvectors of $\mathbf{A}'$ are found by solving the equations

$$\mathbf{A}'\mathbf{m}_1' = \frac{1}{\tau_1}\mathbf{m}_1' \tag{7.82}$$

$$\mathbf{A}'\mathbf{m}_2' = \frac{1}{\tau_2}\mathbf{m}_2' \tag{7.83}$$

where the eigenvectors $\mathbf{m}_1'$ and $\mathbf{m}_2'$ are defined as

$$\mathbf{m}_1' = \begin{pmatrix} m_{11} \\ m_{12} \end{pmatrix}, \qquad \mathbf{m}_2' = \begin{pmatrix} m_{21} \\ m_{22} \end{pmatrix}$$

Note that the eigenvalues of $\mathbf{A}'$ are the same as those for $\mathbf{A}$ since transposing $\mathbf{A}$ has no effect on the determinantal equations 3.45 or 7.78. Writing Eq. 7.82 in explicit terms

$$\begin{pmatrix} a_{11} & a_{21} \\ a_{12} & a_{22} \end{pmatrix}\begin{pmatrix} m_{11} \\ m_{12} \end{pmatrix} = \frac{1}{\tau_1}\begin{pmatrix} m_{11} \\ m_{12} \end{pmatrix} \tag{7.84}$$

leads to the two equations

$$(a_{11} - 1/\tau_1)m_{11} + a_{21}m_{12} = 0 \tag{7.85}$$

$$a_{12}m_{11} + (a_{22} - 1/\tau_1)m_{12} = 0 \tag{7.86}$$

It should be pointed out that Eqs. 7.85 and 7.86 are not independent of each other so that only the ratio m_{11}/m_{12} can be obtained from either one. If we arbitrarily set

$$m_{11} = 1 \tag{7.87}$$

and, again arbitrarily, choose Eq. 7.86 in order to solve for m_{12}, we obtain

$$m_{12} = -\frac{a_{12}}{a_{22} - 1/\tau_1} \tag{7.88}$$

Similarly, Eq. 7.83 leads to the two equations

$$(a_{11} - 1/\tau_2)m_{21} + a_{21}m_{22} = 0 \tag{7.89}$$

$$a_{12}m_{21} + (a_{22} - 1/\tau_2)m_{22} = 0 \tag{7.90}$$

Setting

$$m_{22} = 1 \tag{7.91}$$

and using Eq. 7.89 we obtain

$$m_{21} = -\frac{a_{21}}{a_{11} - 1/\tau_2} \tag{7.92}$$

We now have the components of $\mathbf{M}'$; transposing gives $\mathbf{M}$:

$$\mathbf{M}' = \begin{pmatrix} 1 & -\dfrac{a_{21}}{a_{11} - 1/\tau_2} \\ -\dfrac{a_{12}}{a_{22} - 1/\tau_1} & 1 \end{pmatrix} \tag{7.93}$$

$$\mathbf{M} = \begin{pmatrix} 1 & -\dfrac{a_{12}}{a_{22} - 1/\tau_1} \\ -\dfrac{a_{21}}{a_{11} - 1/\tau_2} & 1 \end{pmatrix} \tag{7.94}$$

Equations 7.47 and 7.48 now read

$$y_1 = x_1 - \frac{a_{12}}{a_{22} - 1/\tau_1} x_2 \tag{7.95}$$

$$y_2 = -\frac{a_{21}}{a_{11} - 1/\tau_2} x_1 + x_2 \tag{7.96}$$

Two features about this result need to be pointed out.

1. Since the choice of Eqs. 7.86 (instead of 7.85) and 7.89 (instead of 7.90) is arbitrary, and the choice $m_{11} = 1$ (instead of $m_{12} = 1$) and $m_{22} = 1$ (instead of $m_{21} = 1$) is arbitrary, a different combination of choices leads to different sets of final equations; this is the reason for the (to the novice somewhat confusing) fact that different authors come up with different equations for the y_j values. For example, choosing Eqs. 7.85 and 7.90 with $m_{11} = m_{22} = 1$ leads to

$$y_1 = x_1 - \frac{a_{11} - 1/\tau_1}{a_{21}} x_2 \tag{7.97}$$

$$y_2 = -\frac{a_{22} - 1/\tau_2}{a_{12}} x_1 + x_2 \tag{7.98}$$

whereas using the same equations but setting $m_{12} = m_{21} = 1$ affords

$$y_1 = -\frac{a_{21}}{a_{11} - 1/\tau_1} x_1 + x_2 \tag{7.99}$$

$$y_2 = x_1 - \frac{a_{12}}{a_{22} - 1/\tau_2} x_2 \tag{7.100}$$

In the following we shall work with Eqs. 7.95–7.96.

2. Since two of the m_{ij} terms are arbitrarily set equal to one, a "normalization factor" is left open in Eqs. 7.95–7.100. For Eqs. 7.95–7.96 the (unknown) normalization factors are of course m_{11} and m_{22} and thus the strictly correct form of the equations is

$$\frac{y_1}{m_{11}} = y_1' = x_1 - \frac{a_{12}}{a_{22} - 1/\tau_1} x_2 \tag{7.101}$$

$$\frac{y_2}{m_{22}} = y_2' = -\frac{a_{21}}{a_{11} - 1/\tau_2} x_1 + x_2 \tag{7.102}$$

Though there are procedures that allow the determination of these normalization factors, it is usually not necessary to know them, as we shall see.

In some applications one needs to go no further than Eqs. 7.95–7.96. However, if the actual amplitudes are desired (Eqs. 7.49–7.50 or 7.51–7.52, respectively), one has to determine the elements of $\mathbf{M}^{-1}$ so one can solve Eq. 7.57. According to Eq. 7.61 we obtain

$$\mathbf{M}^{-1} = \begin{pmatrix} 1 & \dfrac{a_{12}}{a_{22} - 1/\tau_1} \\ \dfrac{a_{21}}{a_{11} - 1/\tau_2} & 1 \end{pmatrix} \Bigg/ \det \mathbf{M} \tag{7.103}$$

with

$$\det \mathbf{M} = 1 - \frac{a_{12}a_{21}}{(a_{22} - 1/\tau_1)(a_{11} - 1/\tau_2)} \tag{7.104}$$

Hence Eqs. 7.49 and 7.50 now read

$$x_1 = \left(y_1 + \frac{a_{12}}{a_{22} - 1/\tau_1} y_2 \right) \Big/ \det \mathbf{M} \tag{7.105}$$

$$x_2 = \left(\frac{a_{21}}{a_{11} - 1/\tau_2} y_1 + y_2 \right) \Big/ \det \mathbf{M} \tag{7.106}$$

They could of course also be obtained by just solving the simultaneous equations 7.95–7.96 for x_1 and x_2.

Since starting with Eqs. 7.101–7.102 instead of 7.95–7.96 leads to the same equations 7.105–7.106, except that y_1 and y_2 are replaced by y_1' and y_2' respectively, we now see that a knowledge of the normalization factors m_{11} and m_{22} is not necessary; m_{11} and m_{22} simply drop out of the equations.

In the following section we shall see how the formalism developed here is applied to some specific situations.

7.3 Applications of Normal Mode Analysis

7.3.1 Amplitudes and Physical Interpretation of Normal Modes in the A ⇌ M ⇌ B System

There are four interesting and commonly occurring special cases of scheme 7.1 which are now discussed in some detail. They are

$$\begin{aligned} \text{I.} \quad & k_1, k_{-1} \gg k_2, k_{-2} \\ \text{II.} \quad & k_1, k_2 \gg k_{-1}, k_{-2} \\ \text{III.} \quad & k_1, k_2 \ll k_{-1}, k_{-2} \\ \text{IV.} \quad & k_{-2} = k_{-1} \end{aligned}$$

In all these cases the relations between the x_j and y_j terms (Eqs. 7.95–7.96 and 7.105–7.106) become quite simple as shown in Table 7.1; in fact, it is possible

Table 7.1. Relaxation Times and Normal Concentrations in A ⇌ M ⇌ B System

	Case I $k_1, k_{-1} \gg k_2, k_{-2}$	Case II $k_1, k_2 \gg k_{-1}, k_{-2}$	Case III $k_1, k_2 \ll k_{-1}, k_{-2}$	Case IV $k_{-2} = k_{-1}$
$\frac{1}{\tau_1}$	$k_1 + k_{-1}$	$k_1 + k_2$	k_{-1}	$k_1 + k_2 + k_{-1}$
$\frac{1}{\tau_2}$	$\frac{k_2 k_{-1}}{k_1 + k_{-1}} + k_{-2}$	$\frac{k_{-1}k_2}{k_1 + k_2} + \frac{k_{-2}k_1}{k_1 + k_2}$	k_{-2}	k_{-1}
y_1	$\Delta c_A + \frac{k_1}{k_1 + k_{-1}} \Delta c_B$	$\Delta c_A + \Delta c_B = -\Delta c_M$	Δc_A	$\Delta c_A + \Delta c_B = -\Delta c_M$
y_2	Δc_B	$-\frac{k_2}{k_1} \Delta c_A + \Delta c_B$	Δc_B	$-\frac{k_2}{k_1} \Delta c_A + \Delta c_B$
Δc_A	$y_1 - \frac{k_1}{k_1 + k_{-1}} y_2$	$\frac{k_1}{k_1 + k_2} y_1 - \frac{k_1}{k_1 + k_2} y_2$	y_1	$\frac{k_1}{k_1 + k_2} y_1 - \frac{k_1}{k_1 + k_2} y_2$
Δc_B	y_2	$\frac{k_2}{k_1 + k_2} y_1 + \frac{k_1}{k_1 + k_2} y_2$	y_2	$\frac{k_2}{k_1 + k_2} y_1 + \frac{k_1}{k_1 + k_2} y_2$
Δc_M	$-y_1 - \frac{k_{-1}}{k_1 + k_{-1}} y_2$	$-y_1$	$-y_1 - y_2$	$-y_1$

to derive them by intuition and physical reasoning, i.e., without recourse to the mathematical apparatus developed above, as will be shown for some cases.

Case IV. As will become apparent case IV is in many respects the most interesting one and is now discussed in detail. The procedure for obtaining the relaxation amplitudes presented here is equally applicable to the other cases.

Our first aim is to obtain the individual concentration changes associated with a given relaxation time, namely Δc_A^{01}, Δc_A^{02}, Δc_B^{01}, etc. Writing the expressions for y_1, y_2, Δc_A, Δc_B, and Δc_M (from Table 7.1) for time $t = 0$ provides

$$y_1{}^0 = \Delta c_A{}^0 + \Delta c_B{}^0 = -\Delta c_M{}^0 \tag{7.107}$$

$$y_2{}^0 = -\frac{k_2}{k_1}\Delta c_A{}^0 + \Delta c_B{}^0 \tag{7.108}$$

$$\Delta c_A{}^0 = \frac{k_1}{k_1 + k_2} y_1{}^0 - \frac{k_1}{k_1 + k_2} y_2{}^0 \tag{7.109}$$

$$\Delta c_B{}^0 = \frac{k_2}{k_1 + k_2} y_1{}^0 + \frac{k_1}{k_1 + k_2} y_2{}^0 \tag{7.110}$$

$$\Delta c_M{}^0 = -y_1{}^0 \tag{7.111}$$

Recalling that $\Delta c_A{}^0 = \Delta c_A^{01} + \Delta c_A^{02}$ (Eq. 7.12), etc., we can now write

$$\Delta c_A^{01} = \frac{k_1}{k_1 + k_2} y_1{}^0 = \frac{k_1}{k_1 + k_2}(\Delta c_A{}^0 + \Delta c_B{}^0) = -\frac{k_1}{k_1 + k_2}\Delta c_M{}^0 \tag{7.112}$$

$$\Delta c_B^{01} = \frac{k_2}{k_1}\Delta c_A^{01} \tag{7.113}$$

$$\Delta c_M^{01} = -y_1{}^0 = \Delta c_M{}^0 \tag{7.114}$$

$$\Delta c_A^{02} = -\frac{k_1}{k_1 + k_2} y_2{}^0 = \frac{k_2}{k_1 + k_2}\Delta c_A{}^0 - \frac{k_1}{k_1 + k_2}\Delta c_B{}^0 \tag{7.115}$$

$$\Delta c_B^{02} = -\Delta c_A^{02} \tag{7.116}$$

$$\Delta c_M^{02} = 0 \tag{7.117}$$

A. Concentration Changes in Temperature-Jump Experiments

As pointed out earlier, $\Delta c_A{}^0$, $\Delta c_B{}^0$, and $\Delta c_M{}^0$ depend on the method of perturbation. For a temperature-jump experiment we have Eqs. 7.7–7.9. Thus combining Eqs. 7.7–7.9 with Eqs. 7.112–7.117 affords

$$\Delta c_A^{01} = \frac{k_1(K_1\,\Delta K_1/K_1 + K_2\,\Delta K_2/K_2)}{(k_1 + k_2)(1 + K_1 + K_2)^2} c_0 \tag{7.118}$$

$$\Delta c_B^{01} = \frac{k_2}{k_1}\Delta c_A^{01} \tag{7.119}$$

$$\Delta c_M^{01} = -\frac{K_1\,\Delta K_1/K_1 + K_2\,\Delta K_2/K_2}{(1 + K_1 + K_2)^2} c_0 \tag{7.120}$$

$$\Delta c_A^{02} = \frac{K_1 K_2(\Delta K_1/K_1 - \Delta K_2/K_2)}{(K_1 + K_2)(1 + K_1 + K_2)} c_0 \tag{7.121}$$

$$\Delta c_{\mathrm{B}}^{02} = -\Delta c_{\mathrm{A}}^{02} \tag{7.122}$$

$$\Delta c_{\mathrm{M}}^{02} = 0 \tag{7.123}$$

The following features are noteworthy.

1. The second relaxation process does not involve any change in the concentration of M (Eq. 7.123). This is not an idiosyncrasy of the temperature-jump method but is always true when $k_{-2} = k_{-1}$, as seen from Eq. 7.117. Hence τ_2 is not detectable if only c_{M} or a physical property related to c_{M} only is chosen to monitor relaxation.

2. Although we do not have the same problem with c_{A} or c_{B} there are circumstances where $\Delta c_{\mathrm{A}}^{02}$ and $\Delta c_{\mathrm{B}}^{02}$ may be very small or zero, namely when $\Delta K_1/K_1 \approx \Delta K_2/K_2$. This is because systems where $k_{-2} = k_{-1}$ will typically be reactions where A and B are chemically very similar, for example conformational or geometrical isomers, so that the reaction enthalpies ΔH_1 and ΔH_2 are bound to be very similar. An example where this is likely to occur is the reaction of nucleophiles such as HO^- or alkoxide ions with 1,3,5-trinitrobenzene (Scheme 7.124); in a rapid first step a Meisenheimer complex (M) is formed which reacts to form a cis (B) or trans (A) 1:2 complex (*19*). With $[\mathrm{Nu}]_0 \gg [\mathrm{T}]_0$ and the rapid

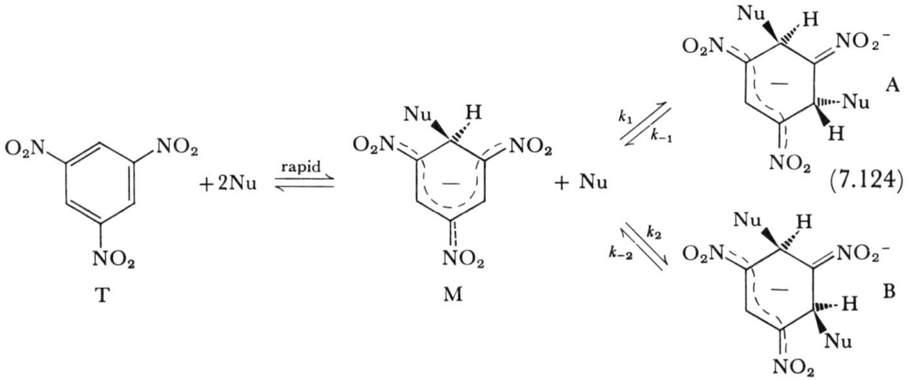

$$\mathrm{T} + 2\mathrm{Nu} \underset{}{\overset{\text{rapid}}{\rightleftharpoons}} \mathrm{M} + \mathrm{Nu} \quad \begin{matrix} \overset{k_1}{\underset{k_{-1}}{\rightleftharpoons}} \mathrm{A} \\ \overset{k_2}{\underset{k_{-2}}{\rightleftharpoons}} \mathrm{B} \end{matrix} \tag{7.124}$$

equilibrium greatly favoring M over T this corresponds exactly to scheme 7.1 with $k_1[\mathrm{N}]_0$ instead of k_1, etc. The question arises whether τ_2 could more easily be detected and measured if another perturbation method is used (vida infra).

3. In general there should be no difficulty in obtaining measurable changes in the concentration of A, B, or M associated with τ_1 except for the usual problems that could arise if the equilibria are too greatly one-sided. In particular there are no cancellations in case $\Delta K_1/K_1 \approx \Delta K_2/K_2$.

B. Concentration Changes in Stopped-Flow Experiments

If the reactions are not too fast, stopped-flow experiments are a viable alternative, as, for example, in the case shown in scheme 7.124. The most obvious type of experiment is to mix T with Nu and monitor the ensuing relaxations. Here $\Delta c_A{}^0$, $\Delta c_B{}^0$, and $\Delta c_M{}^0$ become

$$\Delta c_A{}^0 = -\bar{c}_A = -K_1\bar{c}_M = -\frac{K_1}{1 + K_1 + K_2} c_0 \tag{7.125}$$

$$\Delta c_B{}^0 = -\bar{c}_B = -K_2\bar{c}_M = -\frac{K_2}{1 + K_1 + K_2} c_0 \tag{7.126}$$

$$\Delta c_M{}^0 = c_M{}^0 - \bar{c}_M = c_0 - \bar{c}_M = \frac{K_1 + K_2}{1 + K_1 + K_2} c_0 \tag{7.127}$$

In combination with Eqs. 7.112–7.117 we obtain

$$\Delta c_A^{01} = -\frac{K_1}{1 + K_1 + K_2} c_0 \tag{7.128}$$

$$\Delta c_B^{01} = \frac{k_2}{k_1} \Delta c_A^{01} = \frac{K_2}{K_1} \Delta c_A^{01} \tag{7.129}$$

$$\Delta c_M^{01} = \frac{K_1 + K_2}{1 + K_1 + K_2} c_0 \tag{7.130}$$

$$\Delta c_A^{02} = \Delta c_B^{02} = \Delta c_M^{02} = y_2{}^0 = 0 \tag{7.131}$$

It is apparent that with respect to detecting the second relaxation effect the situation is even worse than with the temperature-jump method since none of the species undergoes any change in concentration! Thus, without knowing from independent evidence that there are indeed two reactions, the experimenter would conclude that there is only one process.

There is, however, a different kind of stopped-flow experiment that would allow τ_2 to be detected. It is based on introducing either A or B into the system, for example if one could rapidly convert an unreactive form (A′) into its active form by mixing *A′* with a reagent (R). Such a possibility is represented in the following scheme:

$$\begin{array}{ccc} & \text{A} \xleftarrow{\text{rapid}} \text{A}' + \text{R} \\ & {\scriptstyle k_1} \nearrow\!\!\swarrow {\scriptstyle k_{-1}} \\ \text{M} & \\ & {\scriptstyle k_{-2}} \nwarrow\!\!\searrow {\scriptstyle k_2} \\ & \text{B} \end{array} \tag{7.132}$$

Whether this is feasible on chemical grounds is a separate question. Though in our example (scheme 7.124) it does not seem practical the possibility certainly is a real one in other systems.

Assume the reaction is initiated by mixing a solution containing A′ (and Nu) with a solution of R. Instead of Eqs. 7.125–7.127 we now have

$$\Delta c_A{}^0 = c_A{}^0 - \bar{c}_A = c_0 - \bar{c}_A = \frac{1 + K_2}{1 + K_1 + K_2} c_0 \tag{7.133}$$

$$\Delta c_B{}^0 = -\bar{c}_B = -\frac{K_2}{1 + K_1 + K_2} c_0 \tag{7.134}$$

$$\Delta c_M{}^0 = -\bar{c}_M = -\frac{1}{1 + K_1 + K_2} c_0 \tag{7.135}$$

Combining these with Eqs. 7.112–7.117 leads to

$$\Delta c_A^{01} = \frac{k_1}{k_1 + k_2} \frac{1}{1 + K_1 + K_2} c_0 \tag{7.136}$$

$$\Delta c_B^{01} = \frac{k_2}{k_1} \Delta c_A^{01} \tag{7.137}$$

$$\Delta c_M^{01} = -\frac{1}{1 + K_1 + K_2} c_0 \tag{7.138}$$

$$\Delta c_A^{02} = \frac{k_2(1 + K_1) + k_1 K_2}{(k_1 + k_2)(1 + K_1 + K_2)} c_0 \tag{7.139}$$

$$\Delta c_B^{02} = -\Delta c_A^{02} \tag{7.140}$$

$$\Delta c_M^{02} = 0 \tag{7.141}$$

In contrast to the stopped-flow experiment initiated by mixing M and Nu, Δc_A^{02} $(= -\Delta c_B^{02})$ is now of appreciable size (unless K_1 is very large and K_2 is very small) and hence the effect in principle is easily detected. On the other hand, it is now τ_1 that may be difficult to measure, at least when $K_1, K_2 > 1$ or $\gg 1$. For example, if $K_1 = K_2 = 2$ (or $= 10$) we obtain $\Delta c_M^{01} = 0.80c_0$ (or $= 0.95c_0$) according to Eq. 7.130, but only $\Delta c_M^{01} = -0.20c_0$ (or $= -0.048c_0$) according to Eq. 7.138.

A perturbation similar to the one occurring through generating A by mixing A′ with R could in principle also be produced if the reaction A′ + R → A were reversible and its equilibrium position were displaced by a temperature-jump. Rapid relaxation of the presumably fast equilibrium A′ + R ⇌ A would produce a concentration jump in A and thus in fact perturb our system in a qualitatively similar manner as does mixing of R and A′. Thus, as far as the amplitudes associated with τ_1 and τ_2 are concerned we must come to the same qualitative conclusions though the mathematical treatment is somewhat more involved and not developed here.

C. Physical Interpretation

The easiest way of understanding the physical meaning of these relationships is to try to derive the expressions for τ_1^{-1}, τ_2^{-1}, y_1, and y_2 on the basis of physical reasoning and intuition. We start by recalling a crucial feature of the normal modes, namely that they equilibrate independently from each other which is expressed mathematically by

$$dy_1/dt = -(1/\tau_1)y_1 \tag{7.142}$$

$$dy_2/dt = -(1/\tau_2)y_2 \tag{7.143}$$

The fact that the second relaxation process never involves any change in the concentration of M ($\Delta c_M^{02} = 0$) must mean that the normal mode or normal reaction associated with it is a process *not* involving M, at least not explicitly. We suspect that this process corresponds to the equilibration between A and B, represented as

$$A \underset{\overleftarrow{k}}{\overset{\overrightarrow{k}}{\rightleftharpoons}} B \tag{7.144}$$

For the other normal reaction we are left with a process that *does* involve M but which must be independent of whether or not A and B are at equilibrium with each other. This process can be represented as

$$M \underset{k_{-1}}{\overset{k_1+k_2}{\rightleftharpoons}} (A, B) \tag{7.145}$$

with no assumption about the ratio $c_A:c_B$.

That these assignments of the normal reactions are in fact correct can be shown by proving that they lead to Eqs. 7.142–7.143 with $1/\tau_1 = k_1 + k_2 + k_{-1}$ and $1/\tau_2 = k_{-1}$. We begin by setting up the rate equation for the normal reaction 7.145

$$d\,\Delta c_M/dt = -(k_1 + k_2)\,\Delta c_M + k_{-1}(\Delta c_A + \Delta c_B) \tag{7.146}$$

After substituting $-\Delta c_M$ for $\Delta c_A + \Delta c_B$ we immediately obtain

$$d\,\Delta c_M/dt = -(k_1 + k_2 + k_{-1})\,\Delta c_M \tag{7.147}$$

which is indeed equivalent to Eq. 7.142.

With respect to the normal reaction 7.144 we proceed as follows. We argue that when equilibrium between A and B is reached

$$c_B - (k_2/k_1)c_A = 0 \tag{7.148}$$

must hold; c_B and c_A may or may not refer to the final equilibrium concentrations $\bar{c}_A$ and $\bar{c}_B$, respectively; this will depend on whether or not the first normal reaction is at equilibrium. This is expressed by

$$\bar{c}_B + \Delta c_B - (k_2/k_1)(\bar{c}_A + \Delta c_A) = 0 \tag{7.149}$$

Combining Eq. 7.148 (now written for $c_A = \bar{c}_A$, $c_B = \bar{c}_B$) with Eq. 7.149 affords

$$\Delta c_B - (k_2/k_1)\,\Delta c_A = 0 \tag{7.150}$$

If equilibrium between A and B is *not* established, Eq. 7.150 is not fulfilled. Thus we suspect that the difference $\Delta c_B - k_2\,\Delta c_A/k_1$ is a measure of how far the second normal reaction is from equilibrium. In fact, setting up the rate equation in terms of $\Delta c_B - k_2\,\Delta c_A/k_1$ leads to

$$\begin{aligned}\frac{d(\Delta c_B - k_2\,\Delta c_A/k_1)}{dt} &= k_2\,\Delta c_M - k_{-1}\,\Delta c_B - k_2\,\Delta c_M + \frac{k_2 k_{-1}\,\Delta c_A}{k_1} \\ &= -k_{-1}\left(\Delta c_B - \frac{k_2\,\Delta c_A}{k_1}\right)\end{aligned} \tag{7.151}$$

which is of the form of Eq. 7.143 with $1/\tau_2 = k_{-1}$.

Note that since for a reaction such as Eq. 7.144 one has $1/\tau_2 = \vec{k} + \overleftarrow{k}$, and its equilibrium constant is $\vec{k}/\overleftarrow{k} = k_2/k_1$, it follows that $\vec{k} = k_{-1}k_2/(k_1 + k_2)$, $\overleftarrow{k} = k_{-1}k_1(k_1 + k_2)$.

We now appreciate the meaning of the fact that in a stopped-flow experiment initiated by mixing T with Nu, $y_2{}^0 = 0$ and with it $\Delta c_A^{02} = \Delta c_B^{02} = \Delta c_M^{02} = 0$. This is a consequence of A and B being formed directly in their final equilibrium ratio (Eq. 7.150), so that equilibration along the second normal reaction becomes redundant. In the alternative stopped-flow experiment (releasing A by reacting A′ with R; thus at $t = 0$, $c_A = c_0$, $c_B = 0$) A and B are very far from their final equilibrium ratio at the beginning; thus $y_2{}^0 \neq 0$ and τ_2 is very much in evidence.

The small amplitude for τ_2 in a temperature-jump experiment is now also easily understood in physical terms. The equilibrium constant for the second normal reaction, Eq. 7.144, is $K_{AB} = K_2/K_1$ and thus the normal enthalpy of reaction $\Delta H_{AB} = \Delta H_2 - \Delta H_1$. For similar ΔH_1 and ΔH_2 we obtain $\Delta H_{AB} \approx 0$ and thus the equilibrium displacement of the normal reaction induced by a temperature jump is very small.

D. Analogy to Vibrating Molecule

The normal reactions of Eqs. 7.144–7.145 are analogous to the following vibrational normal modes of the linear three-atom molecule:

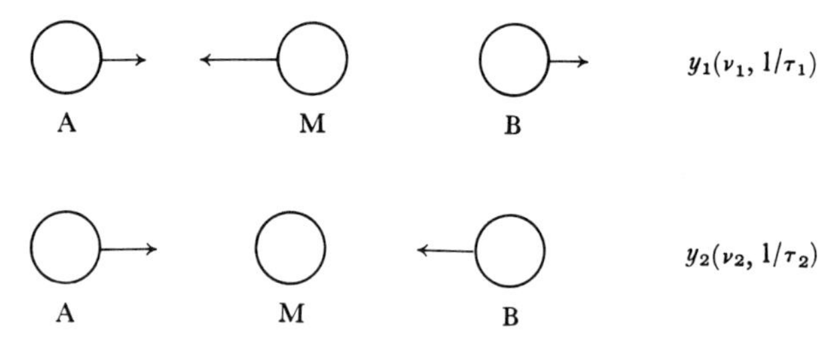

The length of the arrows, which indicate the amplitude of the motion of the respective atoms, corresponds to the Δc_A^{01}, $\Delta c_B^{01}, \ldots$ in the chemical system. Note that in the second normal mode the middle atom does not move (corresponding to $\Delta c_M^{02} = 0$) because of a special relationship between the force constants of the A—M and B—M bonds and the masses of A and B, respectively. (For symmetrical molecules such as CO_2 the masses and force constants are equal, corresponding to $k_2 = k_1$ in the chemical example.)

The problem of detectability of a certain normal mode (of vibration as well as of reaction) is perhaps best illustrated with a discussion of the CO_2 molecule. The first normal mode ($\nu_1 = 2349\ \text{cm}^{-1}$) (*20*) can be excited by infrared light, but not, however, the second normal mode ($\nu_2 = 1340\ \text{cm}^{-1}$) (*20*). This is because only vibrational modes accompanied by a change in the symmetry of charge distribution are infrared active. This situation corresponds to the stopped-flow experiments initiated by mixing M with N, or the temperature-jump experiments when $\Delta H_{AB} = 0$. The second vibrational mode of CO_2 is visible in the Raman spectrum. This corresponds to the second type of stopped-flow experiment or to the temperature-jump experiment where $\Delta H_{AB} \neq 0$. Note that here the analogy is not as perfect because in general both τ_1 and τ_2 will be observable whereas ν_1 for CO_2 is not visible in the Raman spectrum.

E. Spectrophotometric Detection

As shown for single-step systems, the experimental relaxation amplitudes depend of course also on the method of detection. In the case of spectrophotometric detection we write for the two relaxation effects in a cuvette of 1 cm path length

$$\Delta\text{OD}^{01} = \epsilon_A\, \Delta c_A^{01} + \epsilon_B\, \Delta c_B^{01} + \epsilon_M\, \Delta c_M^{01} \tag{7.152}$$

$$\Delta\text{OD}^{02} = \epsilon_A\, \Delta c_A^{02} + \epsilon_B\, \Delta c_B^{02} + \epsilon_M\, \Delta c_M^{02} \tag{7.153}$$

For the sake of simplifying the following equations we now assume $k_2 = k_1$. For the temperature-jump experiment we obtain

$$\Delta\text{OD}^{01} = \tfrac{1}{2}(\epsilon_A + \epsilon_B - 2\epsilon_M)\, \frac{K_1(\Delta K_1/K_1 + \Delta K_2/K_2)}{(1 + 2K_1)^2}\, c_0 \tag{7.154}$$

$$\Delta\text{OD}^{02} = (\epsilon_A - \epsilon_B)\, \frac{K_1(\Delta K_1/K_1 - \Delta K_2/K_2)}{2(1 + 2K_1)}\, c_0 \tag{7.155}$$

For the stopped-flow experiments when mixing M with N

$$\Delta\text{OD}^{01} = (2\epsilon_M - \epsilon_A - \epsilon_B)\, \frac{K_1}{1 + 2K_1}\, c_0 \tag{7.156}$$

$$\Delta\text{OD}^{02} = 0 \tag{7.157}$$

while for the stopped-flow experiment when mixing A′ with R

$$\Delta\text{OD}^{01} = -\tfrac{1}{2}(2\epsilon_M - \epsilon_A - \epsilon_B)\, \frac{1}{1 + 2K_1}\, c_0 \tag{7.158}$$

$$\Delta\text{OD}^{02} = \tfrac{1}{2}(\epsilon_A - \epsilon_B) c_0 \tag{7.159}$$

In general it may be relatively easy to find a spectral region where $(\epsilon_A + \epsilon_B) \neq 2\epsilon_M$ so that ΔOD^{01} is of measurable magnitude. On the other hand, when A and B are chemically very similar, as in reaction 7.124, ϵ_A and ϵ_B may also be similar at most wavelengths, thus creating an additional difficulty in detecting τ_2.

Cases I, II, and III. The concentration changes associated with the individual relaxation processes, as functions of the total concentration changes, are summarized in Table 7.2. Tables 7.3 and 7.4 show the same quantities for two specific perturbation experiments, namely, a temperature-jump experiment (Table 7.3) and a stopped-flow experiment (Table 7.4). In Table 7.5 the normal reactions are identified along with their respective rate laws; the vibrational analog for each normal reaction is also indicated.

The following points are noteworthy.

1. In most but not all cases the normal concentrations found by physical reasoning, i.e., from the rate laws of the normal reactions (Table 7.5), are identical with those found by the mathematical analysis (Table 7.1). Where they are not identical this is because the mathematical analysis provides nonnormalized normal concentration. For example, in case I the mathematical analysis provides

$$y_1 = \Delta c_A + \frac{k_1}{k_1 + k_{-1}} \Delta c_B \tag{7.160}$$

whereas from the rate law of the normal reaction we obtain

$$y_1 = \Delta c_A - K_1 \Delta c_M \tag{7.161}$$

(or $y_1 = \Delta c_A{}^1$ if the reference state is $\bar{c}_A{}^1$ instead of $\bar{c}_A{}^f$). However, substituting $-(\Delta c_A + \Delta c_B)$ for Δc_M in Eq. 7.161 leads to

$$y_1 = (1 + K_1)\left(\Delta c_A + \frac{k_1}{k_1 + k_{-1}} \Delta c_B\right) \tag{7.162}$$

which is the same as Eq. 7.160 except for a normalization factor $(1 + K_1)$. Similarly, in case II the mathematical treatment affords

$$y_2 = -(k_2/k_1)\, \Delta c_A + \Delta c_B \tag{7.163}$$

while from the second normal reaction one concludes

$$y_2 = \Delta c_A \tag{7.164}$$

But since in case II $\tau_1^{-1} \gg \tau_2^{-1}$, for times relevant to the second process we have $y_1 = 0$ and thus $\Delta c_M = 0$, which means $\Delta c_M^{02} = 0$, $\Delta c_B^{02} = -\Delta c_A^{02}$ so that Eq. 7.163 becomes

$$y_2 = -(k_2/k_1 + 1)\, \Delta c_A \tag{7.165}$$

This is the same as Eq. 7.164 except for the normalization factor $-(k_2/k_1 + 1)$.

2. In case II it may be difficult to detect τ_1 in a temperature-jump experiment if K_1 and/or K_2 are very large because the denominator in the expressions for Δc_A^{01}, Δc_B^{01}, and Δc_M^{01} contains $(K_1 + K_2)^2$ while the numerator contains only K_1 and K_2 to the first power. Qualitatively the small amplitudes are understandable because in an equilibrium mixture M is present only at very low or even undetectable concentrations and thus Δc_M is very small. On the other hand, a stopped-flow experiment starting with M as reactant (large perturbation,

$\Delta c_M^{01} = c_0$) allows one to monitor the decay of M very easily. As far as the visibility of τ_2 is concerned, it depends on whether $\Delta K_1/K_1 - \Delta K_2/K_2$ (temperature jump) or $k_1K_2 - k_2K_1$ (stopped flow) is sufficiently different from zero. Here the situation is quite similar to that in case IV; in fact, when $k_1K_2 = k_2K_1$ we have case IV.

3. In case III we note that the two reactions $M \rightleftharpoons A$ and $M \rightleftharpoons B$ equilibrate independently from each other; i.e., they are decoupled by virtue of the very high relative concentration of M.

4. It is interesting to compare the spectrophotometric relaxation amplitudes associated with τ_2 in cases II, III, and IV. For cases II and IV we have

$$\Delta OD^{02} = (\epsilon_A - \epsilon_B)\ \Delta c_A^{02} \tag{7.166}$$

while for case III

$$\Delta OD^{02} = (\epsilon_B - \epsilon_M)\ \Delta c_B{}^0 \tag{7.167}$$

In the common situation where $\epsilon_A \approx \epsilon_B \neq \epsilon_M$, for example in scheme 7.124, ΔOD^{02} becomes very small in cases II and IV regardless of the perturbation method. This does not happen in case III even for $\epsilon_A = \epsilon_B$, because here ΔOD^{02} depends on $\epsilon_B - \epsilon_M$ instead of $\epsilon_A - \epsilon_B$; note also that $\Delta OD^{01} = (\epsilon_A - \epsilon_M)\ \Delta c_A{}^0$. In physical terms this result is again easily understandable: In cases II and IV the second normal reaction corresponds to a formation of B at the expense of A or vice versa, which of course does not produce any OD change if $\epsilon_A = \epsilon_B$; in contrast A and B are produced at the expense of M in case III.

7.3.2 The Degenerate $A + 2S \rightleftharpoons AS + S \rightleftharpoons AS_2$ System

A. Normal Concentrations y_1 and y_2

The derivation of the relaxation times for the system

$$\begin{aligned} A + S &\underset{k_{-1}}{\overset{2k_1}{\rightleftharpoons}} AS \\ AS + S &\underset{2k_{-1}}{\overset{k_1}{\rightleftharpoons}} AS_2 \end{aligned} \tag{7.168}$$

has been discussed in Section 3.3.2. This system and its extensions (see Section 10.1) are of particular interest in the context of relaxation amplitudes and the problem of detectability of relaxation phenomena.

Applying Eqs. 7.95–7.96 in combination with Eqs. 3.82–3.84 ($x_1 = \Delta c_A$, $x_2 = \Delta c_{AS_2}$) affords

$$y_1 = \Delta c_A - \Delta c_{AS_2} = \Delta c_S \tag{7.169}$$

$$y_2 = \frac{k_1\bar{c}_S}{k_{-1}}\ \Delta c_A + \Delta c_{AS_2} \tag{7.170}$$

Table 7.2. Individual Concentration Changes as Functions of Total Concentration Changes for the A $\rightleftharpoons$ M $\rightleftharpoons$ B System

	Case I $k_1, k_{-1} \gg k_2, k_{-2}$	Case II $k_1, k_2 \gg k_{-1}, k_{-2}$	Case III $k_1, k_2 \ll k_{-1}, k_{-2}$	Case IV $k_{-2} = k_{-1}\ (k_2/k_1 = K_2/K_1)$
Δc_A^{01}	$\Delta c_A{}^0 + \frac{k_1}{k_1 + k_{-1}} \Delta c_B{}^0$	$-\frac{k_1}{k_1 + k_2} \Delta c_M{}^0$	$\Delta c_A{}^0$	$-\frac{K_1}{K_1 + K_2} \Delta c_M{}^0$
Δc_B^{01}	0	$-\frac{k_2}{k_1 + k_2} \Delta c_M{}^0$	0	$-\frac{K_2}{K_1 + K_2} \Delta c_M{}^0$
Δc_M^{01}	$-\Delta c_A^{01}$	$\Delta c_M{}^0$	$-\Delta c_A{}^0$	$\Delta c_M{}^0$
Δc_A^{02}	$-\frac{k_1}{k_1 + k_{-1}} \Delta c_B{}^0$	$\frac{k_2}{k_1 + k_2} \Delta c_A{}^0 - \frac{k_1}{k_1 + k_2} \Delta c_B{}^0$	0	$\frac{K_2}{K_1 + K_2} \Delta c_A{}^0 - \frac{K_1}{K_1 + K_2} \Delta c_B{}^0$
Δc_B^{02}	$\Delta c_B{}^0$	$-\frac{k_2}{k_1 + k_2} \Delta c_A{}^0 + \frac{k_1}{k_1 + k_2} \Delta c_B{}^0$	$\Delta c_B{}^0$	$-\frac{K_2}{K_1 + K_2} \Delta c_A{}^0 + \frac{K_1}{K_1 + K_2} \Delta c_B{}^0$
Δc_M^{02}	$-\frac{k_{-1}}{k_1 + k_{-1}} \Delta c_B{}^0$	0	$-\Delta c_B{}^0$	0

Table 7.3. Concentration Changes in a Temperature-Jump Experiment for the A $\rightleftharpoons$ M $\rightleftharpoons$ B System

	Case I $k_1, k_{-1} \gg k_2, k_{-2}$	Case II $k_1, k_2 \gg k_{-1}, k_{-2}$	Case III $k_1, k_2 \ll k_{-1}, k_{-2}$	Case IV $k_{-2} = k_{-1}\,(k_2/k_1 = K_2/K_1)$
Δc_A^{01}	$\dfrac{K_1}{(1+K_1)(1+K_1+K_2)}\dfrac{\Delta K_1}{K_1}c_0$	$\dfrac{k_1}{k_1+k_2}\dfrac{K_1\,\Delta K_1/K_1 + K_2\,\Delta K_2/K_2}{(K_1+K_2)^2}c_0$	$K_1\dfrac{\Delta K_1}{K_1}c_0$	$\dfrac{K_1}{K_1+K_2}\dfrac{K_1\,\Delta K_1/K_1 + K_2\,\Delta K_2/K_2}{(1+K_1+K_2)^2}c_0$
Δc_B^{01}	0	$\dfrac{k_2}{k_1}\Delta c_A^{01}$	0	$\dfrac{K_2}{K_1}\Delta c_A^{01}$
Δc_M^{01}	$-\Delta c_A^{01}$	$-\dfrac{K_1\,\Delta K_1/K_1 + K_2\,\Delta K_2/K_2}{(K_1+K_2)^2}c_0$	$-\Delta c_A^{01}$	$-\dfrac{K_1\,\Delta K_1/K_1 + K_2\,\Delta K_2/K_2}{(1+K_1+K_2)^2}c_0$
Δc_A^{02}	$\dfrac{K_1K_2(K_1\,\Delta K_1/K_1 - \Delta K_2/K_2 - K_1\,\Delta K_2/K_2)}{(1+K_1)(1+K_1+K_2)^2}c_0$	$\dfrac{K_1K_2(\Delta K_1/K_1 - \Delta K_2/K_2)}{(K_1+K_2)^2}c_0$	0	$\dfrac{K_1K_2(\Delta K_1/K_1 - \Delta K_2/K_2)}{(K_1+K_2)(1+K_1+K_2)}c_0$
Δc_B^{02}	$\dfrac{(1+K_1)\,\Delta K_2/K_2 - K_1\,\Delta K_1/K_1}{(1+K_1+K_2)^2}K_2c_0$	$-\Delta c_A^{02}$	$K_2\dfrac{\Delta K_2}{K_2}c_0$	$-\Delta c_A^{02}$
Δc_M^{02}	$\dfrac{\Delta c_A^{02}}{K_1}$	0	$-\Delta c_B^{02}$	0

Table 7.4. Concentration Changes in a Stopped-Flow Experiment (Mixing M with N) for the A $\rightleftharpoons$ M $\rightleftharpoons$ B System

	Case I $k_1, k_{-1} \gg k_2, k_{-2}$	Case II $k_1, k_2 \gg k_{-1}, k_{-2}$	Case III $k_1, k_2 \ll k_{-1}, k_{-2}$	Case IV $k_{-2} = k_{-1}\ (k_2/k_1 = K_2/K_1)$
Δc_A^{01}	$-\frac{K_1}{1 + K_1} c_0$	$-\frac{k_1}{k_1 + k_2} c_0$	$-K_1 c_0$	$-\frac{K_1}{1 + K_1 + K_2} c_0$
Δc_B^{01}	0	$\frac{k_2}{k_1} \Delta c_A^{01}$	0	$\frac{K_2}{K_1} \Delta c_A^{01}$
Δc_M^{01}	$-\Delta c_A^{01}$	c_0	$-\Delta c_A^{01}$	$\frac{K_1 + K_2}{1 + K_1 + K_2} c_0$
Δc_A^{02}	$\frac{K_1 K_2}{(1 + K_1)(1 + K_1 + K_2)} c_0$	$\frac{k_1 K_2 - k_2 K_1}{(k_1 + k_2)(K_1 + K_2)} c_0$	0	0
Δc_B^{02}	$-\frac{K_2}{1 + K_1 + K_2} c_0$	$-\Delta c_A^{02}$	$-K_2 c_0$	0
Δc_M^{02}	$\frac{\Delta c_A^{02}}{K_1}$	0	$-\Delta c_B^{02}$	0

Table 7.5. Normal Reactions in $A \rightleftharpoons M \rightleftharpoons B$ System

Case I $k_1, k_{-1} \gg k_2, k_{-2}$	Case II $k_1, k_2 \gg k_{-1}, k_{-2}$	Case III $k_1, k_2 \ll k_{-1}, k_{-2}$	Case IV $k_{-2} = k_{-1}\ (k_1/k_2 = K_1/K_2)$
First normal reaction			
$A \underset{k_1}{\overset{k_{-1}}{\rightleftharpoons}} M$	$M \xrightarrow{k_1 + k_2} (A, B)$	$A \underset{k_1}{\overset{k_{-1}}{\rightleftharpoons}} M$	$M \underset{k_{-1}}{\overset{k_1 + k_2}{\rightleftharpoons}} (A, B)$
$\frac{d\,\Delta c_A{}^1}{dt} = -(k_1 + k_{-1})\Delta c_A{}^1$ or $\frac{d(\Delta c_A - K_1\,\Delta c_M)}{dt}$ $= -(k_1 + k_{-1})(\Delta c_A - K_1\,\Delta c_M)$	$\frac{d\,\Delta c_M}{dt} = -(k_1 + k_2)\,\Delta c_M$	$\frac{d\,\Delta c_A}{dt} = -(k_{-1} + k_1)\,\Delta c_A$ $\approx -k_{-1}\,\Delta c_A$	$\frac{d\,\Delta c_M}{dt} = -(k_1 + k_2 + k_{-1})\,\Delta c_M$
Analogy to vibrating molecule			
Second normal reaction			
$(A \underset{K_1}{\rightleftharpoons} M) \underset{k_{-2}}{\overset{k_2}{\rightleftharpoons}} B$	$A \underset{\overleftarrow{k}}{\overset{\vec{k}}{\rightleftharpoons}} B$	$M \underset{k_{-2}}{\overset{k_2}{\rightleftharpoons}} B$	$A \underset{\overleftarrow{k}}{\overset{\vec{k}}{\rightleftharpoons}} B$
$\frac{d\,\Delta c_B}{dt} = -\left(\frac{k_2 k_{-1}}{k_1 + k_{-1}} + k_{-2}\right)\Delta c_B$ $(\Delta c_A - K_1\,\Delta c_M = 0)$	$\frac{d\,\Delta c_A}{dt} = -(\vec{k} + \overleftarrow{k})\,\Delta c_A$ $(\Delta c_M = 0)$	$\frac{d\,\Delta c_B}{dt} = -(k_{-2} + k_2)\,\Delta c_B$ $\approx -k_{-2}\,\Delta c_B$	$\frac{d(\Delta c_B - k_2\,\Delta c_A/k_1)}{dt}$ $= -k_{-1}\left(\Delta c_B - k_2\frac{\Delta c_A}{k_1}\right)$ or $\frac{d\,\Delta c_A}{dt} = -(\vec{k} + \overleftarrow{k})\,\Delta c_A$ (for special case when $\Delta c_M = 0$; note $\vec{k} + \overleftarrow{k} = k_{-1}$)
Analogy to vibrating molecule			

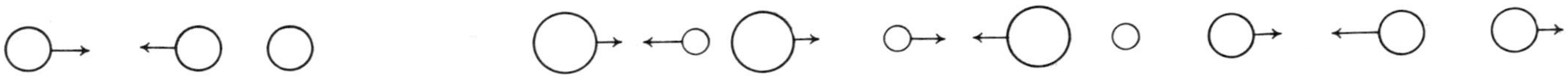

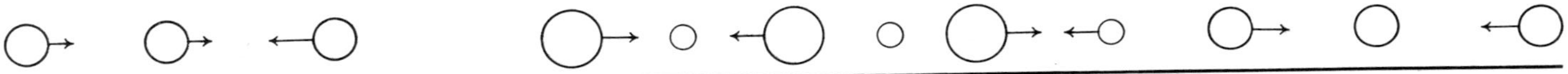

Solving for Δc_A and Δc_{AS_2} (or applying Eqs. 7.105–7.106) leads to

$$\Delta c_A = \frac{k_{-1}}{k_{-1} + k_1\bar{c}_S}(y_1 + y_2) \tag{7.171}$$

$$\Delta c_{AS_2} = -\frac{k_1\bar{c}_S}{k_{-1} + k_1\bar{c}_S}y_1 + \frac{k_{-1}}{k_{-1} + k_1\bar{c}_S}y_2 \tag{7.172}$$

In our earlier discussion we pointed out that the process associated with $\tau_1(y_1)$ is simply the reaction

$$\Phi + S \underset{k_{-1}}{\overset{k_1}{\rightleftharpoons}} \chi \tag{7.173}$$

(which is also reaction 3.85) where Φ and χ stand for free and occupied binding sites, respectively, and $c_\Phi = 2c_A + c_{AS}$, $c_\chi = c_{AS} + 2c_{AS_2}$. For reaction 7.173 the relation

$$\frac{d\,\Delta c_S}{dt} = -\frac{1}{\tau_1}\Delta c_S \tag{7.174}$$

with

$$1/\tau_1 = k_1(2\bar{c}_A + \bar{c}_{AS} + \bar{c}_S) + k_{-1} \tag{7.175}$$

holds which is of the form $dy_1/dt = -(1/\tau_1)y_1$ with y_1 as in Eq. 7.169.

In the language of this chapter reaction 7.173 is the first normal reaction. Since it takes care of all possible net binding processes, the second normal reaction must be associated with a process *not* involving any net binding. It corresponds to the redistribution of S among binding sites, for example

$$AS_2 + A \underset{k_1}{\overset{2k_{-1}}{\rightleftharpoons}} AS + S + A \underset{k_{-1}}{\overset{2k_1}{\rightleftharpoons}} AS + AS \tag{7.176}$$

Scheme 7.176 implies that the mechanism of this redistribution is by dissociation of AS_2 and subsequent new binding. An equivalent mechanism is represented by

$$A + S + AS_2 \underset{k_{-1}}{\overset{2k_1}{\rightleftharpoons}} AS + AS_2 \underset{k_{-1}}{\overset{2k_{-1}}{\rightleftharpoons}} AS + AS + S \tag{7.177}$$

where binding precedes dissociation. There is also a possibility of a direct exchange, $AS_2 + A \rightleftharpoons 2AS$; we shall assume this direct pathway to be negligible.

Let us try to derive the relaxation times for the redistribution processes. Setting up the linearized rate equation in terms of Δc_A for reaction 7.176 (7.177) provides

$$d\,\Delta c_A/dt = -2k_1\bar{c}_A\,\Delta c_S - 2k_1\bar{c}_S\,\Delta c_A + k_{-1}\,\Delta c_{AS} \tag{7.178}$$

We now consider the special situation where the first normal reaction has already reached equilibrium. In this case $y_1 = 0$ and thus $\Delta c_S = 0$, $\Delta c_A = \Delta c_{AS_2}$, and (from Eq. 3.80) $\Delta c_{AS} = -2\,\Delta c_A$. Hence Eq. 7.178 simplifies to

$$d\,\Delta c_A/dt = -2(k_1\bar{c}_S + k_{-1})\,\Delta c_A = -(1/\tau_2)\,\Delta c_A \tag{7.179}$$

Note that τ_2 is of course independent of whether the first relaxation process has reached equilibrium ($y_1 = 0$) or whether it has not ($y_1 \neq 0$) and thus τ_2^{-1} obtained from Eq. 7.179 is generally valid (cf. Eq. 3.84 derived by the determinant procedure). On the other hand, since Eq. 7.179 has been derived under the special conditions where $y_1 = 0$, Δc_A corresponds to y_2 only under these special conditions also, but not for the general case. This is similar to the situation in the $A \rightleftharpoons M \rightleftharpoons B$ system, case IV (Section 7.3.1), where $y_2 = \Delta c_A$ only when $y_1 = 0$. Eigen (*21*) has developed a derivation for y_2 based on physical reasoning for the general case.

B. Detectability of τ_1 and τ_2 in Perturbation Experiments

It is straightforward but tedious to derive the equations for Δc_A^{01}, Δc_A^{02}, $\Delta c_{AS}^{01}, \ldots$ which would prevail in a temperature-jump experiment. For a qualitative understanding this is not necessary. In fact it is immediately obvious that in a temperature-jump experiment only $y_1{}^0 \neq 0$ but $y_2{}^0 = 0$ and thus only τ_1 can be detected. This is because only the first normal reaction (Eq. 7.173) is associated with a finite reaction enthalpy, ΔH_1. The equilibrium constant of the second normal reaction (7.176 or 7.177) is simply equal to the statistical factor 4 (entropy term) and has no finite ΔH associated with it. Hence this equilibrium cannot be perturbed by a change in temperature. For the determination of k_1 and k_{-1} the measurement of τ_1 is sufficient. In fact the analysis of the relaxation curve would be very difficult if τ_2 were also present with an amplitude similar to that of τ_1 because τ_2 is very close to τ_1 (e.g., for $\bar{c}_S \gg 2\bar{c}_A + \bar{c}_{AS}$: $\tau_2^{-1} = 2\tau_1^{-1}$). In other words, the absence of τ_2 is an advantage.

On the other hand, just as in case IV of the $A \rightleftharpoons M \rightleftharpoons B$ system, erroneous conclusions about the reaction mechanism results unless one has independent knowledge of the true mechanism. Thus, trying to make τ_2 visible has to be viewed as a means to prove the mechanism; τ_2 would indeed become detectable in a mixing experiment (if the rates are slow enough to permit such an experiment), for example mixing A with S. Let us show this for the special case where S is used in large excess or is buffered, i.e., $\bar{c}_S = [S]_0$. At $t = 0$, $c_A = [A]_0$, $c_{AS} = c_{AS_2} = 0$ and one obtains

$$\Delta c_A{}^0 = [A]_0 - \bar{c}_A = \frac{2K_1[S]_0 + (K_1[S]_0)^2}{1 + 2K_1[S]_0 + (K_1[S]_0)^2}[A]_0 \qquad (7.180)$$

$$\Delta c_{AS_2}^0 = -\bar{c}_{AS_2} = -\frac{(K_1[S]_0)^2}{1 + 2K_1[S]_0 + (K_1[S]_0)^2}[A]_0 \qquad (7.181)$$

With Eqs. 7.169–7.170 we then have

$$y_1{}^0 = 2K_1[S]_0 \frac{1 + K_1[S]_0}{1 + 2K_1[S]_0 + (K_1[S]_0)^2}[A]_0 \qquad (7.182)$$

$$y_2{}^0 = (K_1[S]_0)^2 \frac{1 + K_1[S]_0}{1 + 2K_1[S]_0 + (K_1[S]_0)^2}[A]_0 \qquad (7.183)$$

Clearly, neither $y_1{}^0$ nor $y_2{}^0$ is zero and thus both relaxation effects should be visible, provided of course that, say, with spectrophotometric detection no cancellation due to special relationships among the extinction coefficients occurs, making ΔOD^{01} or ΔOD^{02} equal to zero.

The same principles discussed here can be applied to extensions of the title system where A has more than two equivalent binding sites. An application to the binding of small molecules to multiunit enzymes is discussed in Section 10.1.

Problems

1. Assume you deal with a system such as Eq. 7.1 where $\Delta H_1 = 4$ kcal/mole and $\Delta H_2 = 6$ kcal/mole. How large are $\Delta c_A^{01}/c_0$, $\Delta c_B^{01}/c_0$, $\Delta c_M^{01}/c_0$, $\Delta c_A^{02}/c_0$, $\Delta c_B^{02}/c_0$, and $\Delta c_M^{02}/c_0$, respectively, for a 10° temperature jump ($T_f = 25°$) for the cases

(a) $K_1 = 10$, $K_2 = 3$, $k_1 = 100k_2$
(b) $K_1 = 100$, $K_2 = 30$, $k_1 = k_2$
(c) $K_1 = 0.01$, $K_2 = 0.03$, $k_1 = k_2$
(d) $K_1 = 1$, $K_2 = 3$, $k_{-2} = k_{-1}$.

Comment on your findings.

2. Assume you are faced with doing a temperature-jump study of the reaction

$$AH \rightleftharpoons A^- + H^+$$

where AH is a carbon acid with $pK_a = 8.0$. Unfortunately the solubility of AH is only 10^{-5} *M* and thus its solution is too dilute to act as its own buffer; consequently, a buffer has to be added for accurate pH control. If you had three different buffers of $pK_a \sim 8$ available, with the acid dissociation constant associated with $\Delta H = +6$, -10, and -4 kcal/mole, respectively, which one would you choose if the dissociation of AH were associated with a $\Delta H = -5$ kcal/mole?

3. Derive the relaxation times which characterize the following reaction scheme:

$$
\begin{array}{ccccc}
 & & B & & \\
 & & k_{-2}\downarrow\uparrow k_{2} & & \\
C & \underset{k_{-3}}{\overset{k_3}{\rightleftharpoons}} & M + N & \underset{k_{-1}}{\overset{k_1}{\rightleftharpoons}} & A \\
 & & k_{-4}\uparrow\downarrow k_{4} & & \\
 & & D & &
\end{array}
$$

Which ones are detectable in an experiment initiated by mixing M + N? (a) Assume $k_{-4} = k_{-3} = k_{-2} = k_{-1}$; (b) assume $k_{-1}, k_{-2}, k_{-3}, k_{-4} \gg k_1, k_2, k_3, k_4$. (*Hint:* First find the normal concentrations by physical reasoning.)

References

1. F. Guillain and D. Thusius, *J. Amer. Chem. Soc.* **92**, 5534 (1970).
2. D. Thusius, *J. Amer. Chem. Soc.* **94**, 356 (1972).
3. R. Winkler, Ph.D. Thesis, Max Planck Inst., Göttingen, and Tech. Univ. Vienna (1969).
4. H. Diebler, M. Eigen, G. Ilgenfritz, G. Maass, and R. Winkler, *Pure Appl. Chem.* **20**, 93 (1969).
5. W. Knoche, C. A. Firth, and D. Hess, *Advan. Mol. Relax. Proc.* **6**, 1 (1974).
6. C. Kalidas, W. Knoche, and D. Papadopoulos, *Ber. Bunsenges. Phys. Chem.* **75**, 106 (1971).
7. C. F. Bernasconi, *J. Org. Chem.* **36**, 1671 (1971).
8. D. Thusius, G. Foucalt, and F. Guillain, *in* "Dynamic Aspects of Conformation Changes in Biological Macromolecules" (C. Sadron, ed.). Reidel, Dordrecht-Holland, 1972.
9. B. H. Robinson, A. Seelig-Löffler, and G. Schwarz, *J. Chem. Soc. Faraday Trans. I* **71**, 815 (1975).
10. R. Zana and J. Lang, *J. Phys. Chem.* **74**, 2734 (1970).
11. M. Hussey and P. D. Edmonds, *J. Phys. Chem.* **75**, 4012 (1971).
12. M. Eigen and L. DeMaeyer, *in* "Technique of Organic Chemistry" (S. L. Friess, E. S. Lewis, and A. Weissberger, eds.), Vol. VIII, part 2, p. 895. Wiley (Intersciençe), New York, 1963.
13. M. Eigen and L. DeMaeyer, *in* "Techniques of Chemistry" (G. G. Hammes, ed.), Vol. VI, part 2, p. 63. Wiley (Interscience), New York, 1973.
14. K. Kustin, D. Shear, and D. Kleitman, *J. Theoret. Biol.* **9**, 186 (1965).
15. G. Schwarz, *Rev. Mod. Phys.* **40**, 206 (1968).
16. H. T. G. Hayman, *Trans. Faraday Soc.* **66**, 1402 (1970).
17. G. G. Hammes and R. Schimmel, *Enzymes* **2**, 67 (1970).
18. M. Eigen and G. Ilgenfritz, unpublished notes.
19. C. F. Bernasconi and R. G. Bergstrom, *J. Amer. Chem. Soc.* **96**, 2397 (1974).
20. G. Herzberg, "Infrared and Raman Spectra of Polyatomic Molecules," p. 272. Van Nostrand-Reinhold, Princeton, New Jersey, 1945.
21. M. Eigen, unpublished Harvard lecture notes (1966).

Chapter 8 | *Complete Solution of the Relaxation Equation*

8.1 Derivation of the Complete Relaxation Equation

In Chapter 1 it was shown that after a single, small perturbation (jump) the approach to the new equilibrium condition of any two-state system can be described by equations of the form

$$d\,\Delta c_j/dt = -(1/\tau)\,\Delta c_j \tag{8.1}$$

$$\Delta c_j = \Delta c_j{}^0 \exp(-t/\tau) \tag{8.2}$$

where $\Delta c_j = c_j(t) - \bar{c}_j{}^{\mathrm{f}}$, $\Delta c_j{}^0 = \bar{c}_j{}^{\mathrm{i}} - \bar{c}_j{}^{\mathrm{f}}$. In deriving these equations it was assumed that the perturbation is instantaneous and that it occurs at $t = 0$.

In reality the jump in the external parameter is not instantaneous, it has a finite rise time. If we want to take this into consideration, we need a somewhat more general approach. This becomes necessary in cases where the relaxation time is very short and comparable to the rise time of the jump.

Another important situation where Eqs. 8.1 and 8.2 cannot apply is in the case of oscillatory perturbations, as in ultrasonic or dielectric relaxation techniques. We shall see that the same extended mathematical formalism can be applied to both situations.

The new element in the more general treatment is to allow the "equilibrium concentration" $\bar{c}_j$ to be possibly time dependent. Let us define

$$\Delta c_j(t) = c_j(t) - c_j{}^* \tag{8.3}$$

$$\Delta \bar{c}_j(t) = \bar{c}_j(t) - c_j{}^* \tag{8.4}$$

where $c_j{}^*$ is a time-*in*dependent reference value the choice of which is arbitrary; most conveniently one selects either $c_j{}^* = \bar{c}_j{}^{\mathrm{f}}$ or $c_j{}^* = \bar{c}_j{}^{\mathrm{i}}$. The concentration $\bar{c}_j(t)$ is then the equilibrium concentration that would establish itself if the system were allowed to relax under the external conditions prevailing at that very moment in time. Thus it is evident that during the initial phases of a jump experiment $\bar{c}_j(t)$ changes with time until the external parameter (e.g., temperature) has reached its final value; at this time $\bar{c}_j(t)$ becomes equal to $\bar{c}_j^{\mathrm{f}}$. In an oscillatory perturbation experiment $\bar{c}_j(t)$ oscillates as long as the perturbation is on but becomes $\bar{c}_j^{\mathrm{f}} = \bar{c}_j^{\mathrm{i}}$ at the time the perturbation is switched off.

Thus, at the end of the jump, or when the oscillating perturbation is turned off, Eqs. 8.3 and 8.4 reduce to

$$\Delta c_j(t) = c_j(t) - \bar{c}_j^{\mathrm{f}} \tag{8.5}$$

$$\Delta \bar{c}_j = \bar{c}_j^{\mathrm{f}} - \bar{c}_j^{\mathrm{f}} = 0 \tag{8.6}$$

if we define $c_j^* = \bar{c}_j^{\mathrm{f}}$, or to

$$\Delta c_j(t) = c_j(t) - \bar{c}_j^{\mathrm{i}} \tag{8.7}$$

$$\Delta \bar{c}_j = \bar{c}_j^{\mathrm{f}} - \bar{c}_j^{\mathrm{i}} = \Delta c_j^{\infty} \tag{8.8}$$

if we define $c_j^* = \bar{c}_j^{\mathrm{i}}$. We of course recognize Eq. 8.5 as being equivalent to 1.9, and Eq. 8.7 as equivalent to 1.27; i.e., our previous treatment is a special case of the general treatment.

Let us now show, again using the A + B $\rightleftharpoons$ C system as an example, that the linearized rate equation becomes

$$d\,\Delta c_j/dt = -(1/\tau)(\Delta c_j - \Delta \bar{c}_j) \tag{8.9}$$

According to Eqs. 8.3 and 8.4 we can express the concentrations of A, B, and C as

$$c_{\mathrm{A}} = c_{\mathrm{A}}^* + \Delta c_{\mathrm{A}} = \bar{c}_{\mathrm{A}} + \Delta c_{\mathrm{A}} - \Delta \bar{c}_{\mathrm{A}}$$

$$c_{\mathrm{B}} = c_{\mathrm{B}}^* + \Delta c_{\mathrm{B}} = \bar{c}_{\mathrm{B}} + \Delta c_{\mathrm{B}} - \Delta \bar{c}_{\mathrm{B}}$$

$$c_{\mathrm{C}} = c_{\mathrm{C}}^* + \Delta c_{\mathrm{C}} = \bar{c}_{\mathrm{C}} + \Delta c_{\mathrm{C}} - \Delta \bar{c}_{\mathrm{C}}$$

From stoichiometry we have

$$\Delta c_{\mathrm{A}} = \Delta c_{\mathrm{B}} = -\Delta c_{\mathrm{C}} = x \tag{8.10}$$

$$\Delta \bar{c}_{\mathrm{A}} = \Delta \bar{c}_{\mathrm{B}} = -\Delta \bar{c}_{\mathrm{C}} = \bar{x} \tag{8.11}$$

The rate equation, in terms of dc_{A}/dt, then becomes (after replacing c_{A} with $c_{\mathrm{A}}^* + x$ in the differential quotient but c_{A} with $\bar{c}_{\mathrm{A}} + x - \bar{x}$, etc., on the right-hand side of the rate equation)

$$dx/dt = -k_1(\bar{c}_{\mathrm{A}} + x - \bar{x})(\bar{c}_{\mathrm{B}} + x - \bar{x}) + k_{-1}(\bar{c}_{\mathrm{C}} - x + \bar{x}) \tag{8.12}$$

In a completely rigorous treatment account would have to be taken of the fact that during the duration of the jump (or of the oscillation) k_1 and k_{-1} may also change with time, e.g., in a temperature-jump experiment the change is from k_1^{i} and k_{-1}^{i} at $t = 0$ to k_1^{f} and k_{-1}^{f} at the time of completion of the jump. However, this would make the mathematics very complicated indeed. Since for small perturbations k_1^{i} and k_{-1}^{i} do not differ much from k_1^{f} and k_{-1}^{f}, the assumption that they are constant throughout the experiment is a reasonable approximation.

After multiplying out and linearization Eq. 8.12 reduces to

$$dx/dt = -(1/\tau)(x - \bar{x}) \tag{8.13}$$

with $1/\tau = k_1(\bar{c}_A + \bar{c}_B) + k_{-1}$. Equation 8.13 is called the complete relaxation equation. It can be integrated by the common method of expressing $x(t)$ as the product of two subfunctions, $x(t) = u(t)\cdot v(t)$, as follows. Rearranging Eq. 8.13 leads to

$$\tau(dx/dt) + x = \bar{x} \tag{8.14}$$

Since $dx/dt = u(dv/dt) + v(du/dt)$, Eq. 8.14 becomes

$$\tau u \frac{dv}{dt} + \tau v \frac{du}{dt} + uv = \bar{x} \tag{8.15}$$

Let $v(t)$ be a function so that

$$\tau \frac{dv}{dt} + v = 0 \tag{8.16}$$

Multiplying Eq. 8.16 with u and subtracting from Eq. 8.15 affords

$$\tau \frac{du}{dt} = \frac{\bar{x}}{v} \tag{8.17}$$

Integration of Eq. 8.16,

$$\int_{v(0)}^{v(t)} \frac{dv}{v} = -(1/\tau) \int_0^t dt \tag{8.18}$$

leads to

$$v(t) = v(0) \exp(-t/\tau) \tag{8.19}$$

We now insert Eq. 8.19 into 8.17

$$\tau \frac{du}{dt} = \frac{\bar{x}(t)}{v(0)} \exp\left(\frac{t}{\tau}\right) \tag{8.20}$$

Integration,

$$\int_{u(0)}^{u(t)} du = u(t) - u(0) = \frac{1}{\tau v(0)} \int_0^t \bar{x}(t) \exp\left(\frac{t}{\tau}\right) dt \tag{8.21}$$

and forming the product $u(t)\cdot v(t)$ affords

$$x(t) = x_0 \exp\left(\frac{-t}{\tau}\right) + \frac{\exp(-t/\tau)}{\tau} \int_0^t \bar{x}(t) \exp\left(\frac{t}{\tau}\right) dt \tag{8.22}$$

Equation 8.22 represents the complete solution of the relaxation equation. The first term is called the transient solution, the second term the forced solution of the relaxation equation; $\bar{x}(t)$ is known as the forcing function.

The same treatment is applicable to multistep mechanisms where the normal concentration y (see Section 7.2) is substituted for x, and $\bar{y}$ for $\bar{x}$. Thus all equations to follow are equally valid for any y_j and $\bar{y}_j$.

8.2 Transient and Forced Solutions as Special Cases of the Complete Solution

In a single jump experiment with an "instantaneous" jump Eq. 8.22 reduces to the familiar

$$x(t) = x_0 \exp(-t/\tau) \tag{8.23}$$

if $c_j{}^* = \bar{c}_j{}^{\mathrm{f}}$ ($\bar{x} = 0$, $x_0 = \bar{c}_j{}^{\mathrm{i}} - \bar{c}_j{}^{\mathrm{f}}$), or to

$$x(t) = x_\infty[1 - \exp(-t/\tau)] \tag{8.24}$$

if $c_j{}^* = \bar{c}_j{}^{\mathrm{i}}$ ($\bar{x} = x_\infty = \bar{c}_j{}^{\mathrm{f}} - \bar{c}_j{}^{\mathrm{i}}$).

If the jump has a finite rise time and is complete at $t = t_1$, we obtain for $t \geq t_1$

$$x(t \geq t_1) = x_0 \exp[-(t + t_1)/\tau] = x(t_1) \exp(-t/\tau) \tag{8.25}$$

For very rapid jumps ($t_1 \ll \tau$), Eqs. 8.25 and 8.23 become equivalent and the complete solution of the relaxation equation is dominated by the transient term which depends only on the initial conditions, x_0, and not on the forcing function, $\bar{x}$.

For $t_1 \sim \tau$ the complete solution of the relaxation equation must be considered if τ is to be determined. This necessitates a knowledge of the forcing function $\bar{x}$ as well as of the initial conditions. Several cases are discussed in Section 8.3. For $t_1 \gg \tau$, $x(t)$ is dominated by the forced solution during the entire useful time and τ cannot be determined from a jump experiment.

When dealing with oscillating perturbations only the forced solution

$$x_{\text{forced}} = \frac{\exp(-t/\tau)}{\tau} \int_0^t \bar{x}(t) \exp\left(\frac{t}{\tau}\right) dt \tag{8.26}$$

has to be considered. This is because after the initial phases of the experiment the transient term vanishes from Eq. 8.22 since at $t \gg \tau$, $x_0 \exp(-t/\tau) \approx 0$. Note that, in contrast to the transient solution, the forced solution depends only on the forcing function, $\bar{x}$, and not on the initial condition, x_0. Explicit solutions of Eq. 8.26 for oscillating $\bar{x}(t)$ are dealt with in Section 8.4.

8.3 Complete Solution for Some Common Step Functions (Transient Relaxation Methods)

We now consider a few common step functions, $\bar{x}$, as they occur in transient relaxation methods and solve the complete relaxation equation.

8.3.1 Rectangular Step Function

A rectangular step function is equivalent to an instantaneous perturbation (zero rise time). If we define $c_j{}^* = \bar{c}_j{}^{\mathrm{f}}$, we have $\bar{x}(t < 0) = x_0$, $\bar{x}(t \geq 0) = 0$, $x(t < 0) = x_0$, and $x(t \geq 0)$ given by Eq. 8.23; with the definition $c_j{}^* = \bar{c}_j{}^{\mathrm{i}}$

we have $\bar{x}(t < 0) = 0$, $\bar{x}(t \geq 0) = x_\infty$, $x(t < 0) = 0$, and $x(t \geq 0)$ given by Eq. 8.24. The latter situation is shown in Fig. 8.1.

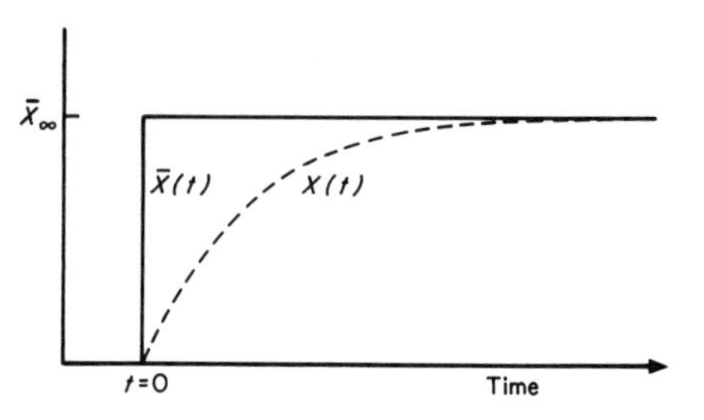

FIGURE 8.1. Relaxational response to rectangular step function.

8.3.2 Step Functions with Linear Increase

The step function with linear increase (and nonzero rise time) is of the form

$$\bar{x} = 0, \qquad \text{for} \quad t < 0 \tag{8.27}$$

$$\bar{x} = x_\infty(t/t_1), \qquad \text{for} \quad t_1 \geq t \geq 0 \tag{8.28}$$

$$\bar{x} = x_\infty, \qquad \text{for} \quad t \geq t_1 \tag{8.29}$$

when the definition $c_j^* = \bar{c}_j^{\,\mathrm{i}}$ (here more convenient than $c_j^* = \bar{c}_j^{\,\mathrm{f}}$) prevails. Thus for x we obtain from Eq. 8.22

$$x = x_\infty\left(\frac{t}{t_1}\right)\left[1 - \frac{\exp(t/\tau) - 1}{t/\tau}\exp\left(\frac{-t}{\tau}\right)\right] \qquad \text{for} \quad t_1 \geq t \geq 0 \tag{8.30}$$

$$x = x_\infty\left[1 - \frac{\exp(t_1/\tau) - 1}{t_1/\tau}\exp\left(\frac{-t}{\tau}\right)\right] \qquad \text{for} \quad t \geq t_1 \tag{8.31}$$

The functions $\bar{x}(t)$ and $x(t)$ are shown in Fig. 8.2. Note that for $t_1 \ll \tau$: $\exp(t_1/\tau) \approx 1 + (t_1/\tau)$ and Eq. 8.31 simplifies to Eq. 8.24.

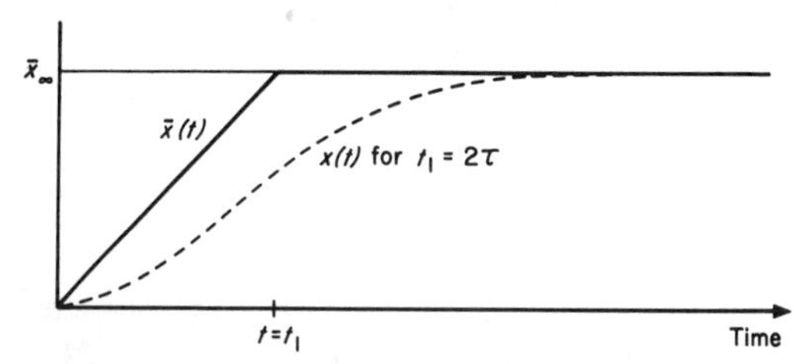

FIGURE 8.2. Relaxational response to a step function with linear increase. [From Eigen and DeMaeyer (*1*), by permission of John Wiley & Sons.]

The temperature jump produced by the discharge of a cable (Section 11.1.1) is an example of a step function with a linear increase.

8.3.3 Step Function with Exponential Increase

The exponential step function is one of the easiest to realize experimentally. It occurs, for example, in the most popular relaxation method, namely the Joule heating temperature-jump technique based on the discharge of a capacitor, as is discussed in more detail in Section 11.1.1. The forcing function (again with the definition $c_j^* = \bar{c}_j^{\,\mathrm{i}}$) is

$$\bar{x} = 0 \qquad \text{for} \quad t < 0 \tag{8.32}$$

$$\bar{x} = x_\infty[1 - \exp(-t/\tau_\mathrm{d})] \qquad \text{for} \quad t \geq 0 \tag{8.33}$$

where τ_d is the time constant for the capacitor discharge. Equation 8.22 then becomes

$$x = x_\infty\left[1 + \frac{\tau_\mathrm{d}}{\tau - \tau_\mathrm{d}}\exp\left(\frac{-t}{\tau_\mathrm{d}}\right) - \frac{\tau}{\tau - \tau_\mathrm{d}}\exp\left(\frac{-t}{\tau}\right)\right] \qquad \text{for} \quad t \geq 0 \tag{8.34}$$

which once more reduces to Eq. 8.24 if $\tau_\mathrm{d} \ll \tau$. Figure 8.3 shows $x(t)$ for two different τ/τ_d ratios.

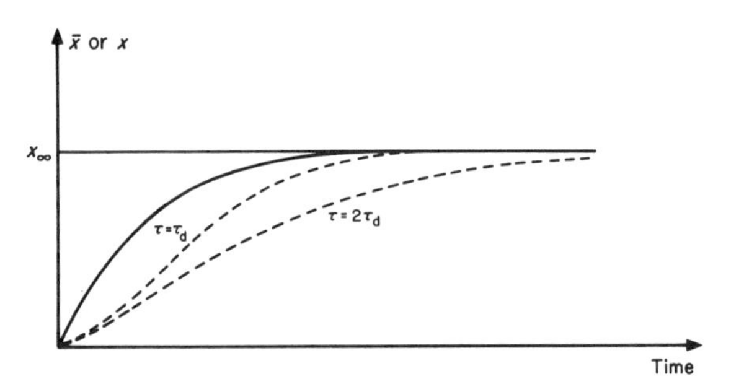

FIGURE 8.3. Relaxational response to a step function with exponential increase; (———) = $\bar{x}(t)$, (– – –) = $x(t)$. [From Eigen and DeMaeyer (*1*), by permission of John Wiley & Sons.]

For the alternative definition $c_j^* = \bar{c}_j^{\,\mathrm{f}}$ the respective functions are

$$\bar{x} = x_0 \qquad \text{for} \quad t < 0 \tag{8.35}$$

$$\bar{x} = x_0 \exp(-t/\tau_\mathrm{d}) \qquad \text{for} \quad t \geq 0 \tag{8.36}$$

$$x = x_0\left[\frac{\tau}{\tau - \tau_\mathrm{d}}\exp\left(\frac{-t}{\tau}\right) - \frac{\tau_\mathrm{d}}{\tau - \tau_\mathrm{d}}\exp\left(\frac{-t}{\tau_\mathrm{d}}\right)\right] \qquad \text{for} \quad t \geq 0 \tag{8.37}$$

We note that the problem of evaluating τ in the event that it is of comparable magnitude to τ_d is the same as that of evaluating two closely spaced relaxation times (see Section 9.2). However, τ_d can usually be determined or calculated independently (see Section 11.1.1) which makes an analysis of τ easier than in the case of two unknown relaxation times.

8.3.4 Rectangular Pulse

A rectangular step function of limited duration is called a rectangular pulse and (with $c_j^* = \bar{c}_j^{\,\mathrm{i}}$) can be described by the following equations:

$$\bar{x} = 0 \qquad \text{for} \quad t < 0 \quad \text{and} \quad t > \theta \tag{8.38}$$

$$\bar{x} = x_\infty \qquad \text{for} \quad 0 \leq t \leq \theta \tag{8.39}$$

θ is the duration of the pulse. For x we then obtain

$$x = 0 \quad \text{for} \quad t < 0 \tag{8.40}$$

$$x = x_\infty[1 - \exp(-t/\tau)] \quad \text{for} \quad 0 \leq t \leq \theta \tag{8.41}$$

$$x = x_\infty[\exp[-(t - \theta)/\tau] - \exp(-t/\tau)] \quad \text{for} \quad t > \theta \tag{8.42}$$

Equation 8.42 is derived as follows. At $t = \theta$ we have

$$x(\theta) = x_\infty[1 - \exp(-\theta/\tau)] \tag{8.43}$$

We now take the point of view that the end of the pulse at $t = \theta$ is equivalent to a new rectangular step function for which

$$\bar{x} = x(\theta) \quad \text{for} \quad t = \theta \tag{8.44}$$

$$\bar{x} = 0 \quad \text{for} \quad t > \theta \tag{8.45}$$

The decay of x is then given by

$$x = x(\theta) \exp[-(t - \theta)/\tau] \quad \text{for} \quad t > \theta \tag{8.46}$$

which combined with Eq. 8.43 gives 8.42. Note that we have $(t - \theta)/\tau$ rather than t/τ in the exponent because the decay starts only at $t = \theta$, whereas t is counted from the beginning.

The forcing function in the shock-wave pressure-jump method can be regarded as a rectangular pulse (Section 12.2); rectangular electric pulses are being used in the electric field-jump method (Section 13.2.1). If the pulse duration is long, $\theta \gg \tau$, the situation is the same as for the rectangular step function and τ can be measured from the transient response. For a pulse duration comparable to τ there is the possibility of determining τ from the ratio $x(\theta)/x_\infty = [1 - \exp(-\theta/\tau)]$; by varying the pulse length and plotting $\log[x_\infty - x(\theta)]$ versus θ one obtains a straight line with a slope of $-1/2.303\tau$. The situation is shown in Fig. 8.4.

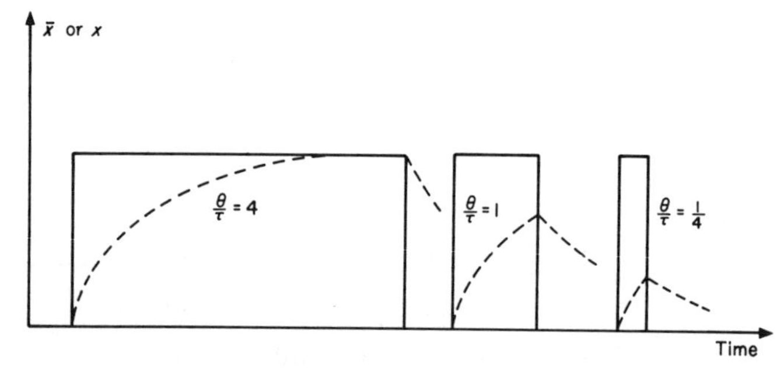

FIGURE 8.4. Relaxational response to rectangular pulses of different duration θ; (———) = $\bar{x}(t)$, (- - -) = $x(t)$. [From Eigen and DeMaeyer (*1*), by permission of John Wiley & Sons.]

8.4 Forced Solution for Oscillating Forcing Function (Stationary Relaxation Methods)

We now consider the forced solution of the relaxation equation for the case where $\bar{x}$ (or $\bar{y}$) is an oscillating forcing function. The situation of highest practical interest is the one of harmonic oscillation, i.e., where $\bar{x}$ is a sinusoidal function of time. This is, for example, the case in an ultrasonic experiment; periodic pressure fluctuations caused by the propagating sound wave induce a periodic fluctuation in the equilibrium constant ($\partial \ln K/\partial p = -\Delta V/RT$), and thus in $\bar{x}$.

A sinusoidal time fluctuation can be expressed as $\bar{x} = A \sin \omega t$ or $A \cos \omega t$ where A is the amplitude (peak values of $\bar{x}$) and $\omega = 2\pi f$ is the angular frequency while f is the frequency in hertz. For the mathematical treatment of harmonic oscillations it is often convenient to add an imaginary part in order to make $\bar{x}$ a complex number

$$\bar{x} = A(\cos \omega t + i \sin \omega t) \tag{8.47}$$

By virtue of the well-known relationship

$$a \pm ib = \rho(\cos \varphi \pm i \sin \varphi) = \rho \exp(\pm i\varphi) \tag{8.48}$$

with $\rho = (a^2 + b^2)^{1/2}$ one can then express $\bar{x}$ as

$$\bar{x} = A \exp(i\omega t) \tag{8.49}$$

which makes the ensuing mathematical manipulations easier; after these mathematical operations one may revert to the trigonometric notation and make use of the real part.

In principle the solution of our problem would now involve the evaluation of Eq. 8.26 with $\bar{x}$ according to Eq. 8.49. This can indeed be done but there is a more convenient procedure based on the following. Qualitatively it is easily appreciated that unless the chemical relaxation time is extremely short compared to the reciprocal frequency of oscillation the equilibrium and thus $x(t)$ cannot adjust itself instantaneously to the value of $\bar{x}(t)$ prevailing at any given time. This leads to a time lag or phase shift between the $x(t)$ and $\bar{x}(t)$. In addition the amplitude of x will be less than that of $\bar{x}$ because there is not enough time to reach the value of x which corresponds to the peak value of $\bar{x}$. In the extreme situation where the oscillation frequency is very much larger than the reciprocal relaxation time, the system cannot adjust at all and the amplitude of the response drops to zero. The various situations are shown in Fig. 8.5.

To find the mathematical description of the curves shown in Fig. 8.5 we proceed as follows. Based on the arguments just presented we describe x by

$$x = A' \cos(\omega t - \varphi') \tag{8.50}$$

where A' is the amplitude and φ' is the phase shift, while the frequency is of course the same as that for $\bar{x}$. Again converting to complex numbers this is equivalent to

$$x = A' \exp(i\omega t - i\varphi') = B \exp(i\omega t) \tag{8.51}$$

with $B = A' \exp(-i\varphi')$. A different way of expressing x is in terms of $\bar{x}$:

$$x = \sigma\bar{x} \tag{8.52}$$

where $\sigma = B/A$ is a time-independent function called the transfer function. It describes how the oscillation is transferred from $\bar{x}$ to x. Specifically, it will contain information about the phase shift (φ') and also about the attenuation of the amplitude (A' compared to A) which arises from absorbing some of the energy by the system. From our qualitative understanding we expect this phase shift and amplitude attenuation to depend on the relation between the relaxation time and the oscillation frequency; hence σ contains all the information necessary for determining the relaxation time.

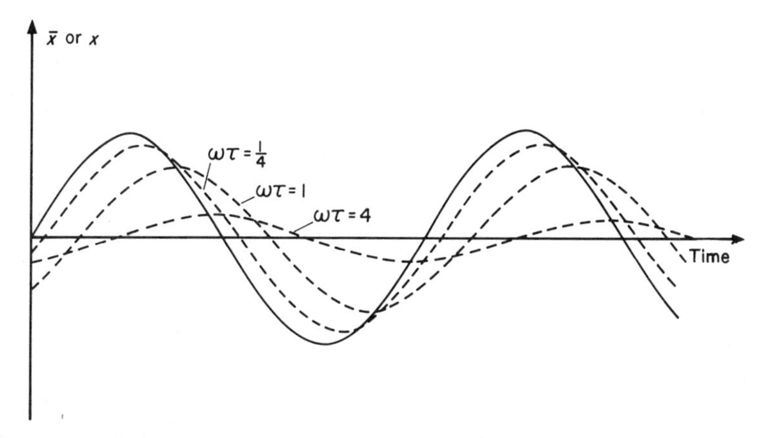

FIGURE 8.5. Relaxational response to a harmonic oscillation; (———) = $\bar{x}(t)$, (- - -) = $x(t)$. [From Eigen and DeMaeyer (*1*), by permission of John Wiley & Sons.]

We find σ easily by substituting $B \exp(i\omega t)$ for x in Eq. 8.13. With $dx/dt = i\omega x$, Eq. 8.13 becomes

$$\tau i\omega x + x = \bar{x} \tag{8.53}$$

or

$$x = \frac{1}{1 + i\omega\tau}\bar{x} = \sigma\bar{x} \tag{8.54}$$

Multiplying both numerator and denominator by $(1 - i\omega\tau)$ yields

$$\sigma = \frac{1}{1 + \omega^2\tau^2} - \frac{i\omega\tau}{1 + \omega^2\tau^2} \tag{8.55}$$

which allows the transfer function to be separated into a real and an imaginary part:

$$\sigma = \sigma_{\text{re}} - i\sigma_{\text{im}} \tag{8.56}$$

with

$$\sigma_{\text{re}} = \frac{1}{1 + \omega^2\tau^2} \tag{8.57}$$

$$\sigma_{\text{im}} = \frac{\omega\tau}{1 + \omega^2\tau^2} \tag{8.58}$$

The two functions σ_{re} and σ_{im} are shown in Fig. 8.6. Note that σ_{re} has the typical shape of a dispersion signal (half-value at $\omega = 1/\tau$), whereas σ_{im} looks like a typical absorption signal with a maximum at $\omega = 1/\tau$.

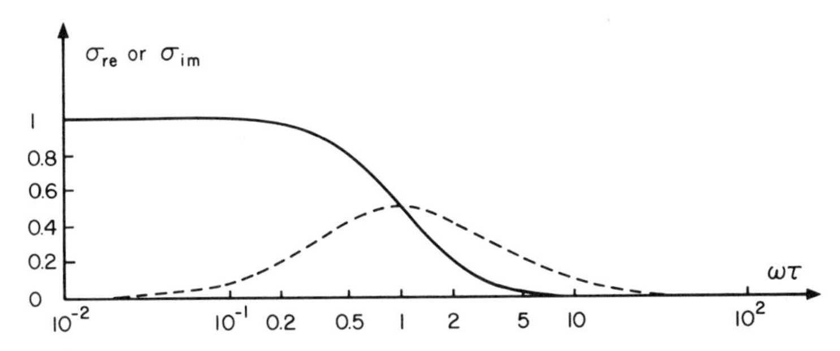

FIGURE 8.6. Frequency dependence of σ_{re} (———) and σ_{im} (– – –). [From Eigen and DeMaeyer (*1*), by permission of John Wiley & Sons.]

The important point is that the frequency dependences of σ_{re} and σ_{im} are directly related to those of measurable physical quantities such as the absorption and velocity dispersion of a sound wave, or the dielectric loss and dispersion of the dielectric constant in a system subjected to an ac electrical field. Relaxation methods based on the measurement of the frequency dependence of sound velocity and sound absorption, as well as of the dielectric constant and the dielectric loss immediately suggest themselves and in fact have been developed (see Chapters 15 and 16).

We can now also find A' and φ' of Eq. 8.51 as functions of $\omega\tau$ as follows. Combining Eqs. 8.49 and 8.52 affords

$$x = \sigma A \exp(i\omega\tau) \tag{8.59}$$

while comparison of Eq. 8.51 with 8.59 leads to

$$A' \exp(-i\varphi') = \sigma A \tag{8.60}$$

Expressing σ as

$$\sigma_{\text{re}} - i\sigma_{\text{im}} = \rho(\cos\varphi - i\sin\varphi) = \rho\exp(-i\varphi) \tag{8.61}$$

with

$$\rho = (\sigma_{\text{re}}^2 + \sigma_{\text{im}}^2)^{1/2} = 1/(1 + \omega^2\tau^2)^{1/2}$$

setting $\varphi' = \varphi$, and combining Eq. 8.60 with 8.61 affords

$$A' = \frac{A}{(1 + \omega^2\tau^2)^{1/2}}$$

Finally, setting $\sigma_{re} = \rho \cos \varphi$ and $\sigma_{im} = \rho \sin \varphi$ leads to

$$\cos \varphi = 1/(1 + \omega^2\tau^2)^{1/2}, \qquad \sin \varphi = \omega\tau/(1 + \omega^2\tau^2)^{1/2}, \qquad \tan \varphi = \sigma_{im}/\sigma_{re} = \omega\tau$$

Thus we see that when $\omega\tau \ll 1$, $A' \rightarrow A$, $\varphi \rightarrow 0$, and

$$x \approx \bar{x} = A \cos \omega\tau$$

For $\omega\tau = 1$, $A' = A/2^{1/2}$, $\varphi = \frac{1}{4}\pi$, making

$$x = \frac{A}{2^{1/2}} \cos(\omega t - \tfrac{1}{4}\pi)$$

while for $\omega\tau \gg 1$, $A' \rightarrow 0$, $\varphi \rightarrow \frac{1}{2}\pi$.

Problems

1. (a) Prove that σ_{im} goes through a maximum when $\omega = \tau^{-1}$.
(b) Where is the inflection point of σ_{re} as a function of $\omega\tau$? Does your result correspond to your intuitive expectation?

2. Calculate the amplitude attenuation and phase shift of the sinusoidal perturbation $\bar{x}$ which occurs by chemical relaxation when $\omega\tau = 0.1$, 0.3, 1.0, 3.0, and 10.0, respectively.

3. After how many τ_d values can you assume that the relaxational response to a step function with exponential increase becomes a pure single exponential? Assume you can tolerate an error of (a) 10%, (b) 5%, (c) 2% in the relaxation time.

Reference

1. M. Eigen and L. DeMaeyer, *in* "Technique of Organic Chemistry" (S. L. Friess, E. S. Lewis, and A. Weissberger, eds.), Vol. VIII, part 2, p. 895. Wiley (Interscience), New York, 1963.

Chapter 9 | Evaluation of Relaxation Times from Experimental Relaxation Curves

In this chapter we deal mainly with transient relaxation techniques. A discussion of how the data are analyzed for the stationary techniques is presented when dealing with the specific techniques.

9.1 One Relaxation Time

The evaluation follows the principles outlined in Section 1.2 (see Fig. 1.1). When fast reaction techniques are used, the relaxation curves are usually displayed on an oscilloscope screen and photographed for evaluation. An example is shown in Fig. 9.1 for a temperature-jump experiment with spectrophotometric

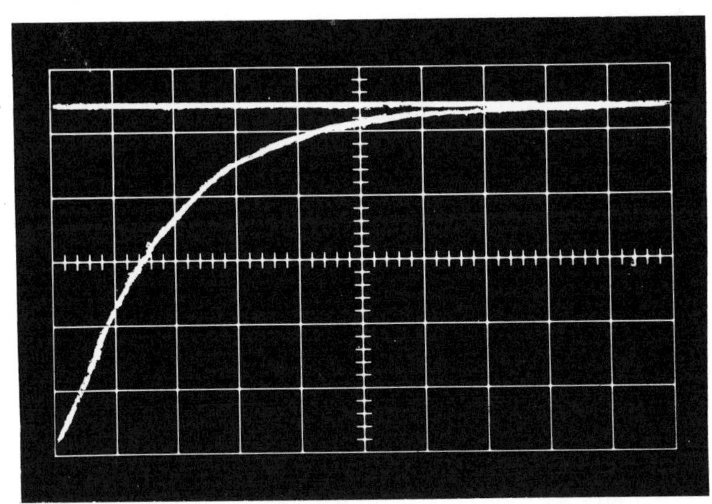

FIGURE 9.1. Temperature-jump oscilloscope trace of an azaviolene redox system (see Section 4.5.1); sweep rate 50 μsec per division.

detection. The vertical axis on the oscilloscope picture corresponds to an electrical signal (millivolts) which is generated by the photomultiplier; the horizontal axis is the time axis. A vertical deflection on the oscilloscope thus corresponds to a change in light intensity; for *small* perturbations this deflection is proportional to ΔOD (see Section 11.2.1).

For the evaluation of τ one typically draws the "infinity line" as indicated in the figure and measures the absolute value of the distance, $|\Delta l|$ (in centimeters), between the relaxation curve and the infinity line at short time intervals. Since Δl is proportional to ΔOD, a plot of $\log|\Delta l|$ versus time provides a straight line with slope $-1/2.303\tau$ as shown in Section 1.2. For more procedural details Yapel and Lumry (*1*) should be consulted.

Evaluation of τ by this method can become tedious when there is a great number of data to be processed. Schuster and Rorabacher [cited by Yapel and Lumry (*1*)] suggest the use of a series of templates which consist of transparent plastic sheets on which standard relaxation curves with known τ-values are drawn. One can then superimpose various templates on an experimental relaxation curve until the one which best approximates the experimental data is found.

Interfacing the photomultiplier output with a computer system is of course an attractive possibility in laboratories producing a lot of routine data. A typical setup used with the temperature-jump technique is to digitize and store the photomultiplier signal by means of a Biomation 802 transient recorder; coupling of the transient recorder to a small computer such as a PDP 11 allows the data to be processed directly. This system can be modified for signal averaging of multiple traces in order to enhance the signal-to-noise ratio and thus the precision with which τ is determined. For more details see Hammes (*2*). Križan and Strehlow (*3*) have described another interfacing system used with their pressure-jump apparatus.

9.2 Two or More Relaxation Times

When there are two relaxation times the relaxation curve is given by the sum of two exponential functions, for example in the case of spectrophotometric detection

$$\Delta\text{OD} = \Delta\text{OD}^{01} \exp(-t/\tau_1) + \Delta\text{OD}^{02} \exp(-t/\tau_2) \tag{9.1}$$

or, since $\Delta\text{OD} = \text{OD} - \overline{\text{OD}}$ with $\overline{\text{OD}}$ referring to $t \gg \tau_2$,

$$\text{OD} = \Delta\text{OD}^{01} \exp(-t/\tau_1) + \Delta\text{OD}^{02} \exp(-t/\tau_2) + \overline{\text{OD}} \tag{9.2}$$

The factors that determine the individual amplitudes ΔOD^{01} and ΔOD^{02}, such as the method of perturbation, the reagent concentrations, the equilibrium and rate constants, and the extinction coefficients of the various species at the monitoring wavelength have been extensively discussed in Chapters 6 and 7.

Equations 9.1 and 9.2 can be generalized for any physical property P (e.g., electrical signal on oscilloscope), used for monitoring relaxation, to

$$\Delta P = \Delta P^{01} \exp(-t/\tau_1) + \Delta P^{02} \exp(-t/\tau_2) \tag{9.3}$$

$$P = \Delta P^{01} \exp(-t/\tau_1) + \Delta P^{02} \exp(-t/\tau_2) + \bar{P} \tag{9.4}$$

9.2.1 Two Widely Separated Relaxation Times

Ideally one would want to determine $1/\tau_1$ under conditions where $\Delta P^{01}/\Delta P^{02} \gg 1$ so that

$$\Delta P \approx \Delta P^{01} \exp(-t/\tau_1) \tag{9.5}$$

and $1/\tau_2$ under conditions where $\Delta P^{01}/\Delta P^{02} \ll 1$ so that

$$\Delta P \approx \Delta P^{02} \exp(-t/\tau_2) \tag{9.6}$$

In this ideal situation the two exponentials do not interfere with each other and plots of log ΔP versus t would simply afford straight lines with slopes of $-1/2.303\tau_1$ or $-1/2.303\tau_2$, respectively.

As long as $1/\tau_1$ and $1/\tau_2$ differ by orders of magnitude this ideal situation is not necessary since the two exponentials are strongly separated and hardly interfere with each other regardless of the value of $\Delta P^{01}/\Delta P^{02}$. For $t = 0$ up to several τ_1 values but $t \ll \tau_2$ we can write Eqs. 9.3 and 9.4 in the form

$$\Delta P \approx \Delta P^{01} \exp(-t/\tau_1) + \Delta P^{02}(1 - t/\tau_2) \tag{9.7}$$

$$P \approx \Delta P^{01} \exp(-t/\tau_1) - \Delta P^{02}(t/\tau_2) + \Delta P^{02} + \bar{P} \tag{9.8}$$

Equations 9.7 and 9.8 describe an exponential function superimposed on a straight line. Figure 9.2 illustrates the situation for the case $\Delta P^{01} = \Delta P^{02}$ and $\tau_2 = 50\tau_1$.

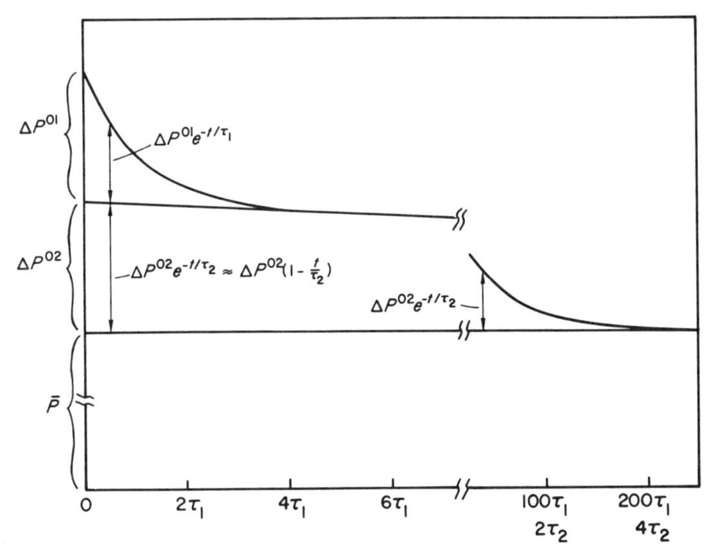

FIGURE 9.2. P according to Eq. 9.4, with $\Delta P^{01} = \Delta P^{02}$ and $\tau_2 = 50\tau_1$. Initial portion of curve approximated by Eq. 9.8.

For the evaluation of $1/\tau_1$ one linearly extrapolates (graphically or numerically) the slightly sloping infinity line of the first part of the relaxation curve to $t = 0$.

This extrapolated line provides the infinity readings for the τ_1 curve as indicated in the figure.

Evaluation of $1/\tau_2$ is even simpler since at $t \gg \tau_1$ Eq. 9.3 simplifies to

$$\Delta P = \Delta P^{02} \exp(-t/\tau_2) \tag{9.9}$$

that is, we are left with a pure exponential.

9.2.2 Two Poorly Separated Relaxation Times

When the two relaxation times are close the precise evaluation of at least one but often of both under any given condition may be quite difficult. In such a situation it frequently pays to spend a considerable effort in first finding conditions where the relaxation curve is dominated by the $\Delta P^{01} \exp(-t/\tau_1)$ term in order to get a relatively good value of τ_1, and then to search for a different set of conditions where the $\Delta P^{02} \exp(-t/\tau_2)$ term dominates for obtaining an accurate τ_2. This is qualitatively obvious; in the following we develop some quantitative relations illustrating the point.

Since only the amplitude *ratios*, $\Delta P^{02}/\Delta P^{01}$, rather than their absolute values are important in these considerations, we can simplify Eq. 9.3 to

$$\Delta Z = \exp(-t/\tau_1) + R \exp(-t/\tau_2) \tag{9.10}$$

where $\Delta Z = \Delta P/\Delta P^{01}$ and $R = \Delta P^{02}/\Delta P^{01}$, without loss in generality.

The properties of the function 9.10 are such that for $\tau_2 \leq 3\tau_1$ and $R > 0$ it becomes very difficult not only to evaluate the relaxation times but even to recognize the presence of two relaxation processes, especially if there is some scatter in the data. That is, the experimenter who does not suspect two relaxation effects but only one and who plots log ΔZ versus t would obtain a reasonably straight line, "confirming" the presence of only one process.

This is illustrated in Fig. 9.3 for the three cases $\tau_2 = 2\tau_1$ with $R = 5$, 1, and 0.2, respectively. The points represent the function of Eq. 9.10 plotted in the range from $\Delta Z = \Delta Z^0$ down to $\Delta Z = 0.05\, \Delta Z^0$ while the lines are the best straight lines through the points, found by least squares analysis. The straight lines correspond to

$$\Delta Z = W \exp(-t/\tau) + V \tag{9.11}$$

with the respective W, V, and τ values indicated in the figure legend; the correlation coefficients are 0.9997, 0.9992, and 0.9995 for $R = 5$, 1, and 0.2, respectively. Note that τ is between τ_1 and τ_2 and closer to the relaxation time associated with the larger amplitude.

The fact that τ depends strongly on the value of R can be exploited in the analysis of such data. Changing some of the conditions, for example measuring the relaxation at a different wavelength or changing the perturbation method, may change R and with it the value of τ, giving to the experimenter a clue that there must be more than one relaxation time. A nice example where the wavelength dependence of τ has provided evidence that at least two relaxation times

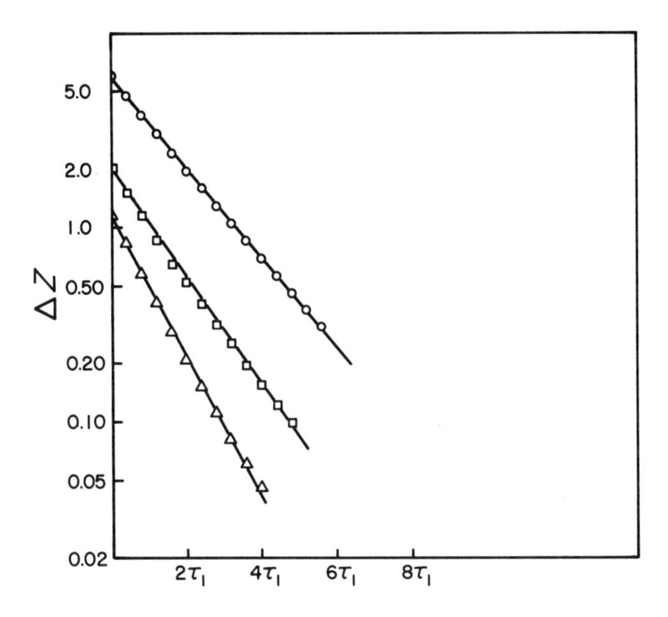

FIGURE 9.3. Logarithmic plots of ΔZ according to Eq. 9.10 with $\tau_2 = 2\tau_1$. Circle (○): $\Delta Z = \exp(-t/\tau_1) + 5\exp(-t/2\tau_1)$; line: $\Delta Z = 5.794\exp(-t/1.89\tau_1) + 0.206$. Square (□): $\Delta Z = \exp(-t/\tau_1) + \exp(-t/2\tau_1)$; line: $\Delta Z = 1.874\exp(-t/1.57\tau_1) + 0.126$. Triangle (△): $\Delta Z = \exp(-t/\tau_1) + 0.2\exp(-t/2\tau_1)$; line: $\Delta Z = 1.147\exp(-t/1.20\tau_1) + 0.053$.

need to be invoked to explain the data is the helix–coil transition of single-stranded, oligo- and polyriboadenylic acid (*4*); though each relaxation curve was "exponential," τ values determined at 285 nm differed by a factor of 2 to 3 from those determined in the range of 240–270 nm, thus implying a mechanism with at least one non-steady-state intermediate rather than a simple "all or none" model.

The situation is more favorable when $R < 0$ because there is a change in the sign of the slope of the relaxation curve (e.g., initial decrease in OD followed by increase in later stages of the curve) which makes it immediately obvious that it is not a single exponential.

Once the fact that there are two relaxation times is established one may try to separate the relaxation times, for example by changing the reagent concentrations (if $1/\tau_1$ and $1/\tau_2$ depend differently on these concentrations) or the temperature (if the activation energies associated with the various rate constants are different). Note that simply trying to fit the data to Eq. 9.10 without further separating τ_1 and τ_2 is fruitless because there will be several sets of τ_1, τ_2, and R which give equally good fits. On the other hand, a ratio $\tau_2/\tau_1 = 5$ may already allow a rough estimation of at least the main relaxation time. Figure 9.4 shows plots of log ΔZ versus t for $\tau_2 = 5\tau_1$ with $R = 5$, 1, and 0.2, respectively. Figure 9.5 shows similar plots for $\tau_2 = 20\tau_1$. All plots at least reveal the fact that there are two relaxation effects.

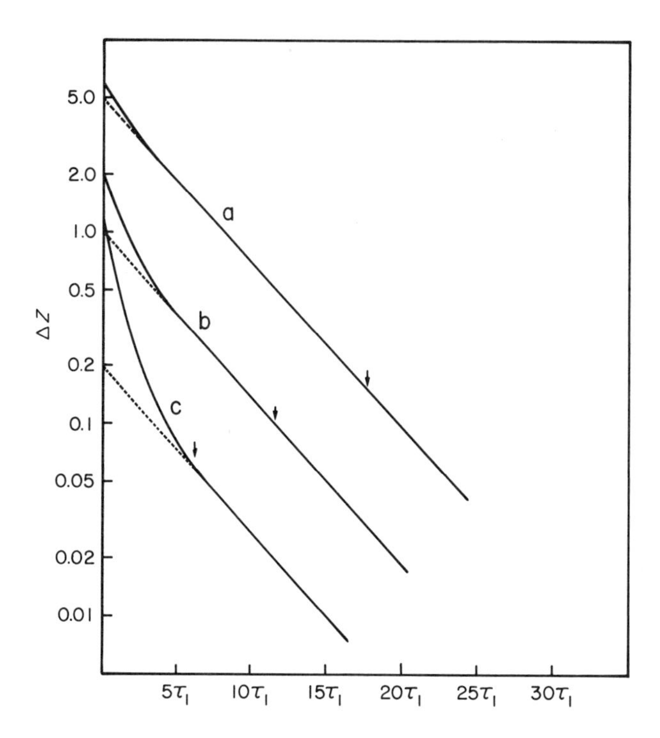

FIGURE 9.4. Logarithmic plots of ΔZ according to Eq. 9.10 with $\tau_2 = 5\tau_1$. Arrows mark point where $\Delta Z = 0.05\, \Delta Z^0$. (a) $\Delta Z = \exp(-t/\tau_1) + 5\exp(-t/5\tau_1)$; (b) $\Delta Z = \exp(-t/\tau_1) + \exp(-t/5\tau_1)$; (c) $\Delta Z = \exp(-t/\tau_1) + 0.2\exp(-t/5\tau_1)$.

In principle such plots allow $1/\tau_1$ and $1/\tau_2$ to be determined as follows. At $t \gg \tau_1$ the relaxation curves are approximated by

$$\Delta Z = R\exp(-t/\tau_2) \tag{9.12}$$

which corresponds to the linear tail in the logarithmic plots. From the slope of this linear portion one obtains $1/\tau_2$, whereas the intercept of the extrapolated straight line at $t = 0$ is equal to $\log R$.

Turning to the evaluation of τ_1^{-1} let us first stress that τ_1^{-1} is *not* equal to $-2.303\cdot$(slope of the initial portion of the log plot) as has been occasionally stated, even by respected scientists. Instead, one uses the extrapolated linear tail portion which affords the values of the function $R\exp(-t/\tau_2)$ at any time (these values can, for example, be directly read off the graph if ΔZ was plotted on semilogarithmic paper) and subtracts $R\exp(-t/\tau_2)$ from ΔZ (graphically or numerically) in order to obtain

$$\exp(-t/\tau_1) = \Delta Z - R\exp(-t/\tau_2) \tag{9.13}$$

Thus a plot of $\log[\Delta Z - R\exp(-t/\tau_2)]$ versus t affords a straight line with slope $-1/2.303\tau_1$.

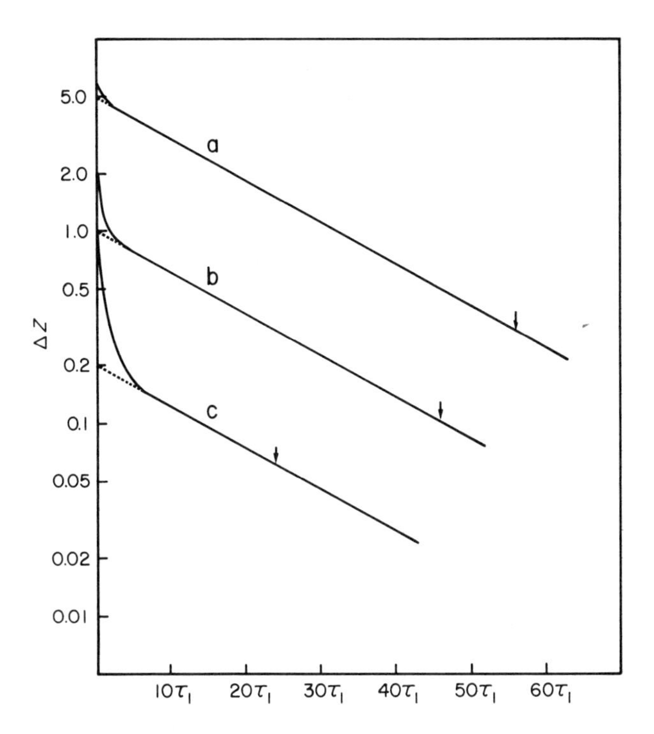

FIGURE 9.5. Logarithmic plots of ΔZ according to Eq. 9.10 with $\tau_2 = 20\tau_1$. Arrows mark points where $\Delta Z = 0.05\,\Delta Z^0$. (a) $\Delta Z = \exp(-t/\tau_1) + 5\exp(-t/20\tau_1)$; (b) $\Delta Z = \exp(-t/\tau_1) + \exp(-t/20\tau_1)$; (c) $\Delta Z = \exp(-t/\tau_1) + 0.2\exp(-t/20\tau_1)$.

As is apparent from the figures the plots are dominated by the $R\exp(-t/\tau_2)$ term except when $R < 1$; that is, the linear portion with slope $-1/2.303\tau_2$ is quite extended and allows a relatively precise determination of $1/\tau_2$ in a time range where the experimental uncertainty in ΔZ (after several τ_2, ΔZ becomes a very small difference between two large numbers) is not yet too high.

In contrast, since the term $[\Delta Z - R\exp(-t/\tau_2)]$ is generally a relatively small difference between two numbers, each with some uncertainty, evaluation of $1/\tau_1$ is intrinsically less precise. For very fast reactions this problem is compounded by the fact that the experimental uncertainty in ΔZ is usually greater in the time range of τ_1 than in the early parts of the time range of τ_2, because of a worse signal-to-noise ratio (see Section 11.2.1). Hence it is particularly important to find experimental conditions where R is as small as possible in order to increase the precision with which τ_1 is evaluated.

Strehlow and Jen (*5*) have discussed the problems considered in this section in some detail and also have proposed some alternative procedures for evaluating the relaxation times. Needless to say, computer fitting procedures can also be applied (*2*, *3*).

9.2.3 Three or More Relaxation Times

The principles of evaluating three or more relaxation times from a relaxation curve are similar to the ones discussed for two relaxation times. However, the uncertainties in the determinations become much more serious if, say, among three relaxation times, all are relatively close to each other (e.g., $\tau_3:\tau_2:\tau_1 = 25:5:1$). Again the situation is largely improved if the relative amplitudes are adjustable to one's wishes. Strehlow and Jen (*5*) again describe some special procedures to determine the main relaxation time in the case of the superposition of several relaxation effects where the amplitude of the main effect is considerably larger than that of all the others.

9.3 Mean Relaxation Times

9.3.1 Definition and Experimental Determination

In a system with n relaxation times, Eq. 9.3 becomes

$$\Delta P = \sum_{j=1}^{n} \Delta P^{0j} \exp(-t/\tau_j) \tag{9.14}$$

Unless the τ_j values are widely separated it is in general not possible to evaluate them separately when $n \geq 3$; when n is very large and all the τ_j values are close, one speaks of a continuous relaxation spectrum. Such relaxation spectra can nevertheless be characterized by several types of "mean relaxation times" which are easily accessible experimentally. Though in general the mean relaxation times cannot give as complete a picture about the reaction system as a determination of all the τ_j values would, they nevertheless can provide some useful information. For small n (say 2 or 3) or for highly degenerate systems with large n a determination of all rate constants from the mean relaxation times is sometimes possible as we shall show.

The concept of mean relaxation times has been introduced by Schwarz (*6*); three types of mean relaxation times, usually symbolized by τ^*, $\tilde{\tau}$, and τ^{**}, have been found most useful. Their definition is best understood with reference to the normalized reaction function, $\Phi(t)$, given by

$$\Phi(t) = \Delta P/\Delta P^0 = \sum_{j=1}^{n} \beta_j \exp(-t/\tau_j) \tag{9.15}$$

with

$$\Delta P^0 = \sum_{j=1}^{n} \Delta P^{0j} \tag{9.16}$$

$$\beta_j = \Delta P^{0j}/\Delta P^0 \tag{9.17}$$

$$\sum_{j=1}^{n} \beta_j = 1 \tag{9.18}$$

From Eq. 9.15 we obtain

$$\frac{d\Phi}{dt} = \frac{1}{\Delta P^0}\frac{d\,\Delta P}{dt} = -\sum_{j=1}^{n}\frac{\beta_j}{\tau_j}\exp\left(\frac{-t}{\tau_j}\right) \tag{9.19}$$

$$\frac{d^2\Phi}{dt^2} = \frac{1}{\Delta P^0}\frac{d^2\,\Delta P}{dt^2} = \sum_{j=1}^{n}\frac{\beta_j}{(\tau_j)^2}\exp\left(\frac{-t}{\tau_j}\right) \tag{9.20}$$

The defining equations for τ^*, $\tilde{\tau}$, and τ^{**} are

$$\frac{1}{\tau^*} = -\left(\frac{d\Phi}{dt}\right)_{t=0} = \sum_{j=1}^{n}\frac{\beta_j}{\tau_j} \tag{9.21}$$

$$\tilde{\tau} = \int_0^\infty \Phi(t)\,dt = \sum_{j=1}^{n}\beta_j\tau_j \tag{9.22}$$

$$\left(\frac{1}{\tau^{**}}\right)^2 = \left(\frac{d^2\Phi}{dt^2}\right)_{t=0} = \sum_{j=1}^{n}\frac{\beta_j}{(\tau_j)^2} \tag{9.23}$$

In all three equations the β_j terms which are the normalized amplitudes associated with the respective τ_j terms (see Chapter 7), can be considered as weight factors.

It is apparent from Eqs. 9.21 and 9.22 that τ^* and $\tilde{\tau}$ are experimentally very easily accessible. $1/\tau^*$ is essentially a normalized initial rate; it is the slope of a tangent to the relaxation curve at $t = 0$, divided by ΔP^0. Simple geometrical considerations show that this tangent intersects the infinity line at $t = \tau^*$ as indicated in Fig. 9.6. $\tilde{\tau}$ corresponds to the shaded area in Fig. 9.6., divided by ΔP^0.

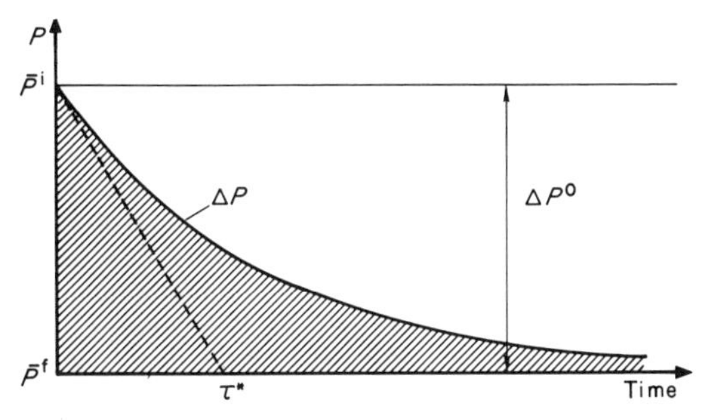

FIGURE 9.6. Determination of $1/\tau^*$ and $\tilde{\tau}$ from the relaxation curve.

In general one would want to determine both τ^* and $\tilde{\tau}$ (for τ^{**} see below). This is because the value of neither τ^* nor $\tilde{\tau}$ alone can provide any insight into the extent or width of the relaxation spectrum, but a comparison of τ^* with $\tilde{\tau}$ gives a measure of this width (see Section 9.3.2).

There is another way to measure the width of the relaxation spectrum for processes where all β_j values are ≥ 0; it is based on determining ρ^*, the relative quadratic deviation of the $1/\tau_j$ terms from $1/\tau^*$, defined as

$$\rho^* = \frac{\sum \beta_j(1/\tau_j - 1/\tau^*)^2}{(1/\tau^*)^2} \tag{9.24}$$

It is easily shown that

$$\rho^* = (\tau^*/\tau^{**})^2 - 1 \tag{9.25}$$

In principle, $(1/\tau^{**})^2$ and thus ρ^* could be determined from the initial curvature of the relaxation curve (Eq. 9.23) but this is not a very accurate procedure because it involves the determination of the slope of a tangent to a curve which itself is made up of points corresponding to the slopes of tangents to the relaxation curve. There is a different, more accurate method that provides ρ^* directly.

From inspection of Eq. 9.24 it is easily appreciated that ρ^* is a measure of how strongly $\Phi(t)$ differs from the exponential function $\exp(-t/\tau^*)$. For example, if the differences $|1/\tau_j - 1/\tau^*|$ are large, particularly for the τ_j values associated with large β_j values, one expects a large difference between $\Phi(t)$ and $\exp(-t/\tau^*)$ and ρ^* is large; for small differences $|1/\tau_j - 1/\tau^*|$ the opposite is true and in the limiting case where all $|1/\tau_j - 1/\tau^*| = 0$ we have of course $\rho^* = 0$ and $\Phi(t) = \exp(-t/\tau^*)$. According to Schwarz (*6*) one can express the deviation, $D(t)$, of the relaxation curve from an exponential as a Taylor series

$$D(t) = \Phi(t) - \exp(-t/\tau^*) = \frac{\rho^*}{2}(t/\tau^*)^2 + \cdots \tag{9.26}$$

as long as $D(t) \ll 1$. Equation 9.26 immediately suggests the following procedure for an experimental evaluation of ρ^*. We plot $\log \Phi(t)$ versus t which typically results in a curve as shown in Fig. 9.7a. The tangent to this curve at $t = 0$

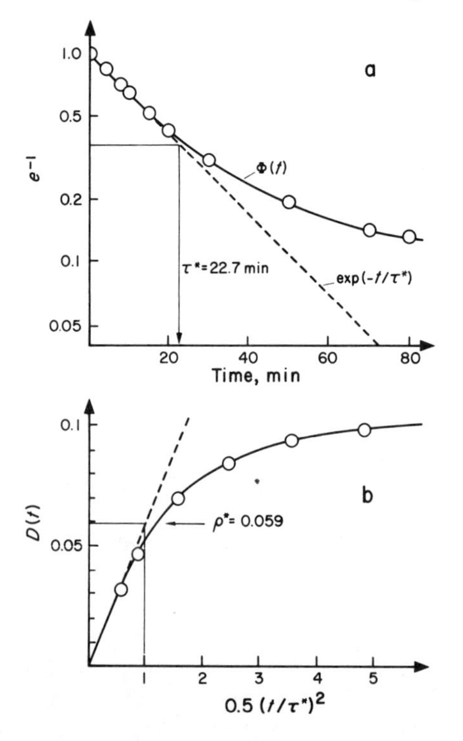

FIGURE 9.7. Determination of ρ^* for the helix I $\rightleftharpoons$ helix II transition of polyproline. [From Schwarz (*7*), by permission of John Wiley & Sons.]

corresponds to log exp$(-t/\tau^*)$. Thus $D(t)$ is easily found from this graph and is in turn plotted versus $0.5(t/\tau^*)^2$ as shown in Fig. 9.7b. The initial slope to this new plot is ρ^*. Incidentally, Fig. 9.7a also shows an alternative method for evaluating τ^*.

Though this procedure is more accurate than the one based on finding $(1/\tau^{**})^2$ from the initial curvature of $\Phi(t)$, a precision of the data higher than that in the determination of $1/\tau^*$ is still required, mainly because $D(t)$ is obtained as the relatively small difference between two relatively large numbers. An application of the use of ρ^* is discussed in Section 10.2.

9.3.2 Information from Mean Relaxation Times

The significance of τ^*, $\tilde{\tau}$, and τ^{**} is best appreciated by considering three extreme situations.

A. All β_j Values Equal

If all β_j terms are equal, we have

$$\beta_1 = \beta_2 = \cdots = \beta_n = 1/n \tag{9.27}$$

and hence

$$1/\tau^* = (1/n) \sum_{j=1}^{n} (1/\tau_j) \tag{9.28}$$

$$\tilde{\tau} = (1/n) \sum_{j=1}^{n} \tau_j \tag{9.29}$$

$$(1/\tau^{**})^2 = (1/n) \sum_{j=1}^{n} (1/\tau_j)^2 \tag{9.30}$$

That is, $1/\tau^*$ is simply the arithmetic mean or average of all $1/\tau_j$ values, $\tilde{\tau}$ is the average of all τ_j values, and $(1/\tau^{**})^2$ is the average of the $(1/\tau_j)^2$ values.

More significantly, comparison of τ^* and $\tilde{\tau}$ gives an indication of the width of the relaxation spectrum, i.e., the larger the difference $(1/\tau^* - 1/\tilde{\tau})$, the wider the spectrum. This is easily appreciated by considering the simple case where $n = 2$. Here we have

$$1/\tau^* = 0.5(1/\tau_1 + 1/\tau_2) \tag{9.31}$$

$$\tilde{\tau} = 0.5(\tau_1 + \tau_2) \tag{9.32}$$

and hence

$$\frac{1}{\tau^*} - \frac{1}{\tilde{\tau}} = \frac{(\tau_1 - \tau_2)^2}{2\tau_1\tau_2(\tau_1 + \tau_2)} \tag{9.33}$$

For a very narrow spectrum ($1/\tau_1 \approx 1/\tau_2$ and thus $1/\tau^* \approx 1/\tau_1 \approx 1/\tilde{\tau}$) we have $[1/\tau^* - 1/\tilde{\tau}] \approx 0$, whereas for a very wide spectrum ($1/\tau_1 \gg 1/\tau_2$ and thus

$1/\tau^* \approx 1/2\tau_1$, $1/\tilde{\tau} \approx 2/\tau_2$) we have $[1/\tau^* - 1/\tilde{\tau}] \approx 1/2\tau_1 \approx 1/\tau^*$. Alternatively, expressing the width of the relaxation spectra by means of ρ^* one obtains, again for $n = 2$,

$$\rho^* = \frac{(1/\tau_1 - 1/\tau_2)^2}{(1/\tau_1 + 1/\tau_2)^2} \tag{9.34}$$

For $1/\tau_1 \approx 1/\tau_2$, $\rho^* \approx 0$ or $\tau^{**} \approx \tau^*$, for $1/\tau_1 \gg 1/\tau_2$, $\rho^* \approx 1$ or $(1/\tau^{**})^2 = 2/(\tau^*)^2$.

These considerations will be qualitatively valid for $n > 2$ as long as the β_j terms are of comparable magnitude; thus from the value of $1/\tau^*$ and a knowledge of the width of the relaxation spectrum considerable insight into the kinetics of the system can be achieved. However, it must be realized that in order to be sure that the β_j terms are of comparable magnitude, the reaction mechanism and the thermodynamic parameters of the system must be known.

B. Relaxation Spectrum Dominated by One Relaxation Time

Frequently some of the relaxation processes dominate the spectrum. In the extreme case of one relaxation process dominating over all the others ($\beta_1 \approx 1 \gg \beta_2, \beta_3, \ldots, \beta_n \approx 0$) Eqs. 9.21–9.23 reduce to $\tau^* = \tilde{\tau} = \tau^{**} = \tau_1$. This situation is of particular interest because it suggests a method by which one could, at least in principle, determine the various τ_j values from measuring τ^* under n different conditions such as (1) $\beta_1 \approx 1$; (2) $\beta_2 \approx 1$; (3) $\beta_3 \approx 1$; etc. This is of course easily recognized as the amplitude problem discussed in Chapter 7. For illustration we therefore go back to our $A \rightleftharpoons M \rightleftharpoons B$ system (Eq. 7.1). Let us take the case where $k_{-2} = k_{-1}$ and assume k_1 and k_2 to be low so that $1/\tau_1 = k_1 + k_2 + k_{-1}$ and $1/\tau_2 = k_{-1}$ are so close that they would be difficult to evaluate if associated with comparable amplitudes. For spectrophotometric detection we have

$$\Delta OD = \Delta OD^{01} \exp(-t/\tau_1) + \Delta OD^{02} \exp(-t/\tau_2) \tag{9.35}$$

with ΔOD^{01} and ΔOD^{02} given by Eqs. 7.154–7.155 in the case of a temperature-jump experiment, Eqs. 7.156–7.157 in an experiment initiated by mixing M with N, or Eqs. 7.158–7.159 in an experiment initiated by mixing (N + A′) with R (Eq. 7.132). Note that $\beta_1 = \Delta OD^{01}/\Delta OD^0$, $\beta_2 = \Delta OD^{02}/\Delta OD^0$.

As discussed in Section 7.3.1 initiation of the reaction by mixing M with N leads automatically to $\beta_2 = 0$ and $\beta_1 = 1$. This case is trivial; the experimental relaxation curve is a true exponential and $\tau^* = \tau_1$. In the temperature-jump experiment we have $\beta_2 < \beta_1$ and perhaps even $\beta_2 \ll \beta_1$. By appropriate choice of the wavelength ($\epsilon_A = \epsilon_B$; see Eq. 7.155), however, β_2 can be made zero so that at least τ_1 can be determined accurately in this experiment.

In the experiment initiated by mixing A′ with R, β_1 and β_2 may or may not be of comparable magnitude. Adjusting the wavelength so that in one experiment $2\epsilon_M = \epsilon_A + \epsilon_B$ and thus $\beta_1 = 0$ (Eq. 7.158), and in another experiment $\epsilon_A = \epsilon_B$ and thus $\beta_2 = 0$ (Eq. 7.159), would allow the determination of both τ_1 and τ_2.

From this analysis it should be clear that in order to vary the experimental conditions intelligently one has to have some knowledge about the reaction mechanism and the spectral characteristics of the species involved. It is also evident that this example is extremely simple. In most other systems which are more complex, either because they are less degenerate ($k_{-2} \neq k_{-1}$) or because they are characterized by more than two relaxation times, it may be practically very difficult to find the n conditions for which one of the β terms at the time is equal to unity.

It should also be pointed out that this analysis is only safe when the τ values are not too widely separated. When they are widely separated, Eqs. 9.21–9.23 may not reduce to $\tau^* = \tilde{\tau} = \tau^{**} = \tau_1$. For example, when $\tau_1 = 1$, $\tau_2 = 10^{-4}$, $\beta_1 = 0.99$, $\beta_2 = 10^{-2}$, one has $\tau^* \approx 0.01$ rather than $\tau^* \approx 1$. The problem is even worse with τ^{**} or ρ^*, as is illustrated with an example in Section 10.2.2 (helix–coil transition).

C. *Information from* $1/\tau^*$ *When* $\Phi(t) = \Delta c_j/\Delta c_j^0$

There exists a somewhat different way of analyzing the relaxation spectrum by means of τ^* which potentially provides the greatest amount of information even in nondegenerate systems. However, it depends on whether it is possible to monitor the relaxation by following the change in concentration of only one species at a time, or of a physical property proportional to it.

According to the rules of differential calculus, for a function $F = F(u_1, u_2, \ldots, u_n)$ where each u_j in turn is a function of t, we can write

$$\frac{dF}{dt} = \sum_{j=1}^{n} \left(\frac{\partial F}{\partial u_j}\right)\left(\frac{\partial u_j}{dt}\right) \tag{9.36}$$

Applying this to ΔP we obtain

$$\frac{1}{\tau^*} = -\frac{1}{\Delta P^0}\left(\frac{d\,\Delta P}{dt}\right)_{t=0} = -\frac{1}{\Delta P^0}\sum_{j=1}^{n}\left(\frac{\partial\,\Delta P}{\partial\,\Delta c_j}\right)\left(\frac{\partial\,\Delta c_j}{dt}\right)_{t=0} \tag{9.37}$$

To avoid confusing the novice we may note that some authors write

$$\frac{1}{\tau^*} = -\frac{1}{\Delta P^0}\left(\frac{dP}{dt}\right)_{t=0} = -\frac{1}{\Delta P^0}\sum_{j=1}^{n}\left(\frac{\partial P}{\partial c_j}\right)_{c_j=\bar{c}_j}\left(\frac{\partial c_j}{\partial t}\right)_{t=0} \tag{9.38}$$

This is of course equivalent because $dP/dt = d\,\Delta P/dt$, $dc_j/dt = d\,\Delta c_j/dt$, and $(\partial P/\partial c_j)_{c_j=\bar{c}_j} = \partial\,\Delta P/\partial\,\Delta c_j$. Let us illustrate the application of Eq. 9.37 to our

$A \rightleftharpoons M \rightleftharpoons B$ system (Eq. 7.1) where no special relation among the four rate constants is stipulated. For spectrophotometric detection Eq. 9.37 becomes

$$\frac{1}{\tau^*} = -\frac{1}{\Delta OD^0}\left(\frac{d\,\Delta OD}{dt}\right)_{t=0}$$

$$= -\frac{1}{\epsilon_A\,\Delta c_A{}^0 + \epsilon_B\,\Delta c_B{}^0 + \epsilon_M\,\Delta c_M{}^0}$$

$$\times\left[\epsilon_A\left(\frac{d\,\Delta c_A}{dt}\right)_{t=0} + \epsilon_B\left(\frac{d\,\Delta c_B}{dt}\right)_{t=0} + \epsilon_M\left(\frac{d\,\Delta c_M}{dt}\right)_{t=0}\right] \quad (9.39)$$

In combination with the rate equations ($d\,\Delta c_A/dt = k_1\,\Delta c_M - k_{-1}\,\Delta c_A$, etc.) we obtain

$$\frac{1}{\tau^*} = -\frac{1}{\Delta OD^0}\left(\frac{d\,\Delta OD}{dt}\right)_{t=0} = -\frac{1}{\epsilon_A\,\Delta c_A{}^0 + \epsilon_B\,\Delta c_B{}^0 + \epsilon_M\,\Delta c_M{}^0}$$

$$\times\,[\epsilon_A(k_1\,\Delta c_M{}^0 - k_{-1}\,\Delta c_A{}^0) + \epsilon_B(k_2\,\Delta c_M{}^0 - k_{-2}\,\Delta c_B{}^0)$$

$$+\,\epsilon_M(k_{-1}\,\Delta c_A{}^0 + k_{-2}\,\Delta c_B{}^0 - k_1\,\Delta c_M{}^0 - k_2\,\Delta c_M{}^0)] \quad (9.40)$$

Assume now that $\epsilon_B = \epsilon_M = 0$ but $\epsilon_A \neq 0$; this reduces Eq. 9.40 to

$$\frac{1}{\tau_A{}^*} = -\frac{1}{\Delta c_A{}^0}\left(\frac{d\,\Delta c_A}{dt}\right)_{t=0} = -\frac{1}{\Delta c_A{}^0}(k_1\,\Delta c_M{}^0 - k_{-1}\,\Delta c_A{}^0) \quad (9.41)$$

and, after substituting $\Delta c_M{}^0 = -\Delta c_A{}^0 - \Delta c_B{}^0$ and rearranging

$$\frac{1}{\tau_A{}^*} = k_1\left(1 + \frac{\Delta c_B{}^0}{\Delta c_A{}^0}\right) + k_{-1} \quad (9.42)$$

Similarly, for the case $\epsilon_A = \epsilon_M = 0$ but $\epsilon_B \neq 0$, we obtain

$$\frac{1}{\tau_B{}^*} = -\frac{1}{\Delta c_B{}^0}\left(\frac{d\,\Delta c_B}{dt}\right)_{t=0} = k_2\left(1 + \frac{\Delta c_A{}^0}{\Delta c_B{}^0}\right) + k_{-2} \quad (9.43)$$

whereas for $\epsilon_A = \epsilon_B = 0$ but $\epsilon_M \neq 0$

$$\frac{1}{\tau_M{}^*} = -\frac{1}{\Delta c_M{}^0}\left(\frac{d\,\Delta c_M}{dt}\right)_{t=0} = k_1 + k_2 + k_{-1}\left(1 + \frac{\Delta c_B{}^0}{\Delta c_M{}^0}\right) + k_{-2}\left(1 + \frac{\Delta c_A{}^0}{\Delta c_M{}^0}\right) \quad (9.44)$$

In case the k_1 and k_2 steps are bimolecular ($M + N \rightarrow A$; $M + N \rightarrow B$) Eqs. 9.42–9.44 allow the determination of at least k_{-1}, k_{-2}, and the sum $k_1 + k_2$ as follows. A plot of $1/\tau_A{}^*$ versus $[N]_0$ (assume $\bar{c}_N \gg \bar{c}_M$) affords a straight line with intercept k_{-1} and slope $k_1[1 + \Delta c_B{}^0/\Delta c_A{}^0]$; from a similar plot of $1/\tau_B{}^*$ one obtains intercept k_{-2} and slope $k_2[1 + \Delta c_B{}^0/\Delta c_A{}^0]$ whereas a plot of $1/\tau_M{}^*$ has a slope $k_1 + k_2$ and an intercept of $k_{-1}[1 + \Delta c_B{}^0/\Delta c_M{}^0] + k_{-2}[1 + \Delta c_A{}^0/\Delta c_M{}^0]$. If ϵ_A and ϵ_B are known, $\Delta c_A{}^0 = \Delta OD^0/\epsilon_A l$ and $\Delta c_B{}^0 = \Delta OD^0/\epsilon_A l$ can also be determined from the respective experiments and thus k_1 and k_2 are obtained individually. In such a case $1/\tau_M{}^*$ does not need to be measured.

D. Other Uses of $1/\tau^*$

A recent application of the relation 9.38 concerns the aggregation kinetics of acridine orange (*8*). Relaxation curves obtained in temperature-jump experiments were definitely nonexponential and τ^* was wavelength dependent. These observations are clear indications that one is dealing with a multistep system. The simplest mechanism of aggregation is one where growth of an aggregate occurs only by adding one monomer unit at a time, such as

$$\begin{array}{llll} A_1 + A_1 \underset{k_{-1}}{\overset{k_1}{\rightleftharpoons}} A_2 & & K_1 = k_1/k_{-1} \\ A_1 + A_2 \underset{k_{-2}}{\overset{k_2}{\rightleftharpoons}} A_3 & & K_2 = k_2/k_{-2} \\ \vdots & & \\ A_1 + A_j \underset{k_{-j}}{\overset{k_j}{\rightleftharpoons}} A_{j+1} & & K_j = k_j/k_{-j} \\ \vdots & & \\ A_1 + A_n \underset{k_{-n}}{\overset{k_n}{\rightleftharpoons}} A_{n+1} & & K_n = k_n/k_{-n} \end{array} \tag{9.45}$$

that is, reactions such as $A_j + A_k \rightleftharpoons A_{j+k}$ are not important.

A number of other assumptions must now be made to make a mathematical treatment feasible. The first is that the optical density of the system can be adequately described by a set of three extinction coefficients defined as follows. ϵ_M refers to a free monomer unit; ϵ_D is the extinction coefficient of a monomer unit which on one side is flanked by the solvent, and on the other side by one or more dye molecules, i.e., sol Ⓐ Asol, solA Ⓐ sol, sol Ⓐ AAsol, solAA Ⓐ sol, etc.; and ϵ_{St} is the extinction coefficient of a monomer unit flanked on both sides by one or more dye molecules, i.e., solA Ⓐ Asol, solA Ⓐ AAsol, solAA Ⓐ A sol, etc. Thus for the total OD we get

$$\mathrm{OD} = l\left\{\epsilon_M c_1 + 2\epsilon_D \sum_{j=2}^{n+1} c_j + \epsilon_{St} \sum_{j=3}^{n+1} (j-2)c_j\right\} \tag{9.46}$$

Applying Eq. 9.38 one obtains

$$\frac{1}{\tau^*} = -\frac{1}{\Delta\mathrm{OD}^0}\left\{\epsilon_M\left(\frac{dc_1}{dt}\right)_{t=0} + 2\epsilon_D \sum_{j=2}^{n+1}\left(\frac{dc_j}{dt}\right)_{t=0} + \epsilon_{St}\sum_{j=3}^{n+1}(j-2)\left(\frac{dc_j}{dt}\right)_{t=0}\right\} \tag{9.47}$$

The rate expressions are

$$\frac{dc_1}{dt} = -2k_1(c_1)^2 + 2k_{-1}c_2 - \sum_{j=2}^{n}(k_j c_1 c_j - k_{-j}c_{j+1})$$

$$\frac{dc_j}{dt} = k_{j-1}c_1c_{j-1} - k_{-(j-1)} + k_{-j}c_{j+1} - k_jc_1c_j$$

Performing a number of straightforward mathematical manipulations (*8*) and assuming $\Delta H_1 = \Delta H_2 = \cdots = \Delta H_j = \cdots = \Delta H_n = \Delta H$ it can be shown that in a temperature-jump experiment

$$\frac{1}{\tau^*} = \frac{1}{\Delta\mathrm{OD}^0} l \left\{ 2(\epsilon_\mathrm{D} - \epsilon_\mathrm{M}) k_1 (\bar{c}_1)^2 + (\epsilon_\mathrm{St} - \epsilon_\mathrm{M}) \bar{c}_1 \sum_{j=2}^{n} k_j \bar{c}_j \right\} \frac{\Delta H}{RT^2} \Delta T \quad (9.48)$$

Making further assumptions Eq. 9.48 can be expressed more explicitly and the kinetic parameters evaluated (*8*). One set of assumptions is that the forward rates are all diffusion controlled, $k_1 = k_2 = \cdots = k_j = \cdots = k_n$ and also all equilibrium constants equal, $K_1 = K_2 = \cdots = K_j = \cdots = K_n$ ("model 2") (*8*); another is that still $k_1 = k_2 = \cdots = k_j = \cdots = k_n$ but $K_j = K_1/j$, i.e., the process is somewhat anticooperative ("model 3") (*8*). It was shown that model 3 gives a better fit with the experimental data than model 2 and thus model 3 was assumed to represent the system adequately.

Problems

1. Evaluate the relaxation time from Fig. 9.1 (disregard the low precision due to the small size of the picture).

2. The following values for Δl were measured from an oscillogram:

Time (msec)	Δl (cm)	Time (msec)	Δl (cm)	Time (msec)	Δl (cm)
0	4.00	0.5	1.42	1.0	0.76
0.1	3.15	0.6	1.24	1.2	0.61
0.2	2.45	0.7	1.08	1.4	0.50
0.3	2.00	0.8	0.95	1.6	0.40
0.4	1.68	0.9	0.85	1.8	0.33

What is the minimum number of relaxation times consistent with the data? Evaluate the relaxation times.

3. The following ΔOD values were measured from the relaxation curve referring to the conformational transition of a bipolymer.

$10^{-2} \times t$ (sec)	ΔOD	$10^{-2} \times t$ (sec)	ΔOD
0	0.100	12.0	0.043
2.5	0.085	18.0	0.031
4.5	0.071	30.0	0.019
6.0	0.063	42.0	0.015
9.0	0.052	48.0	0.013_5

Determine $1/\tau^*$ and ρ^*. Is the relaxation spectrum wide or narrow?

References

1. A. F. Yapel, Jr., and R. Lumry, *Methods Biochem. Anal.* **20**, 169 (1971).
2. G. G. Hammes, *in* "Techniques of Chemistry" (G. G. Hammes, ed.), Vol. VI, part 2, p. 147. Wiley (Interscience), New York, 1973.
3. M. Križan and H. Strehlow, *Chem. Instrum.* **5**, 99 (1973–74).
4. D. Pörschke, *Eur. J. Biochem.* **39**, 117 (1973).
5. H. Strehlow and J. Jen, *Chem. Instrum.* **3**, 47 (1971).
6. G. Schwarz, *Rev. Mod. Phys.* **40**, 206 (1968).
7. G. Schwarz, *Biopolymers* **6**, 873 (1968).
8. B. H. Robinson, A. Seelig-Löffler, and G. Schwarz, *J. Chem. Soc. Faraday Trans. I*, **71**, 815 (1975).

Chapter 10 | *Chemical Relaxation in Complex Systems*

In this chapter a few selected complex systems are discussed, most of which are of biological interest. They are too complex for an exact general mathematical treatment but frequently simplifying assumptions can be made which reduce the problem to a manageable level. Experimental observations usually make these assumptions justifiable. For example, in reactions involving a great many intermediates, a majority of these may never accumulate to detectable levels, thus allowing a steady state treatment. In other systems a number of rate constants may refer to processes that are chemically so similar that they can be assumed to be equal, leading to degeneracies and a great simplification of the relaxation spectrum. In still others the evaluation of the mean relaxation time(s) makes at least a partial analysis feasible.

In selecting the examples to follow we were guided by two main considerations. (1) They should illustrate most of the important principles of chemical relaxation learned in Chapters 1–9, as applied to nontrivial situations; (2) the results and conclusions should be of general scientific interest. This latter criterion is of course somewhat subjective and it is realized that many fine studies could not be included in a book of this scope. However, other applications are mentioned under the various experimental techniques.

The point of view in these discussions is mainly that of the experimenter confronted with analyzing the data; however, referral is made to reviews and papers focusing on the results and conclusions.

10.1 Binding of Small Molecules to Multiunit Enzymes

Consider an enzyme that consists of n equivalent subunits, each with the same intrinsic affinity toward a small molecule. This corresponds to the two-step scheme of case 2 in Table 3.2 expanded to n steps,

$$
\begin{aligned}
A + S &\underset{k_{-1}}{\overset{nk_1}{\rightleftharpoons}} AS \\
AS + S &\underset{2k_{-1}}{\overset{(n-1)k_1}{\rightleftharpoons}} AS_2 \\
&\vdots \\
AS_j + S &\underset{(j+1)k_{-1}}{\overset{(n-j)k_1}{\rightleftharpoons}} AS_{j+1} \\
&\vdots \\
AS_{n-1} + S &\underset{nk_{-1}}{\overset{k_1}{\rightleftharpoons}} AS_n
\end{aligned}
\tag{10.1}
$$

The coefficients in front of the rate constants represent statistical factors. Just as in the two-step system there is only one relaxation time associated with net binding (*1*, *2*); it is easily found by writing the normal reaction corresponding to the binding process

$$\Phi + S \underset{k_{-1}}{\overset{k_1}{\rightleftharpoons}} \chi \tag{10.2}$$

(which is also reaction 7.173) with

$$
\begin{aligned}
1/\tau_1 &= k_1(\bar{c}_\Phi + \bar{c}_S) + k_{-1} \\
\bar{c}_\Phi &= n\bar{c}_A + (n-1)\bar{c}_{AS} + \cdots + \bar{c}_{AS_{n-1}}
\end{aligned}
\tag{10.3}
$$

According to Eigen (*1*, *2*) the remaining $n - 1$ relaxation times are given by

$$
\begin{aligned}
1/\tau_2 &= 2(k_1\bar{c}_S + k_{-1}) \\
1/\tau_3 &= 3(k_1\bar{c}_S + k_{-1}) \\
&\vdots \\
1/\tau_n &= n(k_1\bar{c}_S + k_{-1})
\end{aligned}
\tag{10.4}
$$

They are associated with the various normal modes representing the redistribution of the substrate among the binding sites, just as τ_2 in the two-step system is associated with the normal reaction of redistribution, Eqs. 7.176–7.177.

10.1.1 Allosteric Enzymes

A modification of scheme 10.1 has served as the basis of interpreting the binding of small molecules to so-called allosteric enzymes. The most characteristic feature of allosteric enzymes is that binding is a cooperative process; i.e., saturation curves are sigmoidal rather than hyperbolic (Fig. 10.1). This phenomenon has intrigued scientists for a long time and various theories have been advanced to explain it. We shall discuss only the "allosteric model" proposed by Monod *et al.* (*3*) (MWC model); for more on allosteric enzymes a recent review by Hammes and Wu (*4*) should be consulted.

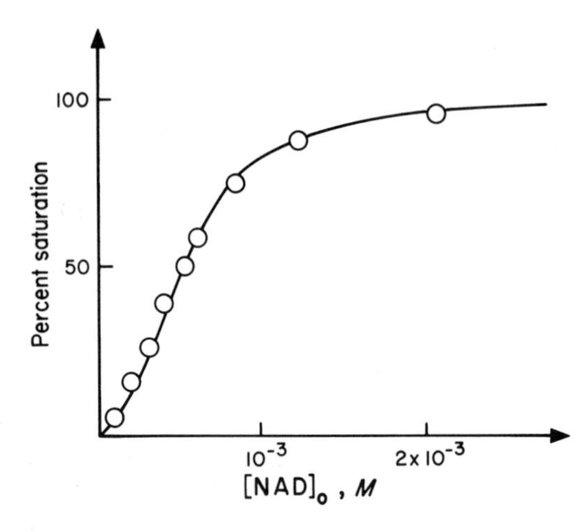

FIGURE 10.1. Binding of nicotinamide adenine dinucleotide (NAD) to D-glyceraldehyde-3-phosphate dehydrogenase (GAPDH) at 40°, pH 8.5. [From Eigen (*1*), by permission of John Wiley & Sons.]

The basic features of the MWC model are the following:

1. The enzyme is not only made up of n subunits but each unit can be present in two different conformations, called the R form and T form.
2. Binding to either the R or the T form can be described by a scheme such as 10.1, i.e., all subunits within the R form have the same intrinsic affinity to the ligand, and the same is true among all the T subunits. However, the affinities of the R and T forms are different.
3. The change from one conformation into the other is an "all or none" process; i.e., the conformation of all subunits change at the same time. This holds true whether or not any substrate is bound to the protein.

For a four-subunit enzyme the MWC model can be represented by Scheme 10.5

$$
\begin{array}{ccc}
4S + R_0 & \underset{k_{-0}}{\overset{k_0}{\rightleftharpoons}} & T_0 + 4S \\
4k_A{}^R \downarrow\uparrow k_D{}^R & & k_D{}^T \uparrow\downarrow 4k_A{}^T \\
3S + R_1 & \underset{k_{-1}}{\overset{k_1}{\rightleftharpoons}} & T_1 + 3S \\
3k_A{}^R \downarrow\uparrow 2k_D{}^R & & 2k_D{}^R \uparrow\downarrow 3k_A{}^T \\
2S + R_2 & \underset{k_{-2}}{\overset{k_2}{\rightleftharpoons}} & T_2 + 2S \\
2k_A{}^R \downarrow\uparrow 3k_D{}^R & & 3k_D{}^T \uparrow\downarrow 2k_A{}^T \\
S + R_3 & \underset{k_{-3}}{\overset{k_3}{\rightleftharpoons}} & T_3 + S \\
k_A{}^R \downarrow\uparrow 4k_D{}^R & & 4k_D{}^T \uparrow\downarrow k_A{}^T \\
R_4 & \underset{k_{-4}}{\overset{k_4}{\rightleftharpoons}} & T_4
\end{array}
\qquad (10.5)
$$

with
$$K_R = \frac{k_D{}^R}{k_A{}^R} = 4\frac{\bar{c}_{R_0}\bar{c}_S}{\bar{c}_{R_1}} = \frac{3}{2}\frac{\bar{c}_{R_1}\bar{c}_S}{\bar{c}_{R_2}} = \frac{2}{3}\frac{\bar{c}_{R_2}\bar{c}_S}{\bar{c}_{R_3}} = \frac{1}{4}\frac{\bar{c}_{R_3}\bar{c}_S}{\bar{c}_{R_4}} \tag{10.6}$$

$$K_T = \frac{k_D{}^T}{k_A{}^T} = 4\frac{\bar{c}_{T_0}\bar{c}_S}{\bar{c}_{T_1}} = \frac{3}{2}\frac{\bar{c}_{T_1}\bar{c}_S}{\bar{c}_{T_2}} = \frac{2}{3}\frac{\bar{c}_{T_2}\bar{c}_S}{\bar{c}_{T_3}} = \frac{1}{4}\frac{\bar{c}_{T_3}\bar{c}_S}{\bar{c}_{T_4}} \tag{10.7}$$

$$L_0 = \frac{k_0}{k_{-0}} = \frac{\bar{c}_{T_0}}{\bar{c}_{R_0}} \tag{10.8}$$

$$L_j = \frac{k_j}{k_{-j}} = \frac{\bar{c}_{T_j}}{\bar{c}_{R_j}} = L_0\left(\frac{K_R}{K_T}\right)^j \tag{10.9}$$

It should be pointed out that the original MWC model only requires equal intrinsic binding *equilibrium* constants (K_R, K_T); the assumption about equal intrinsic rate constants for binding and dissociation ($k_A{}^R$, $k_A{}^T$, $k_D{}^R$, $k_D{}^T$) was introduced by Eigen (*1*, *2*, *5*).

Without going through a mathematical analysis (*3*) let us illustrate qualitatively how the MWC model can account for a sigmoid binding curve. For concreteness we use some equilibrium parameters which were determined for the binding of nicotinamide adenine dinucleotide (NAD) to D-glyceraldehyde-3-phosphate dehydrogenase (GAPDH) (*5*):

$$K_R = 5.3 \times 10^{-5}\,M, \qquad K_T = 1.5 \times 10^{-4}\,M$$
$$L_0 = 30.5, \qquad K_R/K_T = 0.35$$

Note that K_R and K_T are defined as dissociation constants in keeping with the original authors (*5*).

In the absence of NAD the enzyme is about 97% in the approximately threefold less affine T conformation. Addition of initial small amounts of NAD is thus mainly to the less affine but much more abundant T form. This leads to the formation of some T_1 and T_2 which equilibrate with R_1 and R_2. As the degree of saturation increases the ratio between R and T forms becomes less unfavorable ($L_1 = 10.7$, $L_2 = 3.7$, $L_3 = 1.3$, $L_4 = 0.46$); for example, when in states R_3 and T_3, the relative amounts of each form are about equal so that now it is the more affine R_3 form which is principally responsible for further addition of NAD to the enzyme. Thus the overall affinity of the enzyme is increased with increasing degree of saturation. It should be noted that the cooperativity in this example is rather weak; it would be stronger if K_R/K_T was smaller and L larger.

Let us now turn to the kinetic behavior of scheme 10.5 which was studied for the system NAD–GAPDH (*5*). Three relaxation times were observed: $\tau_I^{-1} \sim 10^3$–$10^4\ \text{sec}^{-1}$, $\tau_{II}^{-1} \sim 10^2$–$10^3\ \text{sec}^{-1}$, and $\tau_{III}^{-1} \sim 0.1$–$10\ \text{sec}^{-1}$. τ_I and τ_{II} were measured by the temperature-jump method, τ_{III} by the stopped-flow method. Their respective concentration dependences are shown in Fig. 10.2. These results are entirely consistent with scheme 10.5 and were interpreted as follows. Assuming that the conformational changes, i.e., the reactions $R_0 \rightleftharpoons T_0$, $R_1 \rightleftharpoons T_1, \ldots,$ are slow compared to the binding reactions we have two parallel schemes like

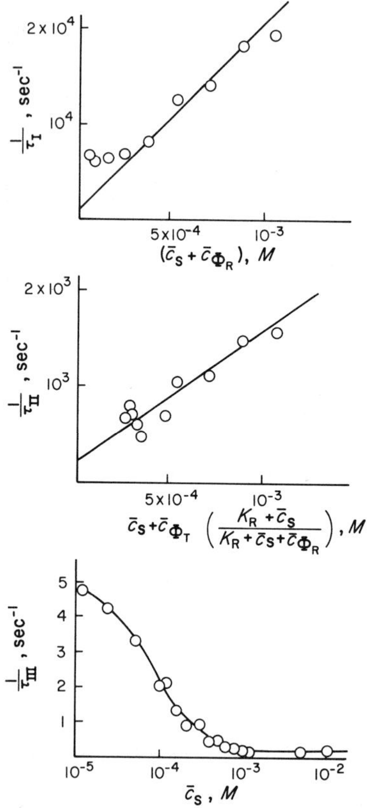

FIGURE 10.2. Concentration dependence of the three relaxation times of the NAD–GAPDH system at 40°, pH 8.5. [From Kirschner *et al.* (*5.*)]

10.1 which by virtue of the common substrate S are coupled to each other. Assuming now further that the binding and dissociation processes in the R states are much faster than in the T states, the equilibration of the part scheme involving only the R states is independent of that of the T states. Thus in analogy to Eqs. 10.3–10.4 we obtain

$$1/\tau_{1R} = k_A{}^R(\bar{c}_S + \bar{c}_{\Phi_R}) + k_D{}^R \tag{10.10}$$

$$1/\tau_{2R} = 2(k_A{}^R\bar{c}_S + k_D{}^R) \tag{10.11}$$

$$1/\tau_{3R} = 3(k_A{}^R\bar{c}_S + k_D{}^R) \tag{10.12}$$

$$1/\tau_{4R} = 4(k_A{}^R\bar{c}_S + k_D{}^R) \tag{10.13}$$

with $\bar{c}_{\Phi_R} = 4\bar{c}_{R_0} + 3\bar{c}_{R_1} + 2\bar{c}_{R_2} + \bar{c}_{R_3}$. Evidently $\tau_{1R} = \tau_I$ (Fig. 10.2a), whereas τ_{2R}, τ_{3R}, and τ_{4R} are undetectable in the temperature-jump experiment.

When considering the binding processes of the T states one can write

$$\chi_R \underset{k_D{}^R}{\overset{k_A{}^R}{\rightleftharpoons}} \Phi_R + S + \Phi_T \underset{k_D{}^T}{\overset{k_A{}^T}{\rightleftharpoons}} \chi_T \tag{10.14}$$

and treat the left side as fast equilibrium. This is analogous to case 6 in Table 3.1 and hence

$$\frac{1}{\tau_{1T}} = k_A{}^T\left(\bar{c}_S + \bar{c}_{\Phi_T}\frac{K_R + \bar{c}_S}{K_R + \bar{c}_S + \bar{c}_{\Phi_R}}\right) + k_D{}^T \tag{10.15}$$

with $\bar{c}_{\Phi_T} = 4\bar{c}_{T_0} + 3\bar{c}_{T_1} + 2\bar{c}_{T_2} + \bar{c}_{T_3}$, while the three relaxation times for substrate redistribution are given by

$$1/\tau_{2T} = 2(k_A{}^T\bar{c}_S + k_D{}^T) \tag{10.16a}$$

$$1/\tau_{3T} = 3(k_A{}^T\bar{c}_S + k_D{}^T) \tag{10.16b}$$

$$1/\tau_{4T} = 4(k_A{}^T\bar{c}_S + k_D{}^T) \tag{10.16c}$$

Again only τ_{1T} is visible in a temperature-jump experiment; it corresponds to τ_{II} (Fig. 10.2b).

We may note at this point that if c_S is kept quasi constant (buffering or large excess) the assumption that the binding–dissociation is much faster in the R states than in the T states is not necessary. In such a situation the two halves of scheme 10.5 become decoupled in very much the same way as in scheme 4.48 when E is quasi constant. (For the reactions in Eq. 10.14 two *independent* rate equations can be written since $\Delta\bar{c}_S = 0$.) For the eight relaxation times the same equations as before result except that Eq. 10.15 simplifies to

$$1/\tau_{1T} = k_A{}^T(\bar{c}_S + \bar{c}_{\Phi_T}) + k_D{}^T$$

Let us turn now to the ninth relaxation time of system 10.5. It must be associated with the conformational change, $\sum R_j \rightleftharpoons \sum T_j$. Its derivation (*1, 2*), although in principle straightforward since all other processes can be assumed to be rapid equilibria, is quite lengthy and not reproduced here. In the case that c_S is quasi constant and further assuming $k_{-0} = k_{-1} = k_{-2} = k_{-3} = k_{-4}$, one obtains (*1, 5*)

$$\frac{1}{\tau_{III}} = k_{-0} + k_0\left[\frac{1 + \bar{c}_S/K_T}{1 + \bar{c}_S/K_R}\right]^4 \tag{10.17}$$

It is easily appreciated that the plot of τ_{III}^{-1} versus $\bar{c}_S$ (Fig. 10.2c) is qualitatively consistent with Eq. 10.17. In the absence of NAD the conformational change is simply the reaction

$$R_0 \underset{k_{-0}}{\overset{k_0}{\rightleftharpoons}} T_0$$

with

$$1/\tau_{III} = k_0 + k_{-0}$$

From $L_0 = 30.5$ it follows that $k_0 = 30.5k_{-0}$ and thus $-\tau_{III}^{-1} = 31.5k_{-0}$. At very high substrate concentrations (saturation) the conformational change is just

$$R_4 \rightleftharpoons T_4$$

with

$$1/\tau_{III} = k_4 + k_{-4}$$

Since $L_4 = 0.46$ and $k_{-4} = k_{-0}$ it follows that $k_4 = 0.46k_{-0}$ and thus $\tau_{\text{III}}^{-1} = 1.46k_{-0}$. Thus there is a dramatic decrease in τ_{III}^{-1} with increasing substrate concentration as is borne out by the experimental data (Fig. 10.2c).

The kinetic parameters obtained from this analysis are

$$k_A{}^R = 1.9 \times 10^7 \, M^{-1}\,\text{sec}^{-1}, \qquad k_D{}^R = 10^3\,\text{sec}^{-1}$$

$$k_A{}^T = 1.37 \times 10^6 \, M^{-1}\,\text{sec}^{-1}, \qquad k_D{}^T = 2.1 \times 10^2\,\text{sec}^{-1}$$

$$k_0 = 5.5\,\text{sec}^{-1}, \qquad k_{-0} = 0.18\,\text{sec}^{-1}$$

Subsequent work by Kirschner (*6, 7*) has led to some refinements without altering the basic conclusions. However, it should be pointed out that although the results of these studies appear to be best explained by the MWC model, other schemes, such as the "induced fit model" of Koshland *et al.* (*8*) may better account for the allosteric behavior of other enzymes (*4*). Also, Loudon and Koshland (*9*) have made a critical analysis of the assumptions made in interpreting the kinetic results of the NAD–GAPDH system and addressed the general problem of the diagnostic relationships of relaxation kinetic data to various allosteric models.

10.2 Cooperative Conformational Transitions of Linear Biopolymers (Helix–Coil Transition of Polypeptides)

Cooperative conformational transitions of biopolymers are of great importance in the understanding of biological processes such as enzyme regulation (see Section 10.1), synthesis and denaturation of nucleic acids and proteins, etc. They have been the focus of numerous theoretical and experimental studies (*10*). These transitions typically proceed in a great number of rapid coupled elementary reactions which lead to complex mechanisms.

In general it is simpler to study these processes on model compounds such as oligonucleotides or polypeptides of well-defined chain length. We shall discuss the theory of helix–coil transitions of polypeptides in some detail.

10.2.1 Equilibrium Theory

Depending on various factors such as temperature, solvent, pH, and salt concentration, polypeptides exist in a state of either random coils or of the more highly ordered α helix. The transitions between the two conformations are usually quite sharp, which is characteristic of cooperative phenomena. Figure 10.3 shows an example where the degree of helicity of polyglutamic acid, measured by the specific optical rotation, is seen to change sharply with changing pH. This behavior is quite different from that of a simple unimolecular reaction, $H \rightleftharpoons C$; it is comparable to a phase transition.

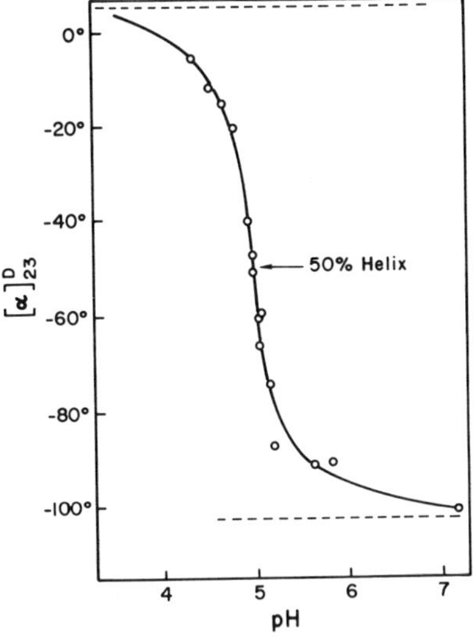

FIGURE 10.3. Helix–coil transition of polyglutamic acid in aqueous solution as measured by the specific optical rotation of the system. [From Schwarz (*12*), by permission of Academic Press.]

According to the statistical mechanical theory of Zimm and Bragg (*11*) such transition curves can, for long polymer chains (no "end effects"), be described by

$$\theta = \frac{\lambda_0 - 1}{\lambda_0 - \lambda_1} = \frac{1}{2}\left[1 + \frac{s - 1}{\lambda_0 - \lambda_1}\right] \tag{10.18}$$

$$\lambda_0, \lambda_1 = \tfrac{1}{2}(1 + s) \pm \tfrac{1}{2}\{(s - 1)^2 + 4\sigma s\}^{1/2}$$

where θ is the fraction of polymer chain in the helical form. The meaning of s and σ is as follows. The theory of Zimm and Bragg is based on a statistical mechanical treatment, i.e., formulated in terms of partition functions. Each conformational state, i.e., each particular sequence of coil and helix segments, contributes a certain term, q_i, to the overall partition function Q; helix and coil segments contribute different statistical weight factors to these q_i terms. The statistical weight factors which depend on the external conditions are defined as

1	for any coil unit
s	for any helix unit following another helix unit
σs	for any helix unit following a coil unit

The contribution to the particular function of a certain sequence of helix and coil units is equal to the products of all statistical factors. It is easily shown that s corresponds to the equilibrium constant for the reaction of adding one helix unit to an already existing helix (ratio of statistical factor $s/1 = s$) whereas σs is the equilibrium constant for the more difficult process of generating a helix unit in a region consisting only of coils. σ is called the "nucleation parameter" and is typically much less than unity; it is a measure of the steepness of the transition curve, i.e., the smaller σ, the steeper the transition. s is strongly dependent on the external conditions; from Eq. 10.18 we see that $s = 1$ defines the transition midpoint ($\theta = 0.5$).

10.2.2 Kinetic Theory

In this section we develop some of the basic features of the kinetic theory, based on the approach of Schwarz (*12–14*). For a full and completely rigorous treatment the original papers should be consulted. The helix–coil transition comes about through a series of elementary steps which represent the transformation of small segments of the polypeptide chain from a helix to a coil unit. A helix unit is an amino acid residue with its dihedral angles characteristic of the helix, within a narrow range of variation; a coil unit is an amino acid residue with one or several of its dihedral angles not those of the helix (*15*).

Each of these elementary steps is a unimolecular reaction

$$\cdots \mathrm{h} \cdots \underset{k_r}{\overset{k_f}{\rightleftharpoons}} \cdots \mathrm{c} \cdots$$

The important point, however, is that k_f and k_r depend on whether the neighboring units are in the helix or coil state. The rate of the macroscopic conformational change is thus dependent on the interplay of all these microscopic processes. In a polypeptide chain with N segments or units there exist up to 2^N different possible sequences or states, leading up to 2^{N-1} relaxation times and therefore giving rise to a virtually continuous relaxation spectrum.

A great simplification can be introduced by assuming that there are essentially only two types of elementary processes:

1. Growth, that is, formation or disintegration of helix units at the beginning or end of a helix region:

$$\begin{matrix} \cdots \mathrm{cch} \cdots \\ \cdots \mathrm{hcc} \cdots \end{matrix} \underset{k_B}{\overset{k_F}{\rightleftharpoons}} \begin{matrix} \cdots \mathrm{chh} \cdots \\ \cdots \mathrm{hhc} \cdots \end{matrix} \tag{10.19}$$

Since reaction 10.19 is the addition of a helix unit to an already existing helix, the equilibrium constant is

$$k_F/k_B = s \tag{10.20}$$

2. Nucleation, that is, formation of a helix unit within a coil region (helix nucleation)

$$\cdots \mathrm{ccc} \cdots \underset{k_{-1}}{\overset{k_1}{\rightleftharpoons}} \cdots \mathrm{chc} \cdots \tag{10.21}$$

or formation of a coil unit within a helix region (coil nucleation)

$$\cdots \mathrm{hch} \cdots \underset{k_{-2}}{\overset{k_2}{\rightleftharpoons}} \cdots \mathrm{hhh} \cdots \tag{10.22}$$

In view of the remarks made in Section 10.2.1 it is easily seen that

$$k_1/k_{-1} = s\sigma, \qquad k_2/k_{-2} = s/\sigma \tag{10.23}$$

It should perhaps be mentioned that these definitions of growth and nucleation steps imply that only the next neighbors have any influence on the ease or difficulty of a given elementary process. Such a model is known as an "Ising model"; it is the simplest model but usually completely satisfactory in interpreting kinetic data. More refined models which take into account possible effects of more distant neighbors do not lead to significant improvements as long as the cooperativity is very large ($\sigma \ll 1$), which is usually the case.

Let us now formulate the rate of the macroscopic helix–coil transition. The polypeptide chain can be considered as being a sequence of triplets as defined in Eqs. 10.19, 10.21, and 10.22. The rate is then

$$\frac{d\theta}{dt} = k_F(f_{cch} + f_{hcc}) - k_B(f_{chh} + f_{hhc}) + k_1 f_{ccc} - k_{-1} f_{chc} + k_2 f_{hch} + k_{-2} f_{hhh} \tag{10.24}$$

where $f_{cch}, f_{hcc}, \ldots$ are the fractions of cch, hcc, ... triplets, respectively; note that by definition we have also

$$f_{cch} + f_{hcc} + f_{chc} + f_{hhc} + f_{ccc} + f_{chh} + f_{hch} + f_{hhh} = 1 \tag{10.25}$$

$$\theta = f_{chh} + f_{hhc} + f_{chc} + f_{hhh} \tag{10.26}$$

Defining further $k_1 = \sigma\gamma_H k_F$, $k_{-1} = \gamma_H k_B$ and $k_2 = \gamma_C k_F$, $k_{-2} = \sigma\gamma_C k_B$ (with $\gamma_H, \gamma_C \geq 1$), and taking into account Eqs. 10.25 and 10.26 leads to

$$\frac{d\theta}{dt} = k_F - (k_F + k_B)\theta - k_F(1 - \gamma_H\sigma)f_{ccc} + k_B(1 - \gamma_C\sigma)f_{hhh} - k_F(1 - \gamma_C)f_{hch} + k_B(1 - \gamma_H)f_{chc} \tag{10.27}$$

We now consider a small perturbation which induces a change from s to $s + \Delta s$ and a subsequent relaxation. According to the principles discussed in Chapter 1 we write

$$\begin{aligned} f_{ccc} &= \bar{f}_{ccc} + \Delta f_{ccc} \\ f_{hhh} &= \bar{f}_{hhh} + \Delta f_{hhh} \\ &\vdots \\ \theta &= \bar{\theta} + \Delta\theta \end{aligned} \tag{10.28}$$

As seen below it is more convenient here to choose the *initial* state as the reference equilibrium, i.e., $\bar{f}_{ccc} = \bar{f}^{i}_{ccc}$, etc. In principle our aim now would be to convert Eq. 10.27 into an expression of the form of Eq. 1.30. Since this is not possible one focuses on the rate of transition at time $t = 0$.

Expressing $k_F/k_B = s + \Delta s$ where s refers to the state before the perturbation, Eq. 10.27 can be written as

$$\frac{1}{k_B}\left(\frac{d\theta}{dt}\right)_{t=0} = s + \Delta s - (s + \Delta s + 1)\bar{\theta} - (s + \Delta s)(1 - \gamma_H\sigma)\bar{f}_{ccc} + (1 - \gamma_C\sigma)\bar{f}_{hhh} - (s + \Delta s)(1 - \gamma_C)\bar{f}_{hch} + (1 - \gamma_H)\bar{f}_{chc} \tag{10.29}$$

The left-hand side of Eq. 10.29 is equivalent to $k_B^{-1}(d\,\Delta\theta/dt)_{t=0}$; the terms on the right-hand side which do not involve Δs add up to $k_B^{-1}(d\bar{\theta}/dt) = 0$. Hence, after making use of Eqs. 10.25 and 10.26 one obtains

$$\frac{1}{k_B}\left(\frac{d\,\Delta\theta}{dt}\right)_{t=0} = (\bar{f}_{\text{cch}} + \bar{f}_{\text{hcc}} + \sigma\gamma_H\bar{f}_{\text{ccc}} + \gamma_C\bar{f}_{\text{hch}})\,\Delta s \qquad (10.30)$$

Expressing the total change in θ as

$$\Delta\theta_\infty = \bar{\theta}^{\text{f}} - \bar{\theta}^{\text{i}} = \frac{\partial\theta}{\partial s}\,\Delta s \qquad (10.31)$$

affords

$$\Delta s = \frac{\Delta\theta_\infty}{\partial\theta/\partial s} \qquad (10.32)$$

while equilibrium theory gives

$$\frac{\partial\theta}{\partial s} = \sigma\,\frac{1+s}{(\lambda_0 - \lambda_1)^3} \qquad (10.33)$$

The equilibrium theory also allows us to express the various f fractions in terms of λ_0 and λ_1 (*12*); after a number of arithmetic manipulations one finally obtains for the mean relaxation time (*12*)

$$\frac{1}{\Delta\theta_\infty}\left(\frac{d\,\Delta\theta}{dt}\right)_{t=0} = \frac{1}{\tau^*} = k_F\,\frac{(s-1)^2 + 4\sigma s}{s\lambda_0} \times \left\{1 + (\gamma_H - 1)\,\frac{\lambda_0 - s}{(1+s)\lambda_0} + (\gamma_C - 1)\,\frac{s(\lambda_0 - 1)}{(1+s)\lambda_0}\right\} \qquad (10.34)$$

In the range near the transition midpoint the factors multiplying $(\gamma_H - 1)$ and $(\gamma_C - 1)$ are very small, and $\lambda_0 \approx s \approx 1$ so that Eq. 10.34 simplifies to

$$\frac{1}{\tau^*} = k_F\{(s-1)^2 + 4\sigma\} = k_F\,\frac{\sigma}{\theta(1-\theta)} \qquad (10.35)$$

Note that Eq. 10.35 could have been obtained more directly as follows. Near the midpoint there are only relatively few places in the chain where nucleation can take place but there is a large number of different growth steps. Thus it is these latter which mainly contribute to the overall rate, i.e., only the k_F and k_B terms in Eq. 10.24 are significant and one can proceed in the same way as before but only considering these two terms. In contrast, when the chain is predominantly in the coil or helix form, nucleation contributes significantly even if it has a slow inherent rate because there are many places in the chain where nucleation can occur, but only a few places where growth can take place.

The most interesting feature about Eq. 10.35 is that it predicts a maximum for τ^* at the midpoint ($\theta = 0.5$ or $s = 1$), given by

$$\tau^*_{\max} = 1/4\sigma k_F \qquad (10.36)$$

This is qualitatively understandable because at this point a maximum number of helix (coil) units are formed and destroyed, respectively, which requires a relatively long time, and also because the growth steps of any helical segment occur sequentially rather than concurrently. As seen in Fig. 10.4 the maximum becomes sharper and τ^*_{max} longer for smaller values of σ.

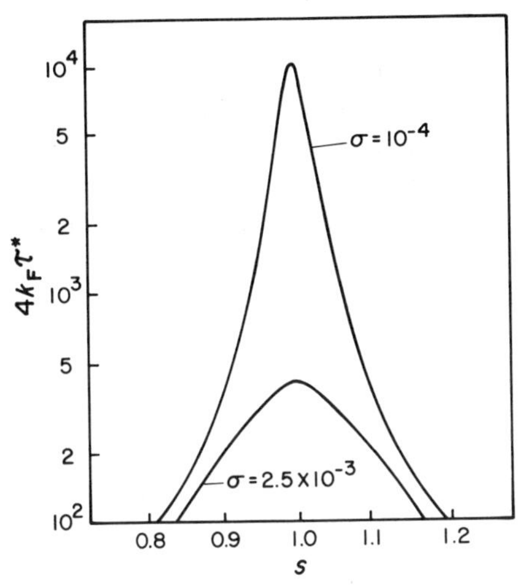

FIGURE 10.4. Mean relaxation time τ^* as a function of s according to Eq. 10.35; τ^* has been multiplied by k_F for a dimensionless plot. [From Schwarz (*12*), by permission of Academic Press.]

An interesting question is how well the relaxation curve can be approximated by a single exponential given by

$$\Delta\theta = \Delta\theta_0 \exp(-t/\tau^*) \tag{10.37}$$

(note here we define $\Delta\theta = \theta - \bar{\theta}^{\text{f}}$, $\Delta\theta_0 = \bar{\theta}^{\text{i}} - \bar{\theta}^{\text{f}} = -\Delta\theta_\infty$). According to Eq. 9.26 we have

$$D(t) = \Phi(t) - \exp(-t/\tau^*) = (\rho^*/2)(t/\tau^*)^2 + \cdots$$

Schwarz (*12, 13*) has shown that under the conditions where Eq. 10.35 is valid, ρ^* is given by

$$\rho^* = 0.5\{(s - 1)^2 + 4\sigma\}^{-1/2} - 1 = 0.5\sigma^{1/2}[\bar{\theta}(1 - \bar{\theta})]^{1/2} - 1 \tag{10.38}$$

It goes through a maximum at $s = 1$, given by

$$\rho^*_{\text{max}} = 1/4\sigma^{1/2} - 1 \tag{10.39}$$

For $\sigma = 10^{-4}$, $\rho^*_{\text{max}} = 24$; for $\sigma = 10^{-2}$, $\rho^*_{\text{max}} = 1.5$. Since ρ^* is a measure of the width of the relaxation spectrum we see that the higher the cooperativity (smaller σ) the wider the spectrum. Schwarz (*12*) has estimated that $\rho^* = 24$ corresponds to a range of relaxation times of about three orders of magnitude, a rather wide spectrum.

For $D(t)$ at the midpoint we obtain (for $\sigma = 10^{-4}$)

$$D(t) = 12(t/\tau^*)^2 + \cdots \tag{10.40}$$

At times $t \leq 0.1\tau^*$, we have $D(t) \leq 0.12$; i.e., the deviation of the relaxation curve from a pure exponential is less than 12%.

As pointed out in Section 9.3, Eq. 9.26 is only valid for $D(t) \ll 1$ and thus cannot be used to estimate $D(t)$ for times longer than $t = 0.1\tau^*$ in the present example. However, Craig and Crothers (*16*), and later Rawlings and Schneider (*17a, b*) approached the problem of helix–coil transition kinetics by computer simulation methods and came to the conclusion that the relaxation curve is very well approximated by an exponential curve over the entire time range; i.e., the relaxation spectrum is rather narrow. This apparent contradiction to the above analysis can be explained by the effect on ρ^* of widely separated relaxation times with strongly differing amplitudes, as discussed in Section 9.3.2 (*18*); as shown by Schwarz (*19*) there are four relaxation times in the spectrum whereby the longest has a large β value and is dominant ($\approx \tau^*$) whereas the others with small β values are extremely short.

10.2.3 Experimental Studies

Recent experimental studies have shown that τ^* in fact goes through a maximum at the midpoint, e.g., in the case of poly(γ-benzyl L-glutamate) in a mixture of ethylene dichloride and dichloroacetic acid (*20*), and for poly(β-benzyl L-aspartate) in *m*-cresol solution (*21*).

The numerical values of $\tau^*_{\max}$ are also "reasonable." Since chemically the conversion of a coil into a helix segment essentially corresponds to the formation of a hydrogen bond its rate can be assumed to be diffusion controlled or nearly so, with $k_F \sim 10^{10}$–10^{11} sec^{-1}. For a typical value of $\sigma \sim 10^{-4}$ one thus expects $\tau^*_{\max} \sim 10^{-7}$ sec. However, in some cases rather high values for $\tau^*_{\max}$ were found, suggesting $k_F < 10^8$ sec^{-1}, e.g., for poly-α,L-glutamic acid in aqueous solution (*22, 23*). This was interpreted as being possibly due to a steric effect making optimal alignment of the CO and NH groups difficult (*23*). Other examples will be mentioned under applications of the temperature-jump, electric field-jump, ultrasonic, and dielectric relaxation methods.

10.3 Cooperative Binding of Small Molecules to Linear Biopolymers

Like conformational transitions, the binding of small molecules such as metal ions, antibiotics, nucleotides, and dyes to biopolymers is of great biological interest. It turns out that there is a great similarity between the theory of cooperative binding and that of cooperative conformational transitions. It is again Schwarz who has made the most significant contributions to the kinetic theory (*24, 25*).

The underlying feature of Schwarz's theory is that the binding comes about through a series of elementary processes of the type

$$A + \cdots u \cdots \rightleftharpoons \cdots a \cdots$$

where A is a monomer molecule, u is an unoccupied binding site of the polymer chain, and a is a site occupied by A. The affinity of u for A is assumed to depend on the state of the nearest neighbors (Ising model), i.e., on whether or not the nearest neighbors are occupied. Just as in the case of conformational transitions it is now assumed that there are only two types of elementary processes, namely growth and nucleation. A growth process is defined as

$$\mathrm{A} + \begin{matrix} \cdots \mathrm{uua} \cdots \\ \cdots \mathrm{auu} \cdots \end{matrix} \underset{k_{\mathrm{D}}}{\overset{k_{\mathrm{R}}}{\rightleftharpoons}} \begin{matrix} \cdots \mathrm{uaa} \cdots \\ \cdots \mathrm{aau} \cdots \end{matrix} \tag{10.41}$$

with

$$k_{\mathrm{R}}/k_{\mathrm{D}} = K = \bar{c}_{\mathrm{uaa}}/\bar{c}_{\mathrm{uua}}\bar{c}_{\mathrm{A}} = \bar{c}_{\mathrm{aau}}/\bar{c}_{\mathrm{auu}}\bar{c}_{\mathrm{A}} \tag{10.42}$$

whereas the two types of nucleation processes are

$$\mathrm{A} + \cdots \mathrm{uuu} \cdots \underset{k_{-1}}{\overset{k_1}{\rightleftharpoons}} \cdots \mathrm{uau} \cdots \tag{10.43}$$

$$\mathrm{A} + \cdots \mathrm{aua} \cdots \underset{k_{-2}}{\overset{k_2}{\rightleftharpoons}} \cdots \mathrm{aaa} \cdots \tag{10.44}$$

with

$$k_1/k_{-1} = K\sigma, \qquad k_2/k_{-2} = K/\sigma \tag{10.45}$$

(That 10.44 is a "nucleation process" becomes obvious when the k_{-2} step is considered.) Note the analogy with Eqs. 10.19–10.23 except that here we deal with a bimolecular process. If we define $Kc_{\mathrm{A}} = s$, the equilibrium formalism describing helix-coil transitions (Eq. 10.18) can be directly applied to the fraction (θ) of occupied binding sites.

The kinetic description of the binding is particularly simple in the range of moderate saturation because here the growth steps are almost exclusively responsible for the overall rate, again in analogy to helix–coil transitions near the midpoint. In this range we can write

$$\frac{dc_{\mathrm{A}}}{dt} = -2k_{\mathrm{R}}c_{\mathrm{A}}c_{\mathrm{uua}} + 2k_{\mathrm{D}}c_{\mathrm{uaa}} \tag{10.46}$$

(note the factor 2 arises because $c_{\mathrm{uua}} = c_{\mathrm{auu}}$, $c_{\mathrm{uaa}} = c_{\mathrm{aau}}$). After introducing $c_{\mathrm{A}} = \bar{c}_{\mathrm{A}} + \Delta c_{\mathrm{A}}$, $c_{\mathrm{uua}} = \bar{c}_{\mathrm{uua}} + \Delta c_{\mathrm{uua}}$, $c_{\mathrm{uaa}} = \bar{c}_{\mathrm{uaa}} + \Delta c_{\mathrm{uaa}}$, and linearizing the rate equation one obtains

$$\frac{d\,\Delta c_{\mathrm{A}}}{dt} = -2k_{\mathrm{R}}\bar{c}_{\mathrm{uua}}\,\Delta c_{\mathrm{A}} - 2k_{\mathrm{R}}\bar{c}_{\mathrm{A}}\,\Delta c_{\mathrm{uua}} + 2k_{\mathrm{D}}\,\Delta c_{\mathrm{uaa}} \tag{10.47}$$

which can also be written as

$$\frac{d\,\Delta c_{\mathrm{A}}}{dt} = -\,2k_{\mathrm{R}}\bar{c}_{\mathrm{uua}}\,\Delta c_{\mathrm{A}}[1 + X(t)] \tag{10.48}$$

with

$$X(t) = \left(\frac{\Delta c_{\mathrm{uua}}}{\bar{c}_{\mathrm{uua}}} - \frac{\Delta c_{\mathrm{uaa}}}{\bar{c}_{\mathrm{uaa}}}\right)\frac{\bar{c}_{\mathrm{A}}}{\Delta c_{\mathrm{A}}}$$

From the equilibrium theory it can be shown that for time $t = 0$, $X(t) \ll 1$. Hence

$$-\frac{1}{\Delta c_A^0}\left(\frac{d\,\Delta c_A}{dt}\right)_{t=0} = \frac{1}{\Delta c_A{}^{\infty}}\left(\frac{d\,\Delta c_A}{dt}\right)_{t=0} = \frac{1}{\tau^*} = 2k_R\bar{c}_{\text{uua}} \qquad (10.49)$$

The equilibrium theory further shows that this is equivalent to

$$\frac{1}{\tau^*} = 2k_R\sigma^{1/2}[\theta(1 - \theta)]^{1/2}gc_P \qquad (10.50)$$

where c_P is the total molar concentration of monomer units, and g is the number of binding sites per monomer unit.

After having stressed the analogy to helix–coil transitions on several occasions it should be pointed out that the result of the kinetic treatment is now quite different; this is mainly because the rate is expressed in terms of changes in c_A rather than θ. The width of the relaxation spectrum is quite narrow, the more so the higher the cooperativity, in complete contrast to the helix–coil transitions. Schwarz (*24*) has shown that for small σ

$$|\rho^*| \ll 1 \qquad (\rho^* \to 0 \quad \text{for} \quad \sigma \to 0)$$

This means that the relaxation curve is very well approximated by a single exponential with $\tau = \tau^*$ over the entire time range.

For small degrees of saturation ($\theta \ll 1$) the situation is much more complex and there are now three relaxation times. The theory has also been developed for this range (*25*).

Experimental studies involving the binding of acridine (*26*) or proflavine (*27*) to poly-L-glutamic acid; of proflavine to polyacrylic acid, poly-α-L-glutamic acid, and polyphosphate (*28*); and toluidine blue to poly-α-L-glutamic acid (*29*) could all be reasonably interpreted by the Schwarz theory. This is equally true for a recent study of the binding of *N*-6,9-dimethyladenine to polyuridylic acid although the results are complicated by the fact that *N*-6,9-dimethyladenine undergoes self-association (*30*).

10.4 Molecular Aggregation Phenomena. Micelle Formation

Molecular aggregation phenomena have become a focus of increasing interest, particularly in biological systems (*31*). The study of the thermodynamics and kinetics of self-association of small molecules is expected to shed some light on similar phenomena in biological systems and thus is of considerable importance. The problems of interpretation of the data and of deriving a reasonable mechanism are very challenging. One particular approach based on the analysis of relaxation amplitudes and mean relaxation times as applied to the self-association of acridine orange has already been discussed in Section 9.3.2.

In this section we shall concentrate on one type of molecular aggregation, namely micelle formation. The mechanism of micellization is a subject of great current interest but is also in a state of considerable confusion (*32*). Part of the problem is that some authors cannot even agree on the experimental facts (*32a*). Several relaxation techniques, including ultrasonics, temperature-jump, pressure-jump, and stopped-flow methods, have been applied to the study of micelle formation. Usually one but sometimes two relaxation times covering a range from about 10^{-2} to 10^{-9} sec have been reported. Most of the relevant literature has been summarized by Rassing *et al.* (*33*) and by Wyn-Jones (*32*).

Several mechanisms have been proposed to account for the experimental data but none has been generally accepted. Some are described here; our discussion does not intend to be comprehensive but is meant to illustrate some of the problems and pitfalls encountered in devising a satisfactory mechanism for a complex system.

10.4.1 The KHDS Model and Variations Thereof

The first relaxation kinetic study of micellization were the temperature-jump experiments on dodecylpyridinium iodide by Kresheck *et al.* (*34*) (KHDS). They found one relaxation time whose reciprocal value was linearly dependent on total surfactant (monomer) concentration, however with a negative ordinate intercept. Such a dependence, which can be described by the general equation

$$1/\tau = k_f c_0 - k_b \tag{10.51}$$

has been observed in many subsequent studies although the time ranges varied greatly. KHDS proposed the following model to account for their data

$$\begin{aligned}
\mathrm{A} + \mathrm{A} &\underset{k_{21}}{\overset{k_{12}}{\rightleftharpoons}} \mathrm{A}_2 \\
\mathrm{A} + \mathrm{A}_2 &\underset{k_{32}}{\overset{k_{23}}{\rightleftharpoons}} \mathrm{A}_3 \\
&\vdots \\
\mathrm{A} + \mathrm{A}_j &\underset{k_{j+1,j}}{\overset{k_{j,j+1}}{\rightleftharpoons}} \mathrm{A}_{j+1} \\
&\vdots \\
\mathrm{A} + \mathrm{A}_{n-1} &\underset{k_{n,n-1}}{\overset{k_{n-1,n}}{\rightleftharpoons}} \mathrm{A}_n
\end{aligned} \tag{10.52}$$

where A is a monomer, A_n is the "finished" micelle, while the other species are intermediates containing various numbers of monomers. The kinetic role of equilibria such as

$$\mathrm{A}_j + \mathrm{A}_k \rightleftharpoons \mathrm{A}_{j+k} \tag{10.53}$$

was assumed to be negligible.

We note that scheme 10.52, except for a somewhat different notation which is the one of the original authors, is the same as that assumed to prevail in the self-association of acridine orange (Eq. 9.45). However, the following assumptions of KHDS are now quite different from those made in the acridine orange case, namely

$$\bar{c}_{\mathrm{A}} \gg \bar{c}_{\mathrm{A}_2} \gg \bar{c}_{\mathrm{A}_3} \gg \cdots \gg \bar{c}_{\mathrm{A}_{n-1}}; \qquad \bar{c}_{\mathrm{A}_n} \gg \bar{c}_{\mathrm{A}_2}$$

which means that the intermediate states are very sparsely populated compared to A and A_n; the second assumption which in fact is implied by the first, is that the last step in scheme 10.52 equilibrates much more slowly than all the others. Thus scheme 10.52 can also be written as

$$(n-1)\mathrm{A} \underset{l'}{\overset{l}{\rightleftharpoons}} \mathrm{A}_{n-1} \qquad \text{(fast)}$$

$$\mathrm{A} + \mathrm{A}_{n-1} \underset{k_{n,n-1}}{\overset{k_{n-1,n}}{\rightleftharpoons}} \mathrm{A}_n \qquad \text{(slow)}$$

For the rate of the slow step one has

$$dc_{\mathrm{A}_n}/dt = k_{n-1,n}c_{\mathrm{A}}c_{\mathrm{A}_{n-1}} - k_{n,n-1}c_{\mathrm{A}_n}$$

Substituting $L(c_{\mathrm{A}})^{n-1}$ for $c_{\mathrm{A}_{n-1}}$ ($L = l/l'$) affords

$$dc_{\mathrm{A}_n}/dt = k_{n-1,n}L(c_{\mathrm{A}})^n - k_{n,n-1}c_{\mathrm{A}_n} \tag{10.54}$$

After linearization we find

$$1/\tau = k_{n-1,n}Ln^2(\bar{c}_{\mathrm{A}})^{n-1} + k_{n,n-1}$$

Since, furthermore,

$$\bar{c}_{\mathrm{A}_n} = L(k_{n-1,n}/k_{n,n-1})(\bar{c}_{\mathrm{A}})^n; \qquad c_0 \approx \bar{c}_{\mathrm{A}} + n\bar{c}_{\mathrm{A}_n}$$

one finally obtains

$$1/\tau = nk_{n,n-1}(c_0/\bar{c}_{\mathrm{A}}) - (n-1)k_{n,n-1} \tag{10.55}$$

Above the critical micelle concentration (cmc), $\bar{c}_{\mathrm{A}}$ remains essentially constant. Hence Eq. 10.55 is indeed of the form of Eq. 10.51.

The KHDS treatment has been criticized mainly on two grounds. As pointed out in Section 5.2 for very large n a rate equation such as 10.54 can only be linearized for extremely small perturbations. In fact Muller (*35*) has shown that the sizes of the temperature jumps used by KHDS were much too large for linearization to be permissible ($n = 87$).

The second criticism is that the KHDS model requires $\bar{c}_{\mathrm{A}_n} \gg \bar{c}_{\mathrm{A}_{n-1}}$ and $k_{n-1,n-2} \gg k_{n,n-1}$; this contradicts the generally accepted notion that the species A_n and A_{n-1} must have very similar properties so that one would rather expect $\bar{c}_{\mathrm{A}_n} \sim \bar{c}_{\mathrm{A}_{n-1}}$ and $k_{n-1,n-2} \sim k_{n,n-1}$ (*35*). Muller has therefore suggested a different model based on the assumption that all $k_{j,j-1}$ are the same for $j = 3, 4, \ldots, n$ and all $k_{j-1,j}$ are the same for $j = 3, 4, \ldots, n$; only the step

$$2\mathrm{A} \underset{k_{21}}{\overset{k_{12}}{\rightleftharpoons}} \mathrm{A}_2$$

is assumed to be much more rapid than the others.

A derivation of τ is very difficult for this model and has not been attempted; however, Muller (*35*) has discussed the situation which occurs in a concentration-jump experiment where a solution containing mainly A_n is rapidly diluted with a large amount of solvent which leads to an essentially irreversible dissociation of the micelles.

10.4.2 The SWR Models

Another scheme, proposed by Sams *et al.* (*36*) (SWR), is the so-called two-state model. The two states are (1) the monomer, (2) all the A_j with $j \geq 2$, i.e., all associated states are considered to be micelles. It is assumed that the rate at which the monomers associate with micelles is proportional to the concentration of monomers, the concentration of the respective micelles, and to the number of monomers in the respective micelle; thus

$$r_f = k_f c_A \{2c_{A_2} + 3c_{A_3} + \cdots + (n-1)c_{A_{n-1}}\} \tag{10.56}$$

No justification for the third assumption was given in the original paper but later Sams *et al.*, (*37*) argued that the assumption can be justified on the basis of the increased surface area of the larger micelles which offers a larger target to the attacking monomer. The assumption is equivalent to assuming that the rate constants for association are proportional to j, i.e., $k_{23} = 2k_f$, $k_{34} = 3k_f$, etc. (see scheme 10.52).

For the dissociation of a monomer unit from a micelle it is assumed that the rate is proportional to the concentration of the respective micelle and to the number of monomers contained in it; thus

$$r_b = k_b\{2c_{A_2} + 3c_{A_3} + \cdots + n\bar{c}_{A_n}\} \tag{10.57}$$

Since the overall surfactant concentration is given by

$$c_0 = \sum_{j=1}^{n} jc_{A_j} \tag{10.58}$$

the terms in the braces of Eqs. 10.56 and 10.57 are (with the implicit assumption that $n \to \infty$) equal, and given by $c_0 - c_A$. Hence one obtains

$$-dc_A/dt = k_f c_A(c_0 - c_A) - k_b(c_0 - c_A) \tag{10.59}$$

At equilibrium $dc_A/dt = 0$ and thus from Eq. 10.59 $k_f\bar{c}_A = k_b$. Linearization of Eq. 10.59 affords

$$1/\tau = k_f c_0 + k_b - 2k_f\bar{c}_A = k_f c_0 - k_b \tag{10.60}$$

(which is also Eq. 10.51) which again is consistent with the experimental observations.

However, the SWR model is not without flaws. When the assumption that the rate of association of a monomer to A_j is proportional to j is viewed in terms of the equivalent assumption that $k_{23} = 2k_f$, $k_{34} = 3k_f, \ldots$ it becomes apparent that expression 10.56 neglects the inclusion of the dimerization step. In fact, as has already been pointed out by others (*38*, *39*), inclusion of the dimerization step would lead to a sign reversal in Eq. 10.60(!), i.e.,

$$1/\tau = k_f c_0 + k_b$$

We might add that the implicit assumption that $n \rightarrow \infty$ seems hardly justified either.

An improved version of the SWR two-state model is based on a treatment of the addition of a monomer to a micelle as an adsorption phenomenon (*33*). Defining a_0 as the area that a monomer occupies on the micelle surface (on a molar basis, i.e., a_0 is the area covered by 1 mole of monomer), α as the fraction of the total micelle surface area S occupied by monomers, one has for S

$$S = \sum_{j=2}^{n} j c_{A_j}(a_0/\alpha) \tag{10.61}$$

and in combination with Eq. 10.58

$$S = (a_0/\alpha)(c_0 - c_A)$$

Following the principles of the Langmuir adsorption theory, according to which the rate of adsorption is proportional to the free surface area, and the rate of desorption is proportional to the occupied surface area, we can write

$$dc_A/dt = -k_1 c_A(1 - \alpha)S + k_{-1}\alpha S \tag{10.62}$$

where k_1 is the rate constant for the condensation of monomers on the micellar surfaces and k_{-1} is the rate constant for the desorption process. Linearization of Eq. 10.62 and combination with Eq. 10.61 affords

$$1/\tau = k_1 a_0[(1 - \alpha)/\alpha]c_0 - k_{-1}a_0 \tag{10.63}$$

Equation 10.63 is equivalent to Eq. 10.51 if we set

$$k_f = k_1 a_0(1 - \alpha)/\alpha, \qquad k_b = k_{-1}a_0 \tag{10.64}$$

We note that according to Eq. 10.64 k_f if not a true rate constant but depends on α, a "packing factor." Inasmuch as α depends on the temperature one expects that the temperature dependence of k_f is not necessarily following the Arrhenius equation, while k_b should have an Arrhenius dependence. These expectations have been borne out by some recent experiments (*40*).

References

1. M. Eigen, *in Nobel Sympos.*, *5* (S. Claesson, ed.), p. 333. Wiley (Interscience), New York, 1967.
2. M. Eigen, *Quart. Rev. Biophys.* **1**, 3 (1968).
3. J. Monod, J. Wymann, and P. Changeux, *J. Mol. Biol.* **12**, 88 (1965).
4. G. G. Hammes and C.-W. Wu, *Ann. Rev. Biophys. Bioeng.* **3**, 1 (1974).
5. K. Kirschner, M. Eigen, R. Bittman, and B. Voigt, *Proc. Nat. Acad. Sci. U.S.* **56**, 1661 (1966).

6. K. Kirschner, E. Gallego, I. Schuster, and D. Goodall, *J. Mol. Biol.* **58**, 29 (1971).
7. K. Kirschner, *J. Mol. Biol.* **58**, 51 (1971).
8. D. E. Koshland, Jr., G. Nemethy, and D. Filmer, *Biochemistry* **5**, 365 (1966).
9. G. M. Loudon and D. E. Koshland, Jr., *Biochemistry* **11**, 229 (1972).
10. D. Poland and H. A. Scheraga, "Theory of Helix-Coil Transitions in Biopolymers," Academic Press, New York, 1970.
11. B. H. Zimm and J. K. Bragg, *J. Chem. Phys.* **31**, 526 (1959).
12. G. Schwarz, *J. Mol. Biol.* **11**, 64 (1965).
13. G. Schwarz, *Biopolymers* **6**, 873 (1968).
14. G. Schwarz and J. Engel, *Angew. Chem. Int. Ed.* **11**, 568 (1972).
15. D. Poland and H. A. Scheraga, "Theory of Helix-Coil Transitions in Biopolymers," p. 5. Academic Press, New York, 1970.
16. M. E. Craig and D. M. Crothers, *Biopolymers* **6**, 385 (1968).
17a. P. K. Rawlings and F. M. Schneider, *Ber. Bunsenges. Phys. Chem.* **77**, 237 (1973).
17b. P. K. Rawlings and F. M. Schneider, *Ber. Bunsenges. Phys. Chem.* **78**, 773 (1974).
18. G. Schwarz, personal communication (1975).
19. G. Schwarz, *J. Theor. Biol.* **36**, 569 (1972).
20. G. Schwarz and J. Seelig, *Biopolymers* **6**, 1263 (1968).
21. A. Wada, *Chem. Phys. Lett.* **8**, 211 (1971).
22. A. D. Barksdale and J. Stuehr, *J. Amer. Chem. Soc.* **94**, 3334 (1972).
23. T. Yasunaga, Y. Tsuji, T. Sano, and H. Takenaka, *in* "Chemical and Biological Applications of Relaxation Spectrometry" (E. Wyn-Jones, ed.), p. 493. Reidel, Dordrecht-Holland, 1975.
24. G. Schwarz, *Eur. J. Biochem.* **12**, 442 (1970).
25. G. Schwarz, *Ber. Bunsenges. Phys. Chem.* **76**, 373 (1972).
26. G. Schwarz and W. Balthasar, *Eur. J. Biochem.* **12**, 461 (1970).
27. G. Schwarz, S. Klose, and W. Balthasar, *Eur. J. Biochem.* **12**, 454 (1970).
28. G. Schwarz and S. Klose, *Eur. J. Biochem.* **29**, 249 (1972).
29. T. Yasunaga, H. Takenaka, T. Sano, and Y. Tsuij, *in* "Chemical and Biological Applications of Relaxation Spectrometry" (E. Wyn-Jones, ed.), p. 467. Reidel, Dordrecht-Holland, 1975.
30. G. W. Hoffman and D. Pörschke, *Biopolymers* **12**, 1625 (1973).
31. B. Pullman (ed.), "Molecular Association in Biology." Academic Press, New York, 1968.
32. E. Wyn-Jones (ed.), "Chemical and Biological Applications of Relaxation Spectrometry." Reidel, Dordrecht-Holland, 1975.
32a. H. Hoffmann, "Chemical and Biological Applications of Relaxation Spectrometry" (E. Wyn-Jones, ed.), p. 181. Reidel, Dordrecht-Holland, 1975.
33. J. E. Rassing, P. J. Sams, and E. Wyn-Jones, *J. Chem. Soc. Farad. Trans. II* **70**, 1247 (1974).
34. G. C. Kresheck, E. Hamori, G. Davenport, and H. A. Scheraga, *J. Amer. Chem. Soc.* **88**, 246 (1966).
35. N. Muller, *J. Phys. Chem.* **76**, 3017 (1972).
36. P. J. Sams, E. Wyn-Jones, and J. E. Rassing, *Chem. Phys. Lett.* **13**, 233 (1972).
37. P. J. Sams, J. E. Rassing, and E. Wyn-Jones, *in* "Chemical and Biological Applications of Relaxation Spectrometry" (E. Wyn-Jones, ed.), p. 163. Reidel, Dordrecht-Holland, 1975.
38. U. Hermann and M. Kahlweit, *Ber. Bunsenges. Phys. Chem.* **77**, 1119 (1973).
39. D. A. W. Adair, V. C. Reinsborough, N. Plavac, and J. P. Valleau, *Can. J. Chem.* **52**, 429 (1974).
40. J. E. Rassing, P. J. Sams, and E. Wyn-Jones, *J. Chem. Soc. Faraday II* **69**, 180 (1973).

Part II | Experimental Techniques and Applications

In Part I we have seen that there are two types of experimental techniques available, namely the transient perturbation methods or sudden rise perturbation methods, and the stationary perturbation methods or periodic perturbation methods. Each individual method has its own characteristics, advantages, and disadvantages; some of the basic features of the commonly used techniques are summarized in Table II.1.

Table II.1. Relaxation Methods

Method	Time range (sec)	Method of detection
A. Transient methods		
1. Temperature jump	1–10^{-6} (10^{-8})	Spectrophotometric Fluorimetric Polarimetric
2. Pressure jump	10–5×10^{-5} (mechanical pressure release) 5×10^{-4}–5×10^{-7} (liquid shock wave)	Conductometric Spectrophotometric
3. Electrical field pulse	10^{-4}–10^{-8}	Conductometric Spectrophotometric
4. Concentration jump	10^{8}–10^{2} (conventional) 10^{3}–10^{-3} (stopped flow)	Spectrophotometric Fluorimetric and many others
B. Stationary methods		
5. Sound absorption and dispersion	10^{-5}–10^{-11} (overall time range for different acoustical techniques)	Power loss or frequency change: Resonance or reverberation (10^{4}–10^{6} Hz); light diffraction (10^{6}–10^{8} Hz); impulse echo (10^{6}–5×10^{8} Hz); Brillouin scattering
6. Dielectric dispersion	10^{-3}–10^{-12}	Power loss, capacitance change

In the following chapters the various methods are discussed in some detail. The principles and some practical aspects in their applications are stressed; for technical details regarding the construction of such equipment the reader will be referred to some key papers and reviews. We also give a brief account of the types of reactions that have been studied or which are potential candidates for future investigations by a given technique. We shall be selective; no attempt to be comprehensive is made. Rather we shall frequently take the point of view of the investigator who is confronted with the task of devising an experiment or of interpreting the data. The reader is also reminded that a number of important applications have been discussed in Part I.

The most widely used techniques will receive the most detailed treatment. Among the transient methods this is the temperature-jump technique; among the stationary methods it is the ultrasound absorption technique. Features and problems common to all transient techniques such as the detection of the relaxation effects by optical methods are discussed in the chapter dealing with the temperature-jump technique; features common to all stationary techniques such as the evaluation of the experimental data are discussed in the chapter on ultrasonic methods.

Chapter 11 | *The Temperature-Jump Method*

The temperature-jump method is not only the most popular of the transient methods but also of all the relaxation methods. Among the most important reasons for its wide use are the following:

1. Most chemical equilibria are associated with a finite standard enthalpy of reaction, ΔH, and thus are temperature dependent according to the van't Hoff equation

$$\left(\frac{\partial \ln K}{\partial T}\right)_p = \frac{\Delta H}{RT^2} \tag{11.1}$$

2. The time range accessible to the temperature-jump method ($1\text{–}10^{-8}$ sec) is wide and matches the range of relaxation times of a large number of inorganic, organic, and biochemical equilibrium reactions.
3. In comparison to many other relaxation methods the temperature-jump apparatus is less difficult to build, at least for the time range of $1\text{–}10^{-6}$ sec; for those unwilling to assemble their own equipment commercial products are available.

Let us estimate the amount by which an equilibrium constant changes in a typical temperature-jump experiment. We use

$$\frac{\Delta K}{K} = \frac{\Delta H}{RT^2}\Delta T \tag{11.2}$$

which follows from Eq. 11.1 provided we deal with small changes (see Chapter 6). At 25°C, $(RT^2)^{-1} = 5.67 \times 10^{-6}$ mole cal^{-1} deg^{-2} and thus we have

$$\frac{\Delta K}{K} = 5.67 \times 10^{-6}\,\Delta H \cdot \Delta T \tag{11.3}$$

which states that the equilibrium constant changes by 0.567% per degree when $|\Delta H| = 1000$ cal/mole $= 1$ kcal/mole. It is apparent that unless ΔH is extremely small, a temperature jump of a few degrees will usually produce an equilibrium shift which is sufficient for inducing measurable concentration changes, provided the equilibrium is not too one-sided (Section 6.1). In fact when ΔH is very large, one may have to restrict oneself to a small ΔT so that it is still permissible to linearize the rate equations. For example, if $|\Delta H| = 5$ kcal/mole and $\Delta T = 10°C$, K would change by about 33% (note that here Eq. 11.2 is no longer accurate) which in certain cases can no longer be treated as a small perturbation (see Chapter 5).

If, on the other hand, ΔH is so small that even a large ΔT does not change K sufficiently, there is often the possibility of coupling the reaction under study to a rapid equilibrium (buffer) which has a large ΔH. A temperature jump may then induce a concentration jump on the reaction under study as discussed in Section 7.1.2.

11.1 The Temperature Pulse

In principle any device that will heat up the sample uniformly in a time shorter than the relaxation time to be measured can be used. The special appeal of the temperature-jump method is of course the fact that there exist various methods which allow a sample to be heated up very quickly. However, "slow" temperature jumps have also been found useful in some applications as discussed in Section 11.1.4.

The most widely used technique, which was also the first to be developed, is based on Joule heating. Other methods which have been developed more recently involve dielectric heating by microwave pulses and optical heating by light flashes and laser pulses.

11.1.1 Joule Heating

A. Discharge of a Capacitor

Joule heating by means of discharging a high-voltage capacitor through the reaction solution was first proposed by Eigen in 1954 (*1*). The first operational temperature-jump apparatus based on Joule heating was described by Czerlinski and Eigen in 1959 (*2*) in which an optical detection system was used. Since then the original design has been modified and refined in several laboratories (*3–14*), and occasionally also combined with other detection methods such as fluorimetry (*6, 14–16*), polarimetry (*17*) and light scattering (*18a,b*).

Figure 11.1 shows a schematic diagram of a standard temperature-jump apparatus utilizing absorption spectroscopy for following chemical relaxation, by far the most widely used detection method (see Section 11.2.1). In a typical setup a low-inductance capacitor with a capacitance in the range of 0.01–0.1 μF is charged to say 20 or 30 kV by a high-voltage generator. By triggering a spark gap, the capacitor discharges to ground through the sample cell containing the conducting reaction mixture whereby the dissipated energy heats up the solution by a few

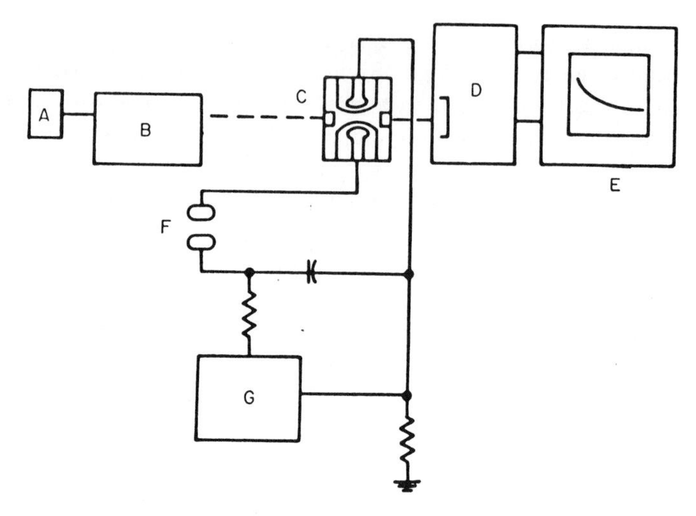

FIGURE 11.1. Schematic diagram of a Joule heating temperature-jump apparatus with spectrophotometric detection. A, light source; B, monochromator; C, observation cell; D, photomultiplier; E, oscilloscope; F, spark gap; G, high-voltage generator. [From French and Hammes (*11*), by permission of Academic Press.]

degrees. Since this energy dissipation among the translational and rotational degrees of freedom of the solvent molecules is extremely rapid (picoseconds), the heating of the solution is quite uniform. Synchronous with the discharge, the oscilloscope sweep is triggered for the observation of the relaxation effect.

There have been numerous different designs for the sample cell (*3–14*). Figure 11.2 shows a cross section through a temperature-jump cell of a commercial apparatus developed by DeMaeyer (*19*) (see Section 11.4.1). The cell shown in the figure is usually made of Lucite or Teflon and contains two platinum- or gold-plated stainless steel electrodes between which the discharge occurs. The electrodes

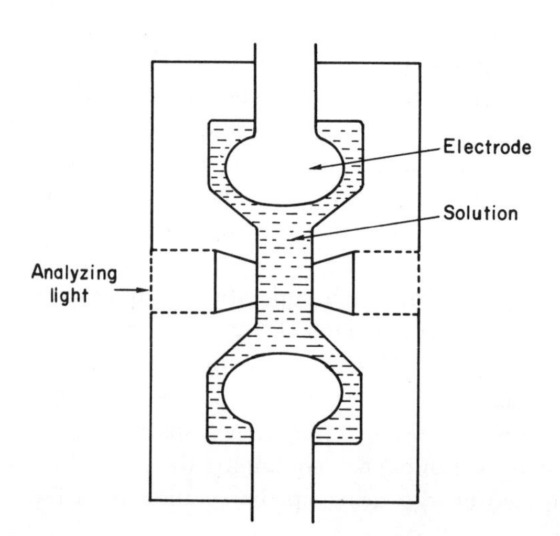

FIGURE 11.2. Cross section through a typical temperature-jump cell. [From Hague (*20*) by permission of John Wiley & Sons.]

are about 1 cm apart and both are immersed in the reaction solution. Perpendicular to the electrode axis there are two quartz windows for optical detection; typically the path length is 1 cm or less. The cell can easily be moved in and out of its metal housing, which is hollowed to provide for thermostating by the circulation of water.

Only the volume between the electrodes and not the entire solution is heated by the temperature jump; this volume is typically about 1 ml, though it can be made smaller if desired. Heat exchange by convection and conduction between the heated volume element and the surrounding liquid starts about 0.5–1 sec after the jump; the larger the jump (ΔT), the earlier the cooling down will occur. This sets an upper limit to the measurable relaxation times at a few hundred milliseconds.

The lower limit for the measurement of relaxation times depends on how rapidly the temperature of the solution can be raised. The discharge of a low-inductance capacitor and thus the heating process follows an exponential function which can be written as

$$\Delta T_t = \Delta T_\infty[1 - \exp(-t/\tau_d)] \tag{11.4}$$

(see also Eq. 8.33), where ΔT_t is the temperature change at time t, ΔT_∞ is the total temperature change produced by the complete discharge of the capacitor, and τ_d is the heating time constant which is equivalent to the discharge time constant of the capacitor. The heating time is given by

$$\tau_d = 0.5RC \tag{11.5}$$

where R is the resistance of the discharge circuit and C is the capacitance of the capacitor. R includes the resistance of the sample cell R_c, and all circuit losses such as skin effects, eddy currents, and radiation losses; R_c is usually by far the largest contributor.

Ideally the heating time is much shorter than the relaxation time to be measured so that the heating process is essentially complete before chemical relaxation has made any significant progress. In this ideal situation the relaxation curve is a pure exponential, at least for single-step equilibria.

When the heating time τ_d is of the same order of magnitude as the relaxation time τ, the relaxation curve becomes distorted and the change of x with time is given by Eqs. 8.34 or 8.37. In such a situation it is very advantageous to know the value of τ_d because this reduces the number of unknowns in Eqs. 8.34 or 8.37 and allows a more accurate evaluation of τ, just as in the case of two poorly separated relaxation times (Section 9.2). By deliberately choosing a very rapid chemical system such as an acid–base indicator in a buffered solution where $\tau \ll \tau_d$, Eq. 8.37 simplifies to

$$x = x_0 \exp(-t/\tau_d) \tag{11.6}$$

from which τ_d is in fact easily evaluated.

Let us now deal with the factors that determine τ_d. According to Eq. 11.5 the heating time can be kept short by decreasing the resistance of the solution and/or by choosing a very low capacitance of the capacitor. The resistance depends to some extent on the design of the sample cell (distance between electrodes, surface area of the electrodes) but is mainly a function of the chemical composition of the solution. A low resistance is usually accomplished by adding an inert electrolyte such as NaCl or KNO_3, typically in concentrations of 0.1–0.5 *M*.

On first sight one might expect that τ_d can be kept short just by choosing a low enough capacitance, thus disposing of the necessity of adding an electrolyte to the solution and with it the necessity of using an ionizing solvent. However, C cannot be chosen below a certain limit because the energy stored on the capacitor, given by

$$Q = \tfrac{1}{2}CV_0^2 \tag{11.7}$$

where V_0 is the voltage, would become too small and with it the size of the temperature jump, ΔT_∞. Assuming that all the energy Q is dissipated in the volume element between the electrodes of the sample cell, the total temperature rise is given by

$$\Delta T_\infty = Q/C_p\rho V \tag{11.8}$$

where C_p is the specific heat at constant pressure, ρ is the density, and V is the volume element of the solution. As an example, $V_0 = 40$ kV is about the highest practical voltage to be achieved on DeMaeyer's apparatus (*19*) before spark breakdown occurs. Thus on a capacitor with $C = 0.01\ \mu$F the energy stored at 40 kV would be $Q = 8$ J which is $8/4.184 \approx 1.9$ cal. With a volume of 1 ml one thus obtains $\Delta T_\infty \approx 1.9°$ for an aqueous solution.

Even with this rather low value of C (0.01 μF) we must keep R below 200 Ω if τ_d is to be kept below 1 μsec (Eq. 11.5). Often it is desirable to produce larger temperature jumps and in fact C is often larger, typically 0.05 μF, which necessitates a corresponding reduction in R by adding more inert salt if no increase in τ_d can be tolerated.

The necessity of using ionizing solvents is the most serious limitation of the Joule heating temperature-jump technique. The almost exclusive application of the method to aqueous solutions described in the early literature could, however, lead to the wrong belief that water is in fact the only practical solvent. This is not true. In recent years the Joule heating method has been used in a variety of other polar solvents such as methanol (*21a–22*), mixtures of water with alcohols (*23*), dioxane (*24*), and dimethylsulfoxide (*25*). The possibility to work in these solvents greatly extends the scope of the Joule heating temperature-jump method. Even if the amount of electrolyte which can be added to such solvents may be limited because of solubility problems, one may still be able to apply the method. This is because many such systems have relaxation times of, say, $\geq 50\ \mu$sec so that somewhat longer heating times (10 μsec) can easily be tolerated. Hence relatively low electrolyte concentrations will be satisfactory in many cases. Furthermore, most organic solvents have a lower specific heat capacity and a lower density than water which leads to a larger ΔT_∞ for a given Q (Eq. 11.8). This offers the possibility of decreasing τ_d by choosing a smaller C.

It is clear from this discussion that in general one must try to achieve as short a heating time as possible. However, it should be noted that extremely rapid discharge of the capacitor should be avoided when it is not necessary. The reason is that the shorter the heating time, the higher the intensity of the shock wave produced by the temperature jump, which frequently leads to cavitation and other disturbances such as oscillations. Shock waves are generated when the liquid is heated more rapidly than it can expand, thus producing local regions of high pressure. The high-pressure front of the shock wave is followed by a low-pressure front which can lead to cavities in the solution. The result is an erratic deflection on the oscilloscope screen which may last from 10 to 50 μsec.

In aqueous solutions such effects can be minimized by working near 4°C where the thermal expansion coefficient is zero, for example by jumping from 2°C to 6°C. When working at higher temperatures buffering of the electrodes with silicone rubber may help in reducing oscillations (*26*). Cavitation problems can also often be minimized by degassing the solution since dissolved gases act as nuclei for cavitation. If the problem persists, one may be forced to settle for a smaller temperature jump, which reduces the intensity of the shock wave.

B. Discharge of a Coaxial Cable

Eigen and DeMaeyer (*3*) have suggested using a coaxial cable instead of a capacitor for the storage of the electrical energy. The principal advantage of a coaxial cable is that of a shorter discharge time. The construction of such a temperature-jump apparatus has recently been described (*27*). Heating times as low as 50 nsec were achieved with maximum jumps of 10°. The apparatus features a very small sample cell (minimum volume 40 μl) which should be of particular interest to biochemists.

11.1.2 Dielectric Heating (Microwave Temperature-Jump Apparatus)

As pointed out above, the Joule heating method is limited to conducting solutions and therefore to ionizing solvents. A different way of producing a temperature jump is to submit the sample solution to a short pulse of microwave radiation. In contrast to Joule heating where the temperature rise is produced by frictional loss of the moving ions, the high-frequency field of a microwave pulse causes dielectric energy losses by dipolar reorientation of the solvent molecules.

Though the first experimental realization of this method by Ertl and Gerischer (*28*) has been applied to aqueous solutions, this technique is not restricted to solvents having a high dielectric constant. Indeed its special appeal lies in the fact that it can be applied to any solvent absorbing microwaves, which is the case for most solvents having a minimal polarity (e.g., chlorobenzene); for aqueous solutions the method is potentially useful when a high electrolyte concentration is undesirable.

The microwave pulse is generated by applying a high-voltage pulse to a magnetron. Temperature rises of about 0.5° within 1 μsec have been achieved in recent designs (*29*, *30*). Technical details referring to the construction of a microwave temperature-jump apparatus have also been presented by French and Hammes (*11*), whereas Czerlinski (*7*) discusses the theoretical aspects important for the design of such equipment.

Apart from the rather complex circuitry involved and the high cost of the microwave pulse generator, the main disadvantage of this technique is that only small temperature jumps can be produced. It appears, however, that substantial improvements on this score can be expected through better design according to DeMaeyer, cited by Yapel and Lumry (*31*).

Both conductometric (*28*, *30*) and spectrophotometric (*29*, *30*) detection systems have been used. The former is more sensitive but less selective than the latter and can only be applied to ion-producing or ion-consuming reactions (see Section 12.1.2). With the spectrophotometric detection system special attention has to be given to an optimal optical design in order to achieve a good signal-to-noise ratio (*29*) (see Section 11.2.1).

On the other hand, the size of the temperature jump is more reproducible with the microwave pulse method than with the Joule heating technique so that the sensitivity in the detection can be increased by computer averaging of a series of replicated experiments. Rüppel and co-workers (*32a,b*) describe a repetitive measuring technique based on the conductometric detection method which allows the signal-to-noise ratio to increase by a factor of 25 with 1200 microwave pulses in a typical experiment. Note that in order to avoid continuous heating of the solution by the repetitive temperature jumps a flow system is necessary where the reaction solution can flow through the conductivity cell in a continuous fashion (*32a*).

11.1.3 Optical Heating

A. Laser Temperature-Jump Method

The possibility of using laser pulses for heating has intrigued a number of research laboratories and several attempts have been made in recent years to develop a fast, sensitive, and convenient laser temperature-jump apparatus (*7*, *33a–40a*).

This method indeed offers several advantages over Joule heating and probably has the greatest potential in future developments, particularly for ultrafast temperature jumps. At the present time, however, there are still various fundamental problems to be solved; for a somewhat more detailed discussion of these problems see Caldin *et al.* (*36*).

In principle the method can be used with any solvent, including nonpolar solvents, and thus can have an even wider range of applicability than dielectric heating. The energy of the laser is either absorbed directly by the solvent or by an appropriate dye which is added to the solution. Commercial lasers which in the Q-switched mode deliver 1.5 J ($\approx$0.36 cal) in 20 nsec are readily available. This is enough to produce a 1° temperature rise in a 0.36 ml aqueous sample or a 5° jump in a 72 μl sample (see Eq. 11.8). Sample cells of such sizes are simple to make.

In practice there are several difficulties to overcome. At the present time the only commercially available lasers with enough power to produce useful temperature jumps are ruby and neodymium which emit at 694 and 1060 nm, respectively. In the case of the ruby laser a dye, for example vanadyl phthalocyanine (*36*), has to be added to the solution since the absorbance of most solvents is negligible at 694 nm. In the *non-Q*-switched mode the light energy can readily and effectively be converted into thermal energy; each dye molecule is repeatedly raised to an excited state and subsequently deactivated by collisions with solvent molecules. In fact jumps of 5° or more can easily be achieved in this manner (*34*, *36*). This is because in the non-*Q*-switched mode the pulse duration is 100 μsec or more (*34*, *36*), long enough compared to the lifetime of the excited states of the dye molecules to allow multiple excitation–deactivation. Unfortunately, such long pulse duration is usually undesirable in a temperature-jump apparatus.

The full potential of the laser is only made use of in the *Q*-switched mode, where the energy is released in a very short time (~20 nsec). However, here the conversion of light into thermal energy is inefficient because of the finite lifetime of the excited dye molecules. Thus typical performances are of the order of 0.1–0.3° temperature jumps with a pulse duration of 1 μsec (*34*, *36*).

The promises of very rapid and efficient heating by lasers can only be fulfilled if the radiation can be directly absorbed by the solvent. Unfortunately, even at 1060 nm, the emitting wavelength of the neodymium laser, light absorption is very small for most solvents. For example, the absorptivity of water at 1060 nm is only 0.067 cm^{-1} so that only a very small fraction of the neodymium radiation is absorbed. As a consequence the largest temperature jumps that can thus far be produced in water by this method are a few tenths of a degree (*34*).

Larger jumps can be produced when the sample cell is designed in such a way as to force the laser beam to traverse the solution a number of times; such a technique has been used with good results in conjunction with a conductivity readout (*34*).

Ideally larger temperature jumps could be achieved if lasers with more suitable emission wavelengths were available, namely between 1100 and 2000 nm. Turner *et al.* (*35a*) make use of the stimulated Raman effect to shift the radiation emitted by the neodymium rod to longer wavelengths. By using liquid nitrogen the radiation of the *Q*-switched laser was shifted from 1060 to 1410 nm and temperature jumps of up to 10° with a heating time of about 25 nsec could be achieved with this method. In fact a study of the triiodide equilibrium, where the shortest measured relaxation time was 30 nsec, has been successfully carried out with this method (*35a,b*).

However, the Raman temperature-jump technique is not without its problems. For example, since heating is by light absorption the sample gets heated most at the front end of the cell and decreasingly so as one moves through the solution. For the relatively large temperature jumps achieved with this method this can have a serious effect since the rate constants in the front may significantly differ from those in the middle and rear portions. Although the problem can be reduced by using cells of very short path length (say 0.1–1 mm) (*35a,b*) this is impractical in many cases if the detection is by absorption spectrometry because of optical densities that are too small. Eyring *et al.* (*39*), who have discussed the above and other problems which plague the laser technique, proposed optical waveguide techniques to detect the absorption changes at the first liquid surface of the sample cell with evanescent photons.

Nevertheless, the future of the laser temperature-jump technique looks fairly promising. It can be expected that the high costs of lasers will be reduced. It is also hoped that high-power rare earth lasers operating at wavelengths between1100 and 2000 nm become commercially available. "Tunable" dye lasers operating between 350 and 1300 nm are already available (*40b*) and with a further increase in their power output they should become very promising indeed.

One of the advantages of laser heating is that the sample cell design can be adapted to a multitude of specific problems. Thus Rigler *et al.* (*41*) have built a micro laser temperature-jump apparatus which permits the study of rapid kinetics not only in solution but in such integrated biological structures as single cells or cell organelles; the authors expect to be able to study the kinetics of interaction of neurotransmitters, antibiotics, antibodies, and viruses with cell surfaces and membrane fragments or the kinetics of cell–cell interactions, and so on.

Laser heating has also been the method of choice in a temperature-jump apparatus designed to study kinetics under pressures of up to 3 kbar (*42*).

B. Heating by Light Flash

Instead of using a laser, optical heating can be achieved by a high-energy light flash. Strehlow and Kalarickal (*43*) have developed a technique in which the temperature of the solution can be raised by 1° within approximately 5 μsec. For weakly absorbing solutions a dye has to be added, preferably one with a large molar extinction coefficient over a broad range of the spectrum. The detection of the concentration changes is by conductometry. Yapel and Lumry (*12*) have discussed some of the advantages and drawbacks of this method, which has not been widely used.

An interesting application of the general idea of flash heating is a simple apparatus for subjecting biological systems to temperature jumps by exposing the outside of a silver-walled sample cell to a 200 J light flash (*44*). However, the heating time is about 400 msec, which is very long by any standards.

11.1.4 The "Slow" Temperature-Jump Technique

Most efforts have been spent in extending the useful time range of the temperature-jump technique by developing faster heating methods. Some applications, however, call for an extension on the slow side of the time spectrum; in a typical

Joule heating apparatus the longest relaxation times that can be measured are about 1 sec in favorable, 0.1 sec in unfavorable cases.

Pohl (*45*) has built a slow temperature-jump apparatus which essentially consists of a standard spectrophotometer equipped with a very small jacketed cuvette allowing a rapid heat exchange between the solution and the outside. The temperature change is brought about by switching the circulating fluids of two thermostats alternating through the jacket of the cuvette. Incidentally this method permits both positive and negative temperature jumps; the process of heating (or cooling) takes about 1 sec. In principle there is no limit to the magnitude of the temperature change and thus the technique can also be used for large perturbations.

The device is simple to build and should be easily adaptable to other detection methods such as fluorimetry, polarimetry, light scattering, or conductometry (*45*).

A similar setup has been described by Hui Bon Hoa and Travers (*46*) which permits work at temperatures as low as −60°C. Spatz and Crothers (*47*) have combined Joule heating with the thermostat switching method and achieved results similar to those of Pohl.

Examples of applications of the slow temperature-jump technique are the kinetics of the reversible denaturation of α-chymotrypsin (*45, 46*), the unwinding of DNA (*47*), and the protein unfolding in ribonuclease A (*48*).

11.2 Detection of Concentration Changes

Concentration changes can be monitored very rapidly and with good sensitivity using optical methods such as spectrophotometry and fluorimetry; polarimetric (*12, 17*) and light scattering (*18a,b*) methods have also been described though their applications have been rather limited. Conductometric detection is being used in conjunction with some microwave heating techniques (*30*); it is not a good method to use in conjunction with Joule heating because of the high background electrolyte concentration. Since conductometric detection has found its widest application in the pressure-jump method it is discussed in Chapter 12. In this section the spectrophotometric and fluorimetric methods are considered.

11.2.1 Spectrophotometric Detection

This method is by far the most commonly used in temperature-jump work and warrants the most detailed discussion. Light from a high-intensity light source passes through a monochromator, set to the desired wavelength, and is partially absorbed by the solution in the sample cell. The concentration change induced by the temperature jump produces a change in the OD of the solution which has an effect on the intensity of the transmitted light. The change in the transmitted light intensity is transformed into an electrical difference signal by a photomultiplier and displayed on an oscilloscope, the time base of which is triggered simultaneously with the temperature jump.

According to the Beer–Lambert law the light intensity, I, transmitted through the solution is given by

$$I = I_0 e^{-E} \tag{11.9}$$

with

$$E = \sum_j \epsilon_j' c_j l \tag{11.10}$$

where I_0 is the light intensity before passage through the cell with an optical path length l, c_j is the molar concentration of species j, and ϵ_j' is the molar extinction coefficient of species j. Note that the commonly used extinction coefficient ϵ_j is defined as $\epsilon_j = \epsilon_j'/2.303$, whereas "absorbance" or "OD" is defined as OD = $E/2.303$, with 2.303 being the conversion factor of natural to decadic logarithms.

For the state after a perturbation we can write

$$\sum_j c_j = \sum_j \bar{c}_j + \sum_j \Delta c_j \tag{11.11}$$

$$E = \bar{E} + \Delta E \tag{11.12}$$

with $\bar{E} = \sum_j \epsilon_j' \bar{c}_j l$ and $\Delta E = \sum_j \epsilon_j' \Delta c_j l$. Thus Eq. 11.9 becomes

$$I = \bar{I} + \Delta I = I_0 e^{-(\bar{E} + \Delta E)} = \bar{I} e^{-\Delta E} \tag{11.13}$$

After rearranging and taking logarithms we obtain

$$\ln(1 + \Delta I/\bar{I}) = -\Delta E \tag{11.14}$$

Expansion in a Taylor series affords

$$\Delta I/\bar{I} = -\Delta E + \Delta E^2/2 + \cdots \tag{11.15}$$

which simplifies to

$$\Delta I/\bar{I} = -\Delta E = -2.303\,\Delta\text{OD} \tag{11.16}$$

when $|\Delta E| \ll 1$. Equation 11.16 shows that for small perturbations the relative change in transmitted light intensity and with it the electrical signal output of the photomultiplier is directly proportional to the change in E and thus proportional to the change in concentration(s). This is no longer true when ΔE is large (Eq. 11.15) but since temperature jumps produce generally quite small perturbations, the condition $\Delta E \ll 1$ is usually fulfilled. When this condition is not met, the photomultiplier output must be converted into absorbance readings, either by calculation or electronically.

A source of erroneous interpretation of a temperature-jump experiment by the inexperienced student is the various factors that contribute to ΔE (ΔOD). Though in most cases the largest contribution to ΔE (ΔOD) is chemical, the temperature jump can also produce concentration changes as a consequence of volume expansion of the solvent, which can be considerable. In rare cases a temperature dependence of ϵ_j' may also contribute to ΔE (ΔOD). These latter (physical) effects are, however, extremely rapid and follow the perturbation virtually instantaneously. They allow the determination of the heating time or the photomultiplier rise time (see Section A below) in very much the same way as do very rapid chemical equilibria such as acid–base indicator systems. It is important to recognize these effects for what they are; many a novice has thought he had observed a nice relaxation time when it turned out on closer inspection that what he observed was merely the heating time of the apparatus. In case of doubt the effect of reducing the resistance of the solution by adding more electrolyte (if the Joule heating method is used) should be investigated. Figures 11.3a,c,d illustrate the situation.

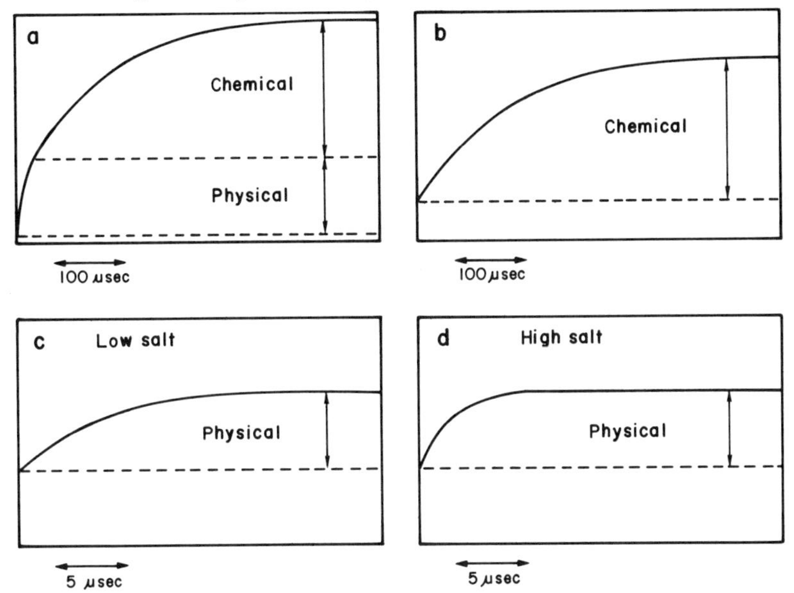

FIGURE 11.3. Physical and chemical relaxation effects. (a) Complete oscillogram; (b) photomultiplier output shorted out during initial 20 μsec; (c) physical effect displayed with faster sweep rate; (d) physical effect after increasing electrolyte concentration.

When performing the measurement of the actual chemical relaxation effect it is common practice to short out the photomultiplier signal for the first few microseconds so that only the chemical effect appears on the oscilloscope screen (Fig. 11.3b). This is also done in the case of systems with two or more (chemical) relaxation times which are widely separated; the fast process is shorted out in the study of the slow one, a particularly useful feature when the amplitude of the fast process is very large.

It was pointed out earlier (see Section 6.2) that in amplitude work it is important to separate the chemical contribution to ΔE (or ΔOD) from the contribution by physical effects.

A. Signal-to-Noise (S/N) Ratios

The precision with which a relaxation effect can be measured depends not so much on the relative change in transmitted light intensity, $\Delta I/\bar{I}$, but on the signal-to-noise ratio (S/N) of the detection system. The main contribution to the noise comes from shot noise (*49*) produced by the continuous background illumination at the photomultiplier cathode.

The S/N is given by

$$\frac{S}{N} = 1.4 \times 10^9 \frac{|\Delta I|}{I} \left(\frac{I_0 A_g S_\lambda}{\Delta f}\right)^{1/2} e^{-E/2} \tag{11.17}$$

where S_λ is the sensitivity of the photocathode at wavelength λ, Δf is the frequency bandwidth of the detection circuitry (reciprocal rise time of the photomultiplier), and A_g is a measure of the efficiency of the optical arrangement (geometrical factor); $I_0 A_g$ is thus the effective light intensity arriving at the photomultiplier cathode in the absence of a sample cell.

From Eq. 11.17 it is apparent that S/N is directly proportional to the relative change in light intensity, $\Delta I/I$, as one would expect intuitively. S/N also increases with the square root of the light intensity I_0 and the photosensitivity of the photomultiplier; it is inversely proportional to the square root of the frequency bandwidth Δf. For relatively long relaxation times, say in the millisecond range, Δf can be chosen small enough to make S/N satisfactory without special attention to the light source, geometrical design, or photomultiplier sensitivity.

On the other hand, when very fast processes are to be measured one needs to work with a large frequency bandwidth which greatly reduces S/N unless the factor $I_0 A_g S_\lambda$ is increased at the same time. Apart from using a highly sensitive photomultiplier, an efficient monochromator with large aperture, and making sure that the light is well focused, the use of high-intensity lamps such as high-pressure mercury arc lamps is often desirable. It is usually also permissible to have the monochromator slits wide open for enhanced light intensity. This is because the distortion of the Beer–Lambert law which occurs for not perfectly monochromatic light (*50*) has no significant effect on the *relative* OD for small OD changes.

A considerable enhancement of S/N can also be achieved by using a photocell rather than a photomultiplier and connecting it directly to the amplifier; this proved of great utility in conjunction with a microwave temperature-jump apparatus (*29*) where only very small jumps (small $\Delta I/I$) can be made.

S/N also depends on E or the OD of the solution. We can write Eq. 11.17 in the form

$$S/N = \text{const}\, E \exp(-E/2) \tag{11.18}$$

since $\Delta I/I = -\Delta E$ must be proportional to E just as Δc_j is proportional to c_j. Figure 11.4 shows how the function $E \exp(-E/2)$ varies with E. It is apparent that the most satisfactory S/N is reached at relatively high absorbance. The optimal S/N can be calculated from

$$\frac{d}{dE}(Ee^{-E/2}) = 0 \tag{11.19}$$

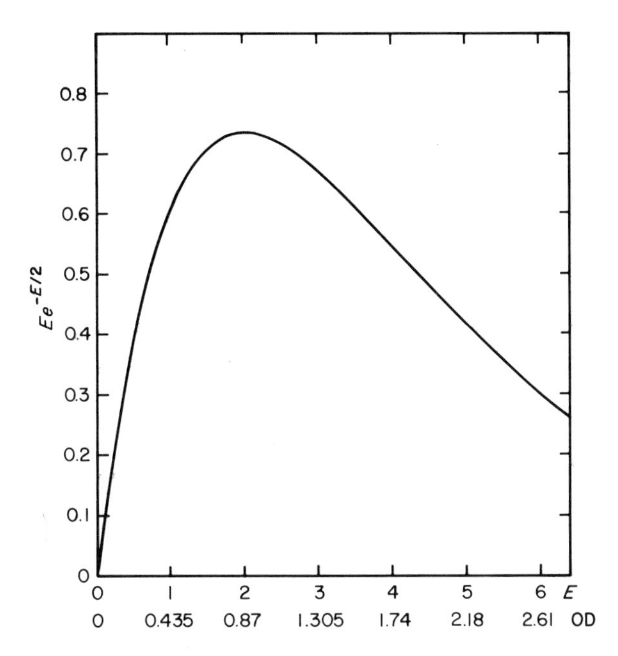

FIGURE 11.4. Function $E \exp(-E/2)$ versus E or OD.

It is reached at $E = 2$ which corresponds to OD $= 0.868$. From Fig. 11.4 it can be seen that for $0.19 \geq \text{OD} \geq 2.34$, which covers an appreciable range of OD values, S/N is at least at half its optimal value.

It should be pointed out, however, that these considerations only apply when the optimal OD can be achieved without having to choose reactant concentrations that may lead to a decrease of Γ/c_0 (Chapter 6). For example, in the A $\rightleftharpoons$ B system, Γ/c_0 is independent of c_0 (Eq. 6.12 or 6.15); the same is true for the A + B $\rightleftharpoons$ C system under pseudo-first-order conditions but not when the concentrations of A and B are comparable (see, e.g., Eq. 6.30; α is dependent on c_0 as seen in Eq. 6.29).

B. Measurement of Relaxation Amplitudes

When measuring amplitudes a very high precision is required; thus it is important to have a large S/N. In constrast to the determination of relaxation *times*, where slight deviations of the Beer–Lambert law are of no consequence, and thus large monochromator slits are permissible, exact values of ΔOD can only be obtained with small slit widths. In fact it is therefore recommended practice to determine the amplitudes at different slit widths and extrapolate to zero slit width. However, small slit widths decrease the S/N. On the other hand, only initial and infinity readings rather than the entire relaxation curve are needed in amplitude determinations so that one can operate the oscilloscope at a slow sweep rate which allows the use of a small Δf with a corresponding enhancement of S/N (see Eq. 11.17). The situation is illustrated in Fig. 11.5.

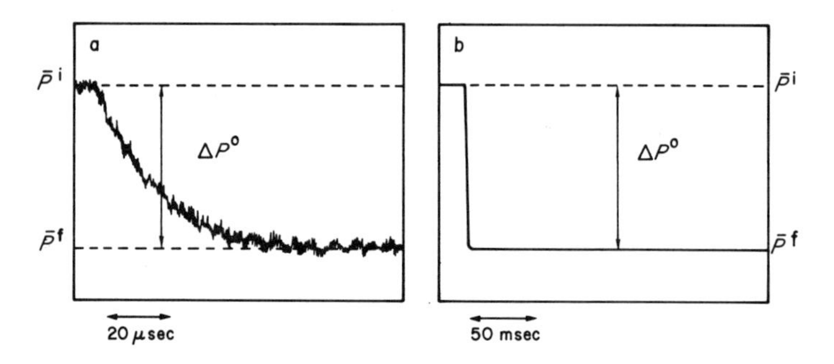

FIGURE 11.5. Determination of relaxation amplitude. (a) Picture taken at high sweep rate with large frequency bandwidth Δf, noisy; (b) picture taken at low sweep rate and small Δf, smooth.

The amplitude ΔOD° can then, for example, be determined as follows. Since the photomultiplier output P (in millivolts) is proportional to the light transmitted through the solution, we have

$$\bar{P}^{i} \propto \bar{I}^{i}, \qquad \bar{P}^{f} \propto \bar{I}^{f}, \qquad \Delta P^{0} = \bar{P}^{i} - \bar{P}^{f} \propto \bar{I}^{i} - \bar{I}^{f} = \Delta I^{0}$$

The proportionality constant is of course the same in the three relations above and thus in combination with Eq. 11.16 one obtains for small perturbations

$$\Delta P^{0}/\bar{P}^{f} = \Delta I^{0}/\bar{I}^{f} = -\Delta E^{0} = -2.303\,\Delta\text{OD}^{0}$$

11.2.2 Fluorimetric Detection

Though of lesser scope than spectrophotometric detection, fluorimetry can offer some advantages over the former method. Owing to its high sensitivity it is particularly suited when very small concentrations have to be employed. For example, in an A + B $\rightleftharpoons$ C system $[1/\tau = k_1(\bar{c}_A + \bar{c}_B) + k_{-1}]$ where the forward reaction is very fast, the term $k_1(\bar{c}_A + \bar{c}_B)$ may be so large as to make τ too short for the temperature-jump method when concentration levels of A and B suitable for spectrophotometric detection are used. With fluorimetric detection the concentrations of A and B may be chosen low enough to reduce the term $k_1(\bar{c}_A + \bar{c}_B)$ sufficiently to make τ measurable.

The counterpart to the Beer–Lambert law for fluorescing light is given by (*3*)

$$I_f = gI_0\{1 - \exp[-(E_{\lambda e} + E_{\lambda f})]\} \tag{11.20}$$

where I_f is the intensity of the fluoresced light at the front end, i.e., the end directed toward the light source of the fluorescing body (*3*); g is a proportionality factor; and $E_{\lambda e}$ and $E_{\lambda f}$ are given by Eq. 11.10 and refer to the wavelengths of the exciting and fluoresced light, respectively.

In analogy to Eq. 11.13 we write

$$I_f = \bar{I}_f + \Delta I_f = gI_0\{1 - \exp[-(\bar{E}_{\lambda e} + \Delta E_{\lambda e} + \bar{E}_{\lambda f} + \Delta E_{\lambda f})]\} \tag{11.21}$$

for the state after equilibrium perturbation. After rearranging and combining Eq. 11.21 with Eq. 11.20 we obtain

$$\frac{\Delta I_f}{\bar{I}_f} = \frac{\exp[-(\bar{E}_{\lambda e} + \bar{E}_{\lambda f})]}{1 - \exp[-(\bar{E}_{\lambda e} + \bar{E}_{\lambda f})]} \{1 - \exp[-(\Delta E_{\lambda e} + \Delta E_{\lambda f})]\} \quad (11.22)$$

For $\Delta E_{\lambda e} + \Delta E_{\lambda f} \ll 1$ this simplifies to

$$\frac{\Delta I_f}{\bar{I}_f} = \frac{\exp[-(\bar{E}_{\lambda e} + \bar{E}_{\lambda f})]}{1 - \exp[-(\bar{E}_{\lambda e} + \bar{E}_{\lambda f})]} (\Delta E_{\lambda e} + \Delta E_{\lambda f}) \quad (11.23)$$

(Taylor series $1 - e^{-x} \approx x$ for $x \ll 1$.).

With respect to Eq. 11.23 two limiting situations are of interest.

1. $\bar{E}_{\lambda e} + \bar{E}_{\lambda f}$ is large, say ≥ 3. Here Eq. 11.23 simplifies to

$$\frac{\Delta I_f}{\bar{I}_f} = \exp[-(\bar{E}_{\lambda e} + \bar{E}_{\lambda f})](\Delta E_{\lambda e} + \Delta E_{\lambda f}) \quad (11.24)$$

The percentage of change in the fluoresced light intensity is proportional to the changes in $E_{\lambda e}$ and $E_{\lambda f}$ and is thus proportional to Δc_j (assuming that only one species is absorbing) just as for spectrophotometric detection. However, there is a strong attenuation by the exponential factor and thus the method is much less sensitive than spectrophotometry at high OD values.

2. When $\bar{E}_{\lambda e} + \bar{E}_{\lambda f} \ll 1$, Eq. 11.23 reduces to

$$\frac{\Delta I_f}{\bar{I}_f} = \frac{\Delta E_{\lambda e} + \Delta E_{\lambda f}}{\bar{E}_{\lambda e} + \bar{E}_{\lambda f}} = \frac{\Delta c_j}{c_j} \quad (11.25)$$

(again assuming only one absorbing species). Here the percentage of change in the light intensity is proportional to the percentage of change in the concentration rather than proportional to the absolute change. Thus for small absolute changes Δc_j at low OD values the fluorimetric method is *more* sensitive than the spectrophotometric method.

Let us illustrate the comparison of the two methods with a numerical example. Assume there is only one absorbing species with $l\epsilon_j' = 10^4$ (spectrophotometry) or $l(\epsilon_{j\lambda e}' + \epsilon_{j\lambda f}') = 10^4$ (fluorimetry), respectively, and the relative change $(\Delta c_j/\bar{c}_j)$ brought about by the temperature jump is 0.1. If $\bar{c}_j = 10^{-5}\ M$, we have $\bar{E} = 0.1$ and $\Delta E = 0.01$. Thus according to Eq. 11.16 we obtain $|\Delta I|/\bar{I} = 0.01$ with spectrophotometric detection. In the case of fluorimetric detection we can use Eq. 11.25 since $\bar{E}_{\lambda e} + \bar{E}_{\lambda f} = 0.1$; hence $\Delta I_f/\bar{I}_f = 0.1$, which shows that fluorimetric sensitivity is ten fold higher than spectrophotometric sensitivity. If, however, $\bar{c}_j = 10^{-4}\ M$, we have $\bar{E} = 1.0$, $\Delta E = 0.1$, and $\Delta I/\bar{I} = 0.1$ with spectrophotometric detection. For the fluorimetric method we can no longer use Eq. 11.25 since $\bar{E}_{\lambda e} + \bar{E}_{\lambda f} = 1.0$ but must use Eq. 11.23 instead. This provides $\Delta I_f/\bar{I}_f = 0.058$; i.e., here fluorimetry is just slightly more than half as sensitive as the spectrophotometric method. Some further aspects of the theory and application of fluorescence relaxation spectrometry have been reviewed recently (*15*).

The first fluorescence temperature-jump apparatus was described by Czerlinski (*6, 16*). A more recent design of high versatility has been developed by Rigler *et al.* (*14*); this apparatus can be used to measure light absorption, nonpolarized or polarized fluorescence, as well as light scattering (for more details see Section 11.4 also). Fluorescence detection is now also part of the standard or accessory equipment of some commercial products (Section 11.4).

11.3 The Combination Stopped-Flow–Temperature-Jump Apparatus

Perhaps the most serious limitation common to all relaxation methods is their exclusive applicability to equilibrium systems; and even if one deals with an equilibrium, a further requirement is that the equilibrium not be too greatly one-sided (see Chapter 6). However, the scope of some relaxation methods can be extended by applying them to stationary or steady states. The consequence of perturbing a stationary state is a chemical relaxation toward a new stationary state just in the same way as a perturbation of an equilibrium state produces a relaxation toward a new equilibrium state. This is best illustrated by an example.

Assume we deal with the reaction

$$\mathrm{A} + \mathrm{B} \underset{k_{-1}}{\overset{k_1}{\rightleftharpoons}} \mathrm{C} \xrightarrow{k_2} \mathrm{D}(+\mathrm{E}) \tag{11.26}$$

which is an extremely common scheme; typically A and B may be two reactants forming the end product(s) D (and E) via a reactive intermediate C, or A may be a substrate with B acting as catalyst (H^+, enzyme, etc.) to form the activated substrate C (e.g., enzyme–substrate complex), or the reaction $\mathrm{C} \rightarrow \mathrm{D}(+\mathrm{E})$ may just be an undesired side reaction, and so on. If the system is set up by mixing A and B and if $k_{-1} \gg k_2$, a stationary state will rapidly develop where C is in equilibrium with A and B, as expressed by

$$c_{\mathrm{C}}^{\mathrm{ss}}/c_{\mathrm{A}}^{\mathrm{ss}}c_{\mathrm{B}}^{\mathrm{ss}} = K_1 \tag{11.27}$$

where ss refers to stationary state. The stationary state concentrations of A, B, and C will be approximately constant for times short on the time scale of the irreversible process (k_2); for longer times they will gradually decrease and eventually approach zero, though still obeying Eq. 11.27 throughout the entire process.

If a relaxation experiment can be performed before most of the starting material has been converted to final products, the relaxation time associated with the fast equilibrium can in principle be measured. The expression for $1/\tau_1$ is given by

$$1/\tau_1 = k_1(c_{\mathrm{A}}^{\mathrm{ss}} + c_{\mathrm{B}}^{\mathrm{ss}}) + k_{-1} \tag{11.28}$$

Since $c_{\mathrm{A}}^{\mathrm{ss}}$ and $c_{\mathrm{B}}^{\mathrm{ss}}$ are time dependent there may be some practical difficulties in determining the concentration dependence of τ_1^{-1}. Whenever possible it will be very advantageous to use pseudo-first-order conditions, for example $[\mathrm{B}]_0 \gg [\mathrm{A}]_0$, so that $c_{\mathrm{A}}^{\mathrm{ss}} + c_{\mathrm{B}}^{\mathrm{ss}} \approx [\mathrm{B}]_0$.

How soon after mixing A and B the relaxation experiment has to be performed depends on the lifetime of the stationary state. With reference to the scheme of

Eq. 11.26 and assuming $[B]_0 \gg [A]_0$ the stationary state decays with the rate at which D (and E) forms, i.e.,

$$-\frac{d(c_A^{ss} + c_C^{ss})}{dt} = \frac{dc_D}{dt} = \frac{k_2 K_1 [B]_0}{1 + K_1[B]_0} (c_A^{ss} + c_C^{ss}) \tag{11.29}$$

and thus the half-life of the stationary state is

$$t_{1/2}^{ss} = \ln 2 \frac{1 + K_1[B]_0}{k_2 K_1 [B]_0} \approx 0.69 \frac{1 + K_1[B]_0}{k_2 K_1[B]_0} \tag{11.30}$$

By reducing $[B]_0$ the half-life can be increased if necessary. Whether this is practical depends on the circumstances since a reduction in $[B]_0$ also reduces the stationary state concentration of C and may lead to difficulties in detecting τ_1 in the relaxation experiment.

When a stationary state persists for a long time, say $t_{1/2}^{ss} \geq 15$ min, rapid handling of freshly prepared solution in conjunction with ordinary relaxation equipment will usually be satisfactory. However, short-lived stationary states must be generated in rapid mixing devices such as the stopped-flow apparatus or by continuous-flow techniques (see Chapter 14). Although the idea of combining flow methods with the temperature-jump technique has been around for some time (*7*, *51*, *52*), its technical realization has been quite difficult. Erman and Hammes (*53*) described the first workable combination stopped-flow–temperature-jump apparatus, which has now been refined (*54*). Another setup based on DeMaeyer's (*19*) temperature-jump apparatus has been developed by Veil (*55*); there is also commercial equipment available (see Section 11.4).

Figure 11.6 shows a schematic diagram of the Erman–Hammes stopped-flow–

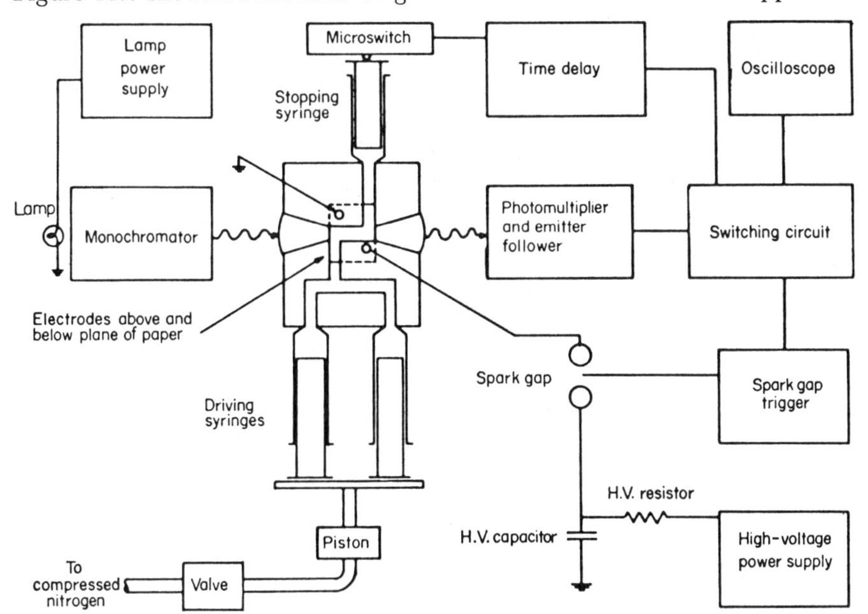

FIGURE 11.6. Schematic diagram of the Erman–Hammes stopped-flow–temperature-jump apparatus. [From Erman and Hammes (*53*), by permission of the American Institute of Physics.]

temperature-jump apparatus. By means of two driving syringes the two reactant solutions are pushed into a mixing chamber and from there into the sample cell. The rapid flow is stopped by a stopping syringe just as in any ordinary stopped-flow apparatus (*56*). However, the mixing or "dead" time is about 10 msec compared to 1 msec for simple stopped-flow devices (for more details on the stopped-flow technique see Section 14.2). A temperature jump can be applied within 1 msec after the flow stops, i.e., about 11 msec after the start of the mixing process. Hence kinetics on the rapid equilibrium part of stationary systems that have a $t_{1/2}^{ss}$ as low as about 10 msec can be measured with this apparatus. For more details see French and Hammes (*11*), Yapel and Lumry (*12*), Erman and Hammes (*53*), and Faeder (*54*).

Thus far the stopped-flow–temperature-jump method has proved particularly useful in the study of the elementary processes of enzyme reactions (*53, 57a–c*). Recent applications to organic reactions involved the measurement of the elementary steps in nucleophilic aromatic substitution reactions (*58*) and are discussed in Section 11.5.4.

11.4 Temperature-Jump Equipment and Its Operation; Commercial Products

Many investigators build their own equipment; detailed descriptions on how to build a temperature-jump apparatus are available elsewhere (*7, 11, 12*) and are not duplicated here. Yapel and Lumry (*12*) offer also the most detailed instructions on how to operate the equipment, calibration procedures, and many useful hints for practical work.

Some investigators have developed temperature-jump apparatuses which allow measurements under unusual conditions, e.g., under very high pressure thus allowing the determination of activation volumes (*13, 59–61*).

The chemist who is not willing to invest the time and effort in building his own machine can choose among three different commercial instruments which will in most cases meet his needs adequately. All three products are based on the Joule heating method. The standard versions are all equipped with spectrophotometric detection systems. We summarize their most important features here; for more technical details Yapel and Lumry (*12*) or Schelly and Eyring (*62*) should be consulted. However, it should be mentioned that one of the instruments described earlier (*12, 62*) is no longer on the market.† On the other hand, a new instrument (Garching Instrumente) has recently become available.

Messanlagen Studiengesellschaft (Göttingen, Germany). The Messanlagen instrument (Fig. 11.7), which was designed by DeMaeyer and was the first on the market, is relatively inexpensive and well suited for very rapid reactions because of the use of low-capacitance energy storage capacitors in conjunction with high-discharge voltages. For added flexibility three interchangeable high-voltage capacitors are provided (0.01, 0.02, and 0.05 μF). Any discharge voltage up to 50 kV can be

† Information from American Instrument Co., Silver Spring, Maryland 20910.

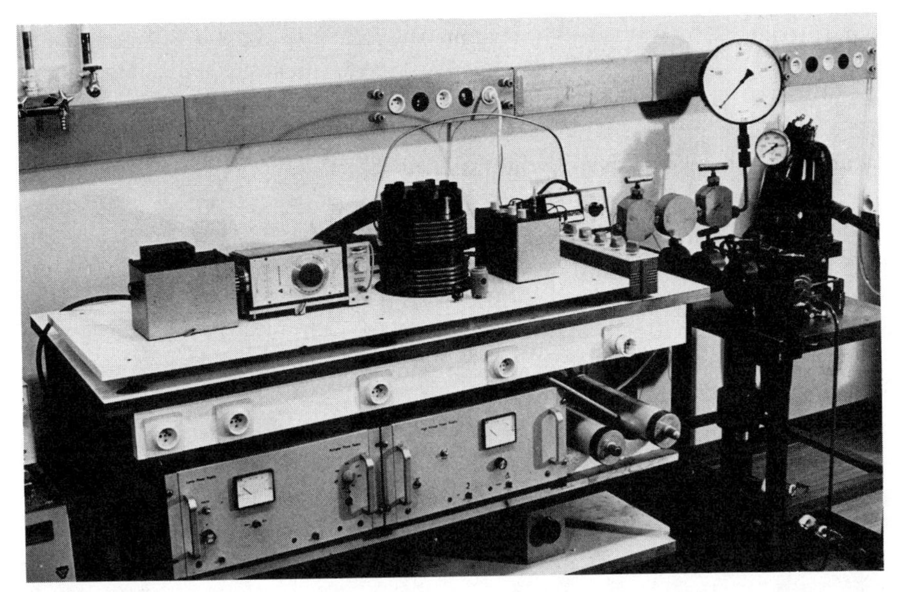

FIGURE 11.7. Temperature-jump apparatus by Messanlagen Gmbh (courtesy of L. De Maeyer).

selected. With the 0.01 μF capacitor a heating time as low as 0.25 μsec can be achieved with a sample resistance of 50 Ω ($\sim$0.2 *M* NaCl; Eq. 11.5).

Cells made out of different materials such as polyacetal, polycarbonate, polyacrylate, epoxy resin, or Teflon are available; they are equipped with gold or platinum electrodes and quartz windows; for a schematic design see Fig. 11.2. Different sample volumes can be chosen; standard size is 6 ml with a 1 cm light path, whereas the "microcell" holds about 1 ml with a 0.7 cm light path. The heated volume of the microcell is small enough to obtain jumps of several degrees even with the 0.01 μF capacitor when high voltages are used. The cell is located in a thermostated pressure compartment which allows the relaxation times to be measured under static pressures of up to 5000 atm if desired.

Different light sources are available: a tungsten–iodide lamp with a quartz envelope for the visible spectrum and high-pressure mercury or xenon arc lamps for both the uv and visible parts of the spectrum. By virtue of a double beam arrangement lamp fluctuation can be minimized, a definite advantage when the arc lamps are used. A Bausch and Lomb high-intensity monochromator is provided with uv (180–400 nm) and visible (350–800 nm) gratings. Signal detection in both the sample and the reference beam is with RCA 1-P28 photomultiplier tubes. No oscilloscope is provided by the manufacturer; a Tektronix Type 549 storage oscilloscope or the equivalent is most suitable. The current price is about \$15,000.†

Durrum Instrument Corporation (*Palo Alto, California*). Durrum is marketing a temperature-jump accessory, called the Model D-150, which can be fitted with

† For details, write to Messanlagen Studiengesellschaft m.b.H., Rudolf Winkel Strasse 9, 34 Göttingen, West Germany.

their Durrum–Gibson stopped-flow apparatus; the combined unit is marketed as the "Durrum Model D-115 Rapid Kinetics Spectrophotometer" and is shown in Fig. 11.8. The most noteworthy feature of this instrument is the provision to use it in the combined stopped-flow–temperature-jump mode; a detailed description of the apparatus has been given by Stewart and Lum (*63*).

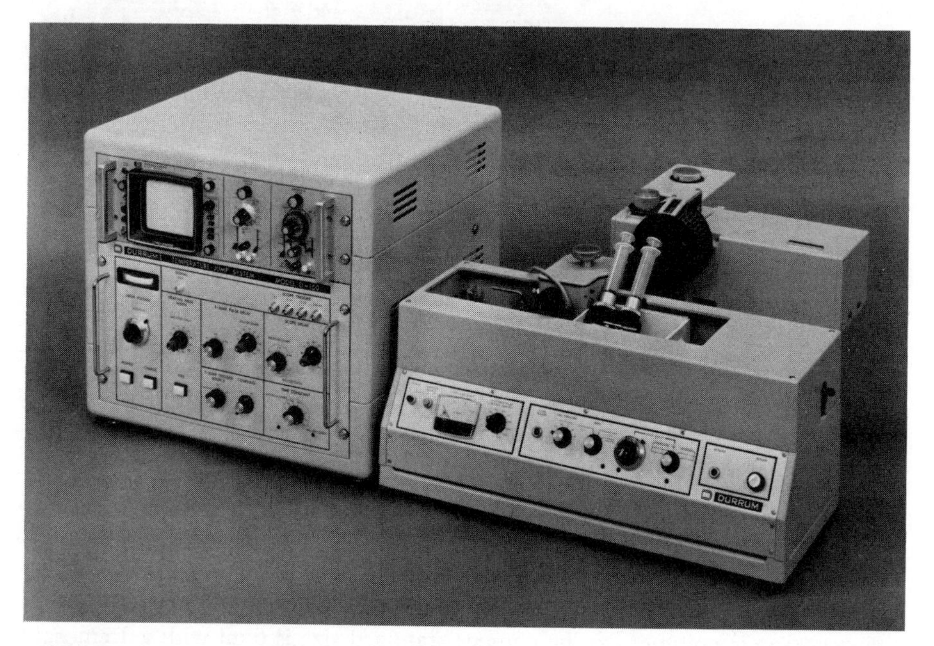

FIGURE 11.8. Temperature-jump apparatus by Durrum Instrument Corporation (through their courtesy).

The standard cell (stainless steel or Kel-F) has a light path of 2 cm with a volume of 200 μl though much larger volumes are needed to fill the cell through a system of syringes. Discharge voltages of 2–5 kV stored on a 1 μF capacitor allow maximum temperature jumps of about 10° or slightly more in aqueous solutions. Because of the high capacitance of the capacitor the heating time is relatively long; according to Eq. 11.5, $\tau_d = 25$ μsec with a 50 Ω sample cell resistance which corresponds to an approximately 0.2 *M* KCl solution. However, there is a provision to shorten the duration of the heating pulse by terminating the discharge of the capacitor after a selected time period. This is achieved by firing a second spark gap that bypasses the sample cell and rapidly discharges the capacitor. The principal drawback of this method is the waste of part of the energy available which may considerably reduce the size of the temperature-jump. If, for example, the discharge is terminated after 10 μsec, only 33% of the energy stored on the capacitor is available for heating with a 50 Ω sample; for a 100 Ω sample, the amount of available energy would be as low as 18%.

The temperature jump can be triggered manually in the simple temperature-jump mode or automatically following the mixing of the reagents. The mixing time of the instrument is about 10 msec with the standard temperature-jump cell. In principle this permits a jump to be performed about 10 msec after initiating the mixing process; in our experience much better results (smoother oscilloscope traces) are obtained with delays of at least 20 msec.

A tungsten–quartz–iodide and deuterium lamp with quartz envelope, and a prism-grating monochromator (200–800 nm) are standard equipment. The instrument includes a storage oscilloscope. The current price of the basic Model D-115 is about $25,000.

Durrum also markets various accessories. Their instrument can also be used in conjunction with a digital recorder instead of the usual oscilloscope.†

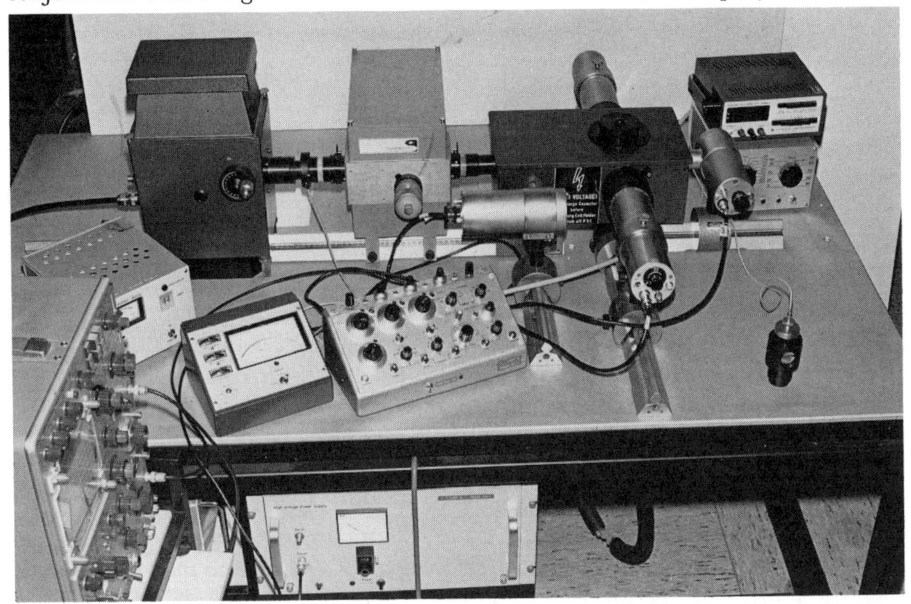

FIGURE 11.9. Temperature-jump apparatus with fluorescence detection by Garching Instrumente. Photograph shows prototype developed at the Max Planck Institute (courtesy of C. R. Rabl).

Garching Instrumente (*Garching, Germany*). This newest development, which at the time of this writing is scheduled to appear on the market in mid-1975, promises to be an outstanding all-purpose instrument for those able to afford it (estimated price about $50,000). The instrument is shown in Fig. 11.9; it is based on the design of Rigler *et al.* (*14*) referred to earlier and has been described in considerable detail (*14*). The time resolution is comparable to that of the Messanlagen instrument; it also has exchangeable capacitors (0.01, 0.02, and 0.05 μF) which can be charged up to 50 kV. The optical arrangement, shown schematically in Fig. 11.10, is carefully thought out, it is the result of long-standing experimental and constructional experience at the birth center of the relaxation techniques (Max Planck Institute für Biophysikalische Chemie, formerly Physikalische Chemie, of

† For details, write to Durrum Instrument Corporation, 3950 Fabian Way, Palo Alto, California 94303.

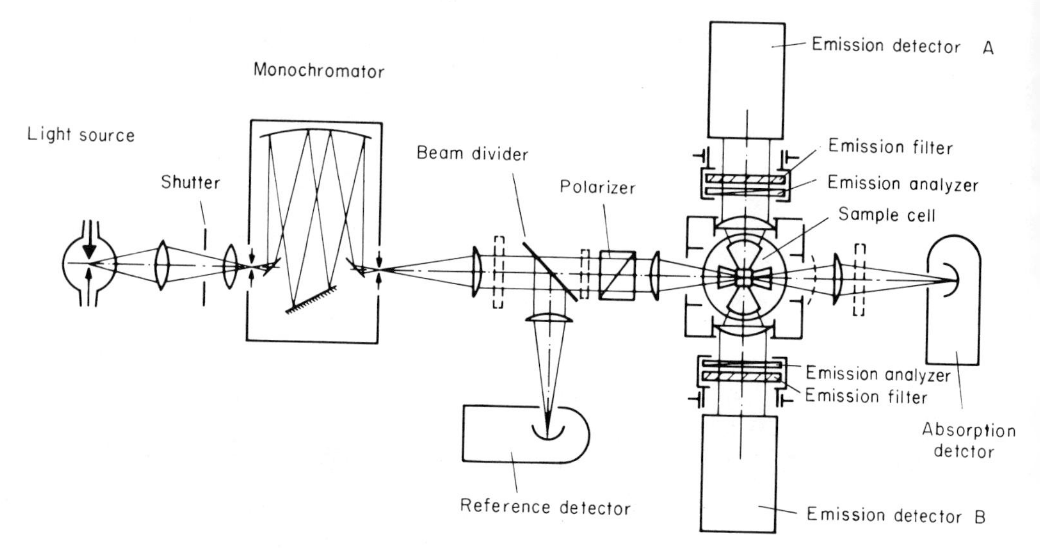

FIGURE 11.10. Schematic diagram of optical arrangement of fluorescence temperature-jump apparatus by Garching Instrumente (courtesy of C. R. Rabl).

Göttingen). It allows absorption measurements with a resolution of $\sim 10^{-4}$ OD units at very short times, $\sim 10^{-5}$ OD units for times larger than 100 μsec. It also permits highly sensitive fluorescence measurements, with concentrations as low as 10^{-8} *M*, and measurements of fluorescence polarization. The design also permits static measurements ("titration") to be made very easily and thus to exploit the high spectrophotometric resolution of the instrument.

A number of accessories are available or in preparation, such as temperature-jump cells of various path lengths from 0.2 mm up to 15 mm, cells allowing measurements up to 90°C or at high pressures up to 130 atm, and most significantly the capability of performing stopped-flow–temperature-jump combination experiments. There is also a provision (optional) to sample the data with digital recorders.†

11.5 Applications of the Temperature-Jump Method

At the time Eigen and DeMaeyer (*3*) wrote their landmark article (1963) there were four main types of reactions that had been explored by the temperature-jump method, namely (1) proton transfer reactions, (2) metal complex formations, (3) electron transfer reactions, and (4) hydration (e.g., of CO_2) and hydrolysis (e.g., of halogens) reactions. Among these studies those on proton transfer and metal complexation were by far the most numerous; they still attract a great deal of attention. In a recent review Hammes (*64*) lists the following eight additional areas in which this technique has now been successfully applied: organic

† For details write to Garching Instrumente, Freisinger Landstr. 25, D-8046 Garching, West Germany.

substitutions, micelle reactions, phospholipid dispersions, protein and enzyme–small-molecule interactions, enzyme catalysis, enzyme regulation, polynucleotide–small-molecule interactions, and polynucleotide interactions.

Any attempt to give a full account of all these applications would of course be a hopeless task; we estimate that studies on more than a thousand different systems have been published. Thus we shall restrict ourselves to a few remarks and mainly refer to reviews or some selected representative papers. Applications of temperature-jump *amplitude* measurements have been discussed in Chapters 6 and 7.

11.5.1 Proton Transfer Process

The general schemes which apply to proton transfer reactions have been extensively discussed in Section 4.4. The results and conclusions from the pioneering relaxation studies on proton transfer reactions have been excellently reviewed by Eigen *et al.* (*3, 65, 66*) and have also been discussed in numerous secondary reviews, monographs, and textbooks (*20, 67–69*). Hence no attempt to review these results again is made here except for the following brief remarks. A reaction of a general acid with OH^-, or of a general base with H_3O^+, can be written as

$$AH + OH^- \underset{k_{-1}}{\overset{k_1}{\rightleftharpoons}} AH\cdots OH^- \underset{k_{-2}}{\overset{k_2}{\rightleftharpoons}} A^-\cdots HOH \tag{11.31}$$

$$B + H_3O^+ \underset{k_{-1}}{\overset{k_1}{\rightleftharpoons}} B\cdots HOH_2^+ \underset{k_{-2}}{\overset{k_2}{\rightleftharpoons}} BH^+\cdots OH_2 \tag{11.32}$$

where $AH\cdots OH^-$ and $B\cdots HOH_2^+$ are encounter complexes, generally assumed to be hydrogen bonded. Under typical reaction conditions these complexes are present at very low concentrations and thus the steady state approximation holds, leading to

$$\vec{k} = k_1k_2/(k_{-1} + k_2), \qquad \overleftarrow{k} = k_{-1}k_{-2}/(k_{-1} + k_2) \tag{11.33}$$

For a diffusion-controlled process we have $k_2 \gg k_{-1}$ and thus

$$\vec{k} = k_1, \qquad \overleftarrow{k} = k_{-1}k_{-2}/k_2 = k_{-1}/K_2 \tag{11.34}$$

When the proton transfer is from or to an atom different from carbon ("normal" acids and bases), these reactions are in fact diffusion controlled in the thermodynamically favored direction, with $k_1 \sim 10^{10}$–$10^{11}\ M^{-1}\ \text{sec}^{-1}$, except when the proton is involved in an intramolecular hydrogen bond. Such intramolecular bonds lower the rate because extra activation energy is required to break it; the rates involving carbon acids and bases are also lower partly because C—H does not easily form a hydrogen bond to the solvent or attacking base.

In the case of a proton transfer from a general acid to a general base the following scheme applies:

$$AH + B \underset{k_{-1}}{\overset{k_1}{\rightleftharpoons}} AH\cdots B \underset{k_{-2}}{\overset{k_2}{\rightleftharpoons}} A^-\cdots HB^+ \underset{k_{-3}}{\overset{k_3}{\rightleftharpoons}} A^- + HB^+ \tag{11.35}$$

Here the proton transfer step is followed by the diffusion of the products, a process that is not required in reactions 11.31 and 11.32 since the products remain hydrogen bonded to the solvent. For the overall rate constants we have

$$\vec{k} = \frac{k_1 k_2 k_3}{k_{-1}k_{-2} + k_2 k_3 + k_{-1}k_3}, \qquad \overleftarrow{k} = \frac{k_{-1}k_{-2}k_{-3}}{k_{-1}k_{-2} + k_2 k_3 + k_{-1}k_3}$$

which for a diffusion-controlled forward process ($k_2 \gg k_{-1}$; $k_2 \gg k_{-2}$) becomes

$$\vec{k} = k_1, \qquad \overleftarrow{k} = k_{-1}k_{-2}k_{-3}/k_2 k_3 = k_{-1}/K_2 K_3$$

Again, for normal acids and bases, the reactions in fact become diffusion controlled for equilibrium constants $>10^2$ to $>10^3$ ($pK_a^{BH} - pK_a^{AH} > 2$ or 3).

Although a fairly large number of proton transfers involving normal acids and bases have been investigated by the temperature-jump method, it is not surprising in view of the above that many systems are beyond the capabilities of this method and have to be studied by faster relaxation techniques such as ultrasonics or the electric field jump. Whether the relaxation time(s) in proton transfer equilibria is (are) in the right range depends primarily on the pK of the acid or base in question. Let us illustrate this with reaction 11.31. Since $\vec{k}/\overleftarrow{k} = K_a^{AH}/K_w$, and assuming $\vec{k} = k_1 \sim 2 \times 10^{10}\ M^{-1}\ \text{sec}^{-1}$ leads to $\overleftarrow{k} = k_1 K_w/K_a^{AH} \sim 2 \times 10^{-4}/K_a^{AH}$. Since $\tau^{-1} = k_1(\bar{c}_{AH} + \bar{c}_{OH}) + k_{-1} \geq k_{-1}$, we have $\tau^{-1} \geq 4 \times 10^5\ \text{sec}^{-1}$ for $K_a \leq 5 \times 10^{-10}$; this is about the limit of the temperature-jump method without pushing it. Similar considerations apply to reaction 11.32. Thus the method is suitable for systems with pK values ranging from $\sim$4.7 to $\sim$9.3.

In systems where the proton transfer is slowed down for the reasons mentioned earlier, the pK limitations become less severe. In fact the temperature-jump method has been particularly successful with intramolecularly hydrogen-bonded systems (*70–73*) and with carbon acids and bases (*74–76*).

More recently proton transfer in weakly polar solvents such as chlorobenzene has been studied by means of the microwave (*77–81*) and laser pulse heating methods (*82*). There are both similarities and differences between these reactions in hydroxylic and aprotic solvents. For a reaction between a neutral acid (say a phenol) and a neutral base (say an amine) scheme 11.35 applies but without the last step because there can be no dissociation of the ion pair $A^- \cdots HB^+$ into free ions. In systems where the overall equilibrium constant is relatively low ($K < 10^2$; e.g., phenols and pryidine bases) direct evidence for the hydrogen-bonded complex $AH \cdots B$ was actually obtained from equilibrium measurements (*78b*, *82*), substantiating the assumed mechanism.

An interesting question which arose in studying such reactions was whether they are also diffusion controlled under favorable thermodynamic conditions. The data could generally be fitted to an equation such as

$$1/\tau = k_f(\bar{c}_{AH} + \bar{c}_B) + k_b \tag{11.36}$$

which in principle is consistent with

(a) $k_2 \gg k_{-1}$: $k_f = k_1$, $k_b = k_{-1}/K_2$

as well as with

(b) $k_{-1} \gg k_2$: $k_f = K_1 k_2, \quad k_b = k_{-2}$

Since k_f was found to depend little, and k_b to depend strongly on the overall equilibrium constant it was concluded that the first hypothesis is correct (*78a,b*). However, an interesting result is that the k_f values were generally found to be appreciably lower than predicted by diffusion theory (*79, 80*). This has led to the suggestion of an additional intermediate as shown in

$$AH + B \rightleftharpoons (AH, B) \rightleftharpoons AH\cdots B \rightleftharpoons A^{-}\cdots HB^{+} \tag{11.37}$$

where (AH, B) is a non-hydrogen-bonded encounter complex. The low k_f values were explained by assuming that AH···B is formed from (AH, B) after rotation into the correct alignment, a processs strongly dependent on steric factors.

11.5.2 Metal Complexes

The general features of metal complex formation are best discussed with reference to the three-step mechanism introduced by Eigen (*83, 84*).

$$\underset{①}{M^{m+}(sol) + L^{l-}(sol)} \underset{k_{21}}{\overset{k_{12}}{\rightleftharpoons}} \underset{②}{M^{m+}(sol, sol)L^{l-}} \underset{k_{32}}{\overset{k_{23}}{\rightleftharpoons}} \underset{③}{M^{m+}(sol)L^{l-}} \underset{k_{43}}{\overset{k_{34}}{\rightleftharpoons}} \underset{④}{ML^{(m-l)+}} \tag{11.38}$$

In Eq. 11.38 ② is an encounter complex, ③ the "outer sphere" complex, and ④ the "inner sphere" complex. In view of the fact that anion–water interactions are generally weaker than cation-water interactions, it is usually implied that the reaction ② → ③ involves the loss of a solvent molecule from the hydration shell of L, and the reaction ③ → ④ the loss of a solvent molecule from the hydration shell of the cation. Support for this contention comes from numerous observations that k_{34} is virtually independent of the nature of L but approximately equal to the rate of solvent exchange of the cation.

Step ① ⇌ ② is always very rapid (diffusion controlled; $k_{12} \sim 10^{10}\ M^{-1}\ \text{sec}^{-1}$, $k_{21} \sim 10^{9}\ \text{sec}^{-1}$) and so is step ② ⇌ ③ ($k_{23}, k_{32} \sim 10^{8}$–$10^{9}\ \text{sec}^{-1}$). Hence an investigation of these steps is clearly beyond the temperature-jump method but is feasible with ultrasonic techniques as discussed in Section 15.4.2.

Step ③ ⇌ ④ is usually rate limiting so that scheme 11.38 can be (and usually is) written as

$$\underset{①}{M^{m+}(sol) + L^{l-}(sol)} \underset{k_{31}}{\overset{k_{13}}{\rightleftharpoons}} \underset{③}{M^{m+}(sol)L^{l-}} \underset{k_{43}}{\overset{k_{34}}{\rightleftharpoons}} \underset{④}{ML^{(m-l)+}} \tag{11.39}$$

For ion concentrations that are not too high the outer sphere complex does not accumulate to measurable levels; hence Eq. 11.39 can be simplified further to

$$M^{m+}(sol) + L^{l-}(sol) \underset{\overleftarrow{k}}{\overset{\vec{k}}{\rightleftharpoons}} ML^{(m-l)+} \tag{11.40}$$

with

$$\overrightarrow{k} = k_{13}k_{34}/(k_{31} + k_{34}), \qquad \overleftarrow{k} = k_{43}k_{31}/(k_{31} + k_{34})$$

Usually $k_{31} \gg k_{34}$ so that

$$1/\tau = K_{13}k_{34}(\bar{c}_M + \bar{c}_L) + k_{43} \tag{11.41}$$

However, in some cases the concentration of ③ has been found to be appreciable, leading to a curved concentration dependence according to

$$\frac{1}{\tau} = \frac{K_{13}k_{34}(\bar{c}_M + \bar{c}_L)}{1 + K_{13}(\bar{c}_M + \bar{c}_L)} + k_{43} \tag{11.42}$$

Examples are the reaction of Ni^{2+} with methyl phosphate (*85*) or the formation of dichloro-1,1,7,7-tetraethyldiethylenetriaminenickel(II) in acetonitrile (*38*); this latter study is also one of the first examples of the application of the laser pulse temperature-jump method to metal complexations.

Many studies have been carried out on systems that conform to Eq. 11.40; since typical values for $\overrightarrow{k}$ range from 10^3 to $10^8\ M^{-1}\ sec^{-1}$ and those for $\overleftarrow{k}$ are $< 10^5\ sec^{-1}$ for a great many complexation reactions, particularly those involving transition metal ions, the temperature-jump method is obviously an ideal technique. The results and conclusions from these studies have been frequently and authoritatively reviewed (*3, 20, 21, 86–92*).

As far as practical experimental aspects are concerned three points should be briefly mentioned.

1. The ligand frequently is a base for which the equilibrium

$$HL^{(l-1)-} \underset{}{\overset{K_a^{HL}}{\rightleftharpoons}} L^{l-} + H^+ \tag{11.43}$$

has to be taken into consideration; it can usually be treated as a rapid preequilibrium. Thus coupling reactions 11.40 and 11.43 corresponds to case 6 in Table 3.1 (with A = HL, B = H^+, C = L, D = M, E = ML). For a buffered solution the expression for τ^{-1} becomes

$$\frac{1}{\tau} = \overrightarrow{k}\left[\bar{c}_L + \frac{\bar{c}_M}{1 + \bar{c}_H/K_a^{HL}}\right] + \overleftarrow{k}$$

Note that $HL^{(l-1)-}$ can sometimes also act as a ligand, e.g., when L is ethylenediamine (*93*), or the inner sphere complex may be involved in an acid–base equilibrium of its own, e.g., in the case of interaction of Mg^{2+} with inorganic and nucleoside phosphates (*94*).

2. Many complex formation reactions are not accompanied by significant spectral changes and thus cannot be directly observed. However, since complexation affects the acid–base equilibrium 11.43, a pH indicator can often be used to monitor the relaxation. Note that in order to assure significant pH changes a nonbuffered system has to be used in such cases. Here the relaxation time is given by

$$\frac{1}{\tau} = \overrightarrow{k}\left[\bar{c}_L + \frac{\bar{c}_M}{1 + \alpha}\right] + \overleftarrow{k}$$

with

$$\alpha = \bar{c}_H \Big/ \left[K_a^{HL} + \bar{c}_L \frac{K_a{}^I + \bar{c}_H}{K_a{}^I + \bar{c}_H + \bar{c}_{In}} \right]$$

where $K_a{}^I$ is the acid dissociation constant of the indicator and c_{In} is the concentration of the indicator base. If K_a^{HL} and $K_a{}^I$ are known, which is usually the case, α can be calculated for any given pH. Thus a plot of τ^{-1} versus $\bar{c}_L + \bar{c}_M/(1 + \alpha)$ provides $\vec{k}$ and $\overleftarrow{k}$.

3. Frequently more than one complex is formed by consecutive reactions such as

$$ML_{i-1} + L \underset{k_{-i}}{\overset{k_i}{\rightleftharpoons}} ML_i, \qquad i = 1, 2, \ldots, n$$

which can lead to relaxation spectra with several relaxation times, though only one τ may be detectable. Examples where this occurs are the complexation of Ni^{2+} or Co^{2+} with glycine (*95*), glycyl-L-leucine (*96*) and other ligands; these examples also illustrate the use of pH indicators as discussed above. Some further aspects of metal complex formation are discussed in Section 15.4.2.

11.5.3 Enzyme–Substrate Interactions

Relaxation methods and in particular the temperature-jump technique have had a great impact on the study of enzyme reactions (see also Section 10.1). The field of enzyme kinetics began more than 60 years ago with Michaelis and Menten's (*97*) well-known mechanism

$$E + S \rightleftharpoons ES \longrightarrow P + E \tag{11.44}$$

Before the development of rapid reaction techniques investigators had to work at very low enzyme concentrations in order to keep the reactions slow. As a consequence ES and intermediates likely to follow ES could not accumulate beyond a very low steady state concentration and thus information regarding the various elementary steps coming from the numerous classical kinetic studies was rather limited (*98*).

It is now generally accepted that the minimum mechanism of a one-substrate enzyme-catalyzed reaction is represented by (*99, 100*)

$$\underset{①}{E + S} \rightleftharpoons \underset{②}{ES} \rightleftharpoons \underset{③}{EP} \rightleftharpoons \underset{④}{P + E} \tag{11.45}$$

The relaxation times characterizing scheme 11.45 have been discussed in Section 4.3.

In practice, even the minimum schemes are usually more complex than Eq.

11.45 because both the free enzyme and the intermediates are present in different protonated states, e.g.,

$$\begin{array}{ccccccc}
E + S & \rightleftharpoons & ES & \rightleftharpoons & EP & \rightleftharpoons & P + E \\
\updownarrows & & \updownarrows & & \updownarrows & & \updownarrows \\
EH^+ + S & \rightleftharpoons & EHS^+ & \rightleftharpoons & EHP^+ & \rightleftharpoons & P + EH^+ \\
\updownarrows & & \updownarrows & & \updownarrows & & \updownarrows \\
EH_2^{2+} + S & \rightleftharpoons & EH_2S^{2+} & \rightleftharpoons & EH_2P^{2+} & \rightleftharpoons & P + EH_2^{2+}
\end{array} \tag{11.46}$$

A further complication arises if S and/or P are also involved in acid–base equilibria. The kinetic equations for such schemes have been discussed by Hammes and Schimmel (*101–103*); if one treats the proton transfers as rapid equilibria, the problems in deriving the relaxation times are relatively straightforward.

Other extensions of scheme 11.45 occur for additional intermediates, e.g., when ES undergoes a conformational change to ES′:

$$E + S \rightleftharpoons ES \rightleftharpoons ES' \rightleftharpoons EP \rightleftharpoons P + E \tag{11.47}$$

Provided all intermediates are present at detectable levels a scheme such as 11.47 is characterized by four relaxation times which may lead to serious difficulties in evaluating the data, as discussed in Chapter 9.

Another experimental problem is that the step EP $\rightarrow$ P + E is often virtually irreversible, making temperature-jump experiments impractical except when coupled with the stopped-flow technique (see below). For these reasons one often employs model substrates, i.e., a compound that is structurally similar to the real substrate so that it has (we hope) the same binding characteristics, however without being transformed into a final product. This reduces schemes 11.45 and 11.47 to the more manageable schemes

$$E + S \underset{k_{-1}}{\overset{k_1}{\rightleftharpoons}} ES \tag{11.48}$$

$$E + S \underset{k_{-1}}{\overset{k_1}{\rightleftharpoons}} ES \underset{k_{-2}}{\overset{k_2}{\rightleftharpoons}} ES' \tag{11.49}$$

A general result from numerous such studies (*20, 64, 103–105*) is that for a great many enzymes and model substrates there is no great variation in k_1; typical values are 10^6–$10^8\ M^{-1}\ \text{sec}^{-1}$. The variations in k_{-1} are much larger and span a range from 10^{-4} to $>10^6\ \text{sec}^{-1}$, reflecting essentially the variations in the equilibrium constants; most often k_{-1} is found to be in the range of 10–$10^4\ \text{sec}^{-1}$.

Kinetic evidence (two relaxation times) for the isomerization step ES $\rightleftharpoons$ ES′ has been reported for several systems and summarized by Hammes and Schimmel (*103*). Typical values for k_2 and k_{-2} are 10^3–$10^4\ \text{sec}^{-1}$.

If one wants to study enzyme reactions with real substrates, the stopped-flow–temperature-jump method is often the only appropriate tool. An example where the first two steps of mechanism 11.47 were successfully studied is the reaction of ribonuclease with uridine-2′,3′-cyclic phosphate (*106*); another example for the reader who likes complexity is the interaction of erythro-β-hydroxyaspartic acid with aspartate aminotransferase where eight (!) relaxation times were found and analyzed (*107*).

The detection method used in the majority of enzyme reactions has been absorption spectrometry. One may study absorbance changes related to the enzyme, at wavelengths around 260 nm (*108*); more often pH indicators are used, exploiting the fact that the enzyme is usually involved in an acid–base equilibrium. In cases where the substrate has a characteristic and strong chromophore its absorption changes may be used to monitor the reaction. The rapid binding of proflavine to enzymes has also been exploited to indicate relaxation of the interaction of nonchromophoric substrates with enzymes (*109*). In a similar way bromophenol blue has been used as an indicator to study the conformational changes in β-lactoglobulin (*110*).

In some systems fluorimetric detection has been used, e.g., in the binding of reduced diphosphopyridine nucleotide to yeast alcohol dehydrogenase (*111*), or of DPNH to malate dehydrogenase (*112*); for an application in conjunction with the stopped-flow–temperature-jump technique see del Rosario and Hammes (*113*).

11.5.4 Organic Reactions

With a few exceptions such as the study of keto–enol tautomerism by Eigen *et al.* (*114*) or the hydration of pyruvic acid (*115*), applications to organic reactions are conspicuously absent from the early literature. It is only relatively recently that the temperature-jump technique has been applied to mechanistic and reactivity problems in organic reactions, and only a fraction of these studies have been done by scientists who were originally trained to be (physical) organic chemists. With the present availability of efficient, versatile, and easy to use commercial equipment it is to be expected that many more physical organic chemists will find the method extremely useful.

Most investigations have hitherto focused on the addition of nucleophiles to carbonyl and aromatic carbon.

A. Carbonyl Addition Reactions

After the first report by Eigen *et al.* (*115*) several other studies of the hydration of aldehydes and ketones appeared: glyoxalate (*116*), mesoxalate (*116*), pyruvic acid ethyl ester (*117*), formaldehyde (*117*), 2-methylbutyraldehyde (*118*), isobutyraldehyde (*119*), D-glyceraldehyde 3-phosphate (*120*), and vitamin B_6 and derivatives (*121*).

Usually the relaxation is monitored spectrophotometrically around 290 nm.

In basic solution the reactions can be described by a scheme such as 11.50 (*119*)

$$\begin{array}{ccc} \rangle C=O \;(\mathbf{1}) & \underset{k_{-1}(-H_2O)}{\overset{k_1(+H_2O)}{\rightleftharpoons}} & \rangle C(OH)_2 \;(\mathbf{2}) \\ {\scriptstyle k_{-2}(-OH^-)}\nwarrow\searrow{\scriptstyle k_2(+OH^-)} & & \nearrow\swarrow {\scriptstyle K} \\ & \rangle C(OH)O^- \;(\mathbf{3}) & \end{array} \tag{11.50}$$

where **2** and **3** are in rapid equilibrium. There is one *observable* relaxation time; under pseudo-first-order conditions it is given by

$$\frac{1}{\tau} = k_1 + k_2\bar{c}_{\mathrm{OH}} + \frac{k_{-1}}{1 + K\bar{c}_{\mathrm{OH}}} + \frac{k_{-2}K\bar{c}_{\mathrm{OH}}}{1 + K\bar{c}_{\mathrm{OH}}} \tag{11.51}$$

(see Section 4.4.1) with $K = \bar{c}_3/\bar{c}_2\bar{c}_{\mathrm{OH}}$. Most of the rate constants can be determined from the concentration dependence of τ^{-1} (*116*, *119*). In acidic solution there is also an acid-catalyzed pathway that needs to be considered (*117–119*); furthermore, in the presence of buffers accelerations by general acid–base catalysis have been observed.

In the case of isobutyraldehyde, aldol condensation in basic solution was rapid enough to necessitate the use of the stopped-flow–temperature-jump technique (*119*). For the hydration of 2-methylbutyraldehyde the reaction enthalpy was determined from relaxation amplitudes (*118*). A study that is of particular interest from the point of view of analyzing a complex system is that of the hydration of vitamin B_6 and related compounds (*121*), because the substrates can exist in various protonated states, each being able to add a water molecule.

For some aldehydes the hydration equilibrium establishes itself quite slowly so that the slow temperature-jump technique (Section 11.1.4) can be used for the rate study. In fact such an example (propionaldehyde) has been exploited for an undergraduate physical chemistry experiment at the University of California at Berkeley (*122*).

Carbinolamine formation by addition of amines to aldehydes has also been studied by the temperature-jump method (*123*, *124*). It was concluded that the reaction with secondary aliphatic amines can be described by

$$\rangle C{=}O + HN\langle \;\underset{k_{-1}}{\overset{k_1}{\rightleftharpoons}}\; -\underset{|}{\overset{O^-}{\overset{|}{C}}}-\overset{+}{\underset{\diagdown}{\overset{H}{\overset{|}{N}}}}\diagup \;\underset{k_{-2}}{\overset{k_2}{\rightleftharpoons}}\; -\underset{|}{\overset{OH}{\overset{|}{C}}}-N\langle \tag{11.52}$$

with the second equilibrium strongly favoring the neutral carbinolamine over the zwitterion. The data further show evidence of general base catalysis in the forward direction which led to the interesting conclusion that proton transfer (in the second step) rather than nucleophilic attack is rate limiting, i.e., $k_{-1} \gg k_2$. The proposed mechanism for the second step is

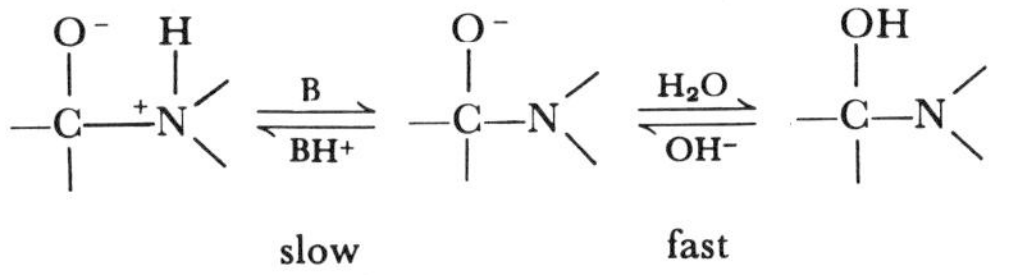

$$\begin{array}{c}\mathrm{O^-}\ \ \mathrm{H}\\ |\ \ \ \ |\\ -\mathrm{C}-\overset{+}{\mathrm{N}}\!<\\ |\end{array} \underset{\mathrm{BH^+}}{\overset{\mathrm{B}}{\rightleftharpoons}} \begin{array}{c}\mathrm{O^-}\\ |\\ -\mathrm{C}-\mathrm{N}\!<\\ |\end{array} \underset{\mathrm{OH^-}}{\overset{\mathrm{H_2O}}{\rightleftharpoons}} \begin{array}{c}\mathrm{OH}\\ |\\ -\mathrm{C}-\mathrm{N}\!<\\ |\end{array} \qquad (11.53)$$

slow fast

A further type of carbonyl addition studied by the temperature-jump method is the formation of a hemimercaptal by addition of 2-mercaptoethanol to the aldehydic group of pyridoxal 5′-phosphate (*125*).

B. Meisenheimer Complexes

Reactions of nucleophiles with aromatic nitro compounds to form anionic σ complexes or "Meisenheimer" complexes (*126–128*) are well suited for temperature-jump investigations (*129*). Some of these reactions have served as models for the various bond-forming and bond-breaking processes which occur in nucleophilic aromatic substitution reactions (*128, 129*).

A mechanistically interesting case which is at the same time instructive for the practicing student is shown in Eq. 11.54 (*130, 131*)

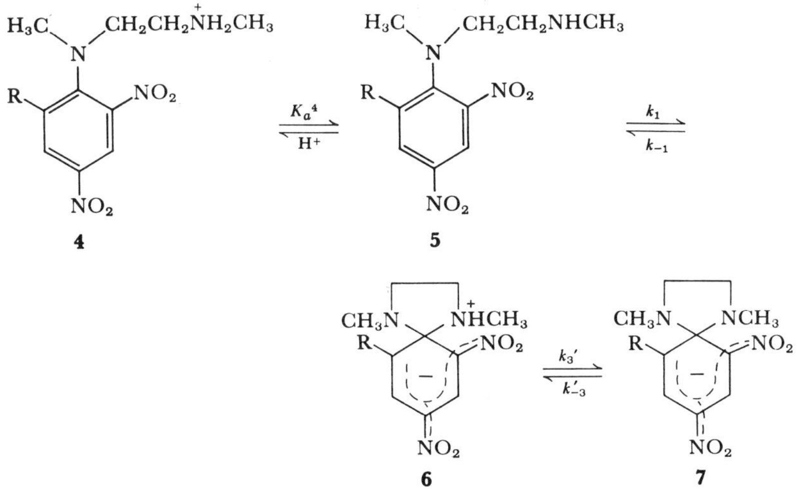

H_3C, $CH_2CH_2\overset{+}{N}H_2CH_3$, N, R, NO_2, NO_2 — **4** $\underset{H^+}{\overset{K_a^4}{\rightleftharpoons}}$ H_3C, $CH_2CH_2NHCH_3$, N, R, NO_2, NO_2 — **5** $\underset{k_{-1}}{\overset{k_1}{\rightleftharpoons}}$

CH_3N, $\overset{+}{N}HCH_3$, R, NO_2, NO_2 — **6** $\underset{k'_{-3}}{\overset{k_3'}{\rightleftharpoons}}$ CH_3N, NCH_3, R, NO_2, NO_2 — **7** (11.54)

R = H or NO_2

with

$$k_3' = k_3 + k_3^{OH}\bar{c}_{OH} + \sum_j k_3^{B_j}\bar{c}_{B_j}; \qquad k'_{-3} = k_{-3}\bar{c}_H + k_{-3}^{OH} + \sum_j k_{-3}^{B_j}\bar{c}_{BH_j}$$

where k_3, k_3^{OH}, and $k_3^{B_j}$ refer to deprotonation rate constants of **6** by the solvent, OH^-, and general bases, respectively, and k_{-3}, k_{-3}^{OH}, and $k_{-3}^{B_j}$ refer to protonation of **7** by H_3O^+, the solvent, and general acids, respectively.

One relaxation time was observed and studied in the presence of buffers at different pH values, always using pseudo-first-order conditions. At relatively low pH (e.g., pH < 9.5 for R = NO_2) (*130*) τ^{-1} depends curvilinearly on buffer concentration, as shown in Fig. 11.11, whereas at high pH τ^{-1} is independent of

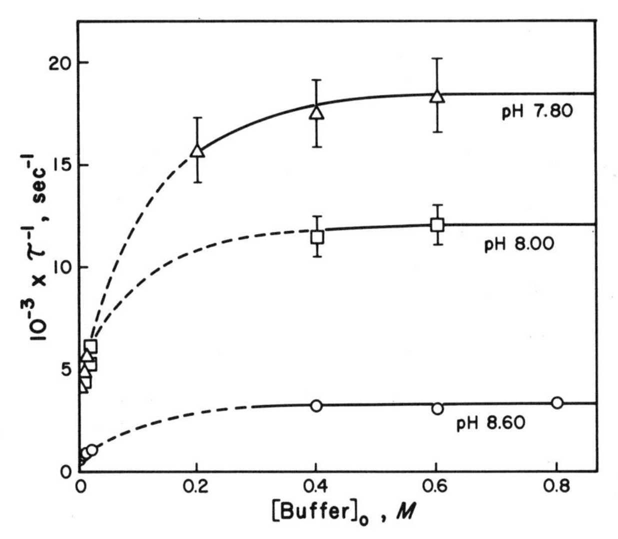

FIGURE 11.11. Buffer dependence of τ^{-1} in system 11.54 with R = NO_2, in aqueous solution 25°C. [From Bernasconi and Gehriger (*130*), by permission of the American Chemical Society.]

buffer concentration. This is consistent with the rate-limiting proton transfer (**6** → **7**) at low buffer concentrations and/or low pH, but the rate-determining nucleophilic attack (**5** → **6**) at high buffer concentration and/or high pH.

Treating the reaction **4** ⇌ **5** as rapid equilibrium and **6** as a steady state intermediate leads to

$$\frac{1}{\tau} = \frac{k_1 k_3'}{k_{-1} + k_3'} \frac{K_a{}^4}{K_a{}^4 + \bar{c}_H} + \frac{k_{-1} k'_{-3}}{k_{-1} + k_3'} \tag{11.55}$$

At high buffer concentration and/or high pH we have $k_3' \gg k_{-1}$ and hence

$$\frac{1}{\tau_{hi}} = k_1 \frac{K_a{}^4}{K_a{}^4 + \bar{c}_H} + \frac{k_{-1}\bar{c}_H}{K_a{}^6} \tag{11.56}$$

where $K_a{}^6$ is the acid dissociation constant of **6**. At zero buffer concentration (intercepts of τ^{-1} versus buffer concentration plots) Eq. 11.55 becomes

$$\frac{1}{\tau_0} = \frac{k_1(k_3 + k_3^{OH}\bar{c}_{OH})}{k_{-1} + k_3 + k_3^{OH}\bar{c}_{OH}} \frac{K_a{}^4}{K_a{}^4 + \bar{c}_H} + \frac{k_{-1}(k_{-3}\bar{c}_H + k_{-3}^{OH})}{k_{-1} + k_3 + k_3^{OH}\bar{c}_{OH}} \tag{11.57}$$

Representative plots of τ_{hi}^{-1} and of τ_0^{-1} versus pH are shown in Fig. 11.12. From these plots coupled with a spectrophotometric determination of $K_a{}^6$, the rate

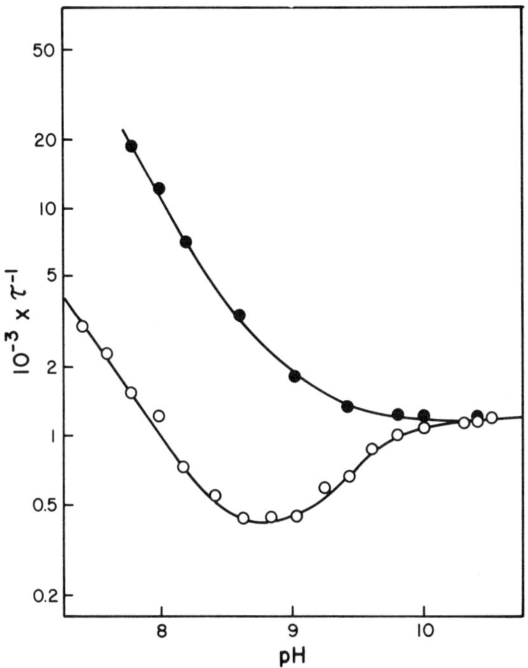

FIGURE 11.12. pH dependence of τ_{hi}^{-1} (●) and τ_0^{-1} (○) for system 11.54 with R = NO_2; the lines are calculated to give best fit with Eqs. 11.56–11.57. [From Bernasconi and Gehriger (*130*).]

constants and $K_a{}^4$ of scheme 11.54 can be evaluated. Furthermore, the initial slopes of the buffer plots, which are given by

$$\text{slope} = \frac{k_1 k_3{}^{\text{B}} \dfrac{K_a{}^4 K_a^{\text{BH}}}{(K_a{}^4 + \bar{c}_{\text{H}})(K_a^{\text{BH}} + \bar{c}_{\text{H}})} + \dfrac{k_{-1} k_3{}^{\text{B}} K_a^{\text{BH}}}{K_a{}^6} \dfrac{\bar{c}_{\text{H}}}{K_a^{\text{BH}} + \bar{c}_{\text{H}}}}{k_{-1} + k_3 + k_3^{\text{OH}} \bar{c}_{\text{OH}}} \tag{11.58}$$

where K_a^{BH} is the acid dissociation constant of BH, allow the evaluation of $k_3{}^{\text{B}}$ and k_{-3}^{B}.

11.5.5 Reactions of Oligonucleotides and Polynucleotides

The temperature-jump method has been applied to a variety of processes involving polynucleotides and oligonucleotides. One type is the cooperative binding of small molecules to the polymer; the theory has been treated in Section 10.3 and several examples have been mentioned. A few other representative studies illustrating some other aspects are the binding of acridine orange or of proflavine to single-stranded polyadenylic acid (*132*), of proflavine to poly A·poly U (double helix) (*133*), and of ethidium bromide to transfer-RNA(tRNA) (*134, 135*). In some cases the interpretation of the results is quite difficult. For example, in the case of poly A with proflavine or acridine orange (*132*) a virtually continuous relaxation spectrum spanning the range of 1 sec–2 × 10^{-5} sec was observed, though with one more or less discrete relaxation effect in the millisecond range. This latter was independent of polymer and dye concentration, indicating an intramolecular process, possibly the stacking of the dyes along the polymer chain, or a conformational change.

Base Pairing in Oligonucleotides

In the remainder of this section we shall discuss the processes of base pairing in oligonucleotides. It is well known that the formation of a double-stranded helix by the combination of two complementary single-stranded oligo- or polynucleotide chains is a cooperative phenomenon; i.e., after a few base pairs have been formed (nucleation) the pairing of the remaining bases becomes easier. The same holds true for *intra*molecular base pairing (loop formation). Experimentally this manifests itself by steep "melting curves" and thus is quite similar to the helix-coil transitions of polypeptides discussed in Section 10.2. The extent of cooperativity increases with increasing chain length. Several mechanistic models have been proposed to account for the cooperative nature of double helix formation, such as the Ising model, the all-or-none model, or the staggering zipper model. The subject has been recently reviewed by Riesner and Römer (*136*).

Since the absorption around 260 nm of the double helix differs significantly from that of the single-stranded random coils, temperature-jump experiments with spectrophotometric detection are easily performed. Such experiments have generally revealed the presence of one relaxation time only, with a concentration dependence typical for a reaction A + B $\rightleftharpoons$ C (*136–138*) in the case of the combination of two single-stranded oligomers. This is consistent with a mechanism as shown in Fig. 11.13 (*136, 138, 139b*) where all intermediates are assumed to be

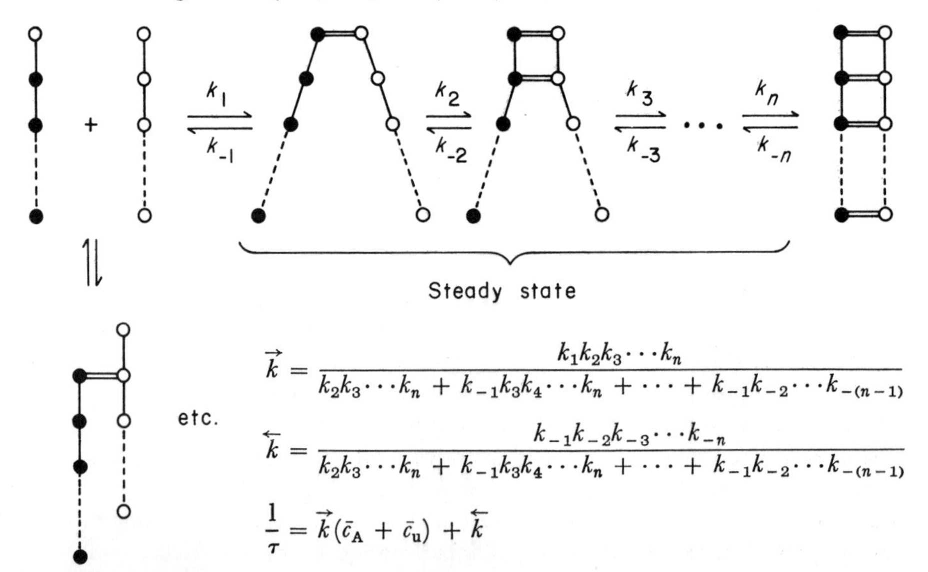

FIGURE 11.13. Mechanism for all-or-none recombination of oligonucleotides involving steady state intermediates. [After Eigen (*139b*).]

in a steady state; this mechanism is usually referred to as the all-or-none model.

An interesting question which is not answered by the rate law regards the rate-determining step. However, two other pieces of kinetic information have been useful in this respect.

1. The bimolecular rate constant $\vec{k}$ is usually on the order of 10^4–10^7 M^{-1} $\sec^{-1}$, i.e., much smaller than the diffusion-controlled rate constants observed in the formation of single base pairs (*139a, 140*). Hence the formation of the first base pair cannot be the rate-determining step.
2. The activation energy for $\vec{k}$ is negative, indicating that $\vec{k}$ cannot refer to an elementary process. Thus a nucleation process involving the formation of two or three base pairs seems more likely (*139a, 140, 141*), i.e., the formation of the second or third base pair is the rate-limiting step followed by a rapid "zippering up" process at a rate of about 10^7 base pairs per second.

Similar observations were made in studying the conformational transitions of naturally occurring tRNA (formation of cloverleaf structure) (*136*). Relaxation amplitudes without kinetic analysis have also been useful in differentiating between processes such as the unstacking of single-stranded regions of tRNA and the actual unfolding of the tRNA molecule (*26*).

A topic of considerable current interest is the kinetics of the binding of base *monomer* to a complementary polynucleotide template. This is interesting in the context of the problem of base recognition in nucleic acid replication. A study mentioned in Section 10.3 refers to the binding of *N*-6,9-dimethyladenine to polyuridylic acid to form a double-stranded helix, investigated by the extrarapid cable discharge temperature-jump method and interpreted in terms of the Schwarz theory (*142*). The kinetics of the dissociation of triple-stranded helices (2:1 polynucleotide–monomer complexes) have also been reported (*143*).

11.5.6 Miscellaneous Reactions

A. Micellization

Temperature-jump studies on detergent solutions undergoing micellization have been reported for dodecylpyridinium iodide (*8*), lauryl sulfate (*144*), dodecylammonium chloride (*145*), sodium dodecylsulfonate (*145*), and octylphenyl polyoxyethylene ether (*146*). The studies of Bennion *et al.* (*145, 146*) are an interesting application of the light scattering detection technique. The results have been generally interpreted along the lines of the KHDS model discussed in Section 10.4.1 and thus at least some need reinterpretation. Also, some later studies by other techniques (*147*) provided results that disagreed with earlier data which could be due to impurities in the earlier samples (*147*).

B. Electron Transfer Reactions

Electron transfer rates in redox equilibria such as

$$\mathrm{Red}_1{}^n + \mathrm{Ox}_2{}^m \underset{k_{-1}}{\overset{k_1}{\rightleftharpoons}} \mathrm{Ox}_1^{n+1} + \mathrm{Red}_2^{m-1} \tag{11.59}$$

can often be studied quite easily; usually there are large spectral differences among the species involved in equilibrium 11.59 with some of the species absorbing in the visible; this makes the absorption spectrophotometric technique ideally suited. A recent study where temperature-jump experiments over a wide pH range provided evidence for three separate electron transfer pathways for azaviolenes has been discussed in Section 4.5.1. Other examples include the classical quinone–semiquinone–hydroquinone system (*148*), the redox reactions of $IrCl_6^{2-}$ with substituted phenanthroline complexes of Fe(II) and Ir(II) (*149*), the reduction of cytochrome *c* by ferrohexacyanide (*150*), and the comproportionation–disproportionation equilibrium of flavin mononucleotide (*151*) and 9,10-phenanthrene-quinone-3-sulfonate (*152*).

C. *Others*

Some recent applications have involved the dimerization kinetics of a zinc porphyrin cation radical (*153*), of the 7,7,8,8-tetracyanoquinodimethane anion radical (*154*), of vanadium(IV) tetrasulfophthalocyanine (*155*), of Congo red (*156*), and of a number of meso-substituted water-soluble porphyrins (*157*), of methylene blue (*39*), and of proflavin (*158*); the last two studies were carried out with the laser Raman temperature-jump method.

In the search for models for enzyme–substrate reactions, the interaction of α-cyclodextrin with a variety of dyes, to form inclusion compounds, has been investigated (*159*).

Phospholipids are being used as model membranes to study the rates and mechanisms of membrane-promoted processes (*160–162*). Since one does not deal with homogeneous solutions but rather with heterogeneous systems special care has to be taken (control experiments!) that no artifacts are being measured and that one has reproducible stock solutions. The detection of the relaxation effect by direct absorption spectrometry is not feasible but fluorescence (*160, 161*) or absorption indicators (*160*) can be used. An alternative is to detect changes in light scattering (*162*). Light scattering and fluorescence detection have also been used in a recent study of the self-association of bovine liver glutamate dehydrogenase (*163*); this also appears to be the first study where relaxation *amplitudes* with light scattering detection have been exploited for mechanistic deduction and determination of thermodynamic parameters.

Problems

1. Calculate ΔT_∞ for a Joule heating temperature-jump experiment in methanol solution when the heated cell volume is 0.8 ml, $C = 2 \times 10^{-8}$ F, $V_0 = 25$ kV, and the starting temperature is 20°C.

2. Since the heated cell volume in a temperature-jump cell is not accurately known it is better to determine ΔT_∞ experimentally (calibration of temperature-jump cell). Suggest methods of calibrating a temperature-jump cell and compare your ideas with the methods described by Yapel and Lumry (*12*).

3. Why is the heating time in the Joule heating technique more than three times longer for a 0.1 M NaCl compared to a 0.1 M HCl solution?

4. Assuming the volume of the reaction solution between the two electrodes of a Joule heating temperature-jump cell corresponds to a cube of 1 cm sides, what is τ_d for a 0.1 M NaCl solution if $C = 5 \times 10^{-8}$ F? (Consult the "Handbook of Chemistry and Physics" for conductivities of electrolyte solutions.)

5. When a relaxation experiment is conducted under pseudo-first-order conditions large perturbations are permitted. Show that unless the photomultiplier output is converted into a signal proportional to OD instead of I, the resulting relaxation curve becomes distorted, and more so the higher the OD. Estimate this distortion in experiments where the OD doubles as a consequence of relaxation, i.e., $\overline{OD}^f = 2\overline{OD}^i$, for the cases (a) $\overline{OD}^f = 0.1$, (b) $\overline{OD}^f = 0.2$, (c) $\overline{OD}^f = 0.5$, (d) $\overline{OD}^f = 1.0$.

References

1. M. Eigen, *Discuss. Faraday Soc.* **17**, 194 (1954).
2. G. H. Czerlinski and M. Eigen, *Z. Elektrochem.* **63**, 652 (1959).
3. M. Eigen and L. DeMaeyer, *in* "Technique of Organic Chemistry" (S. L. Friess, E. S. Lewis, and A. Weissberger, eds.), Vol. VIII, part 2, p. 895. Wiley (Interscience), New York, 1963.
4. H. Diebler, Ph.D. Thesis, Göttingen (1960).
5. G. G. Hammes and P. Fasella, *J. Amer. Chem. Soc.* **84**, 4644 (1962).
6. G. H. Czerlinski, *Rev. Sci. Instrum.* **33**, 1184 (1962).
7. G. H. Czerlinski, "Chemical Relaxation." Dekker, New York, 1966.
8. G. C. Kresheck, E. Hamori, G. Davenport, and H. Scheraga, *J. Amer. Chem. Soc.* **88**, 246 (1966).
9. A. Bewick, M. Fleischmann, J. N. Hiddleston, and L. Wynne-Jones, *Discuss. Faraday Soc.* **39**, 146 (1966).
10. R. F. Pasternak, K. Kustin, L. A. Hughes, and E. Gibbs, *J. Amer. Chem. Soc.* **91**, 4401 (1969).
11. T. C. French and G. G. Hammes, *Methods Enzymol.* **16**, 3 (1969).
12. A. F. Yapel, Jr. and R. Lumry, *Methods Biochem. Anal.* **20**, 169 (1971).
13. A. D. Yu, M. D. Waissbluth, and R. A. Grieger, *Rev. Sci. Instrum.* **44**, 1390 (1973).
14. R. Rigler, C.-R. Rabl, and T. M. Jovin, *Rev. Sci. Instrum.* **45**, 580 (1974).
15. R. Rigler and M. Ehrenberg, *Quart. Rev. Biophys.* **6**, 139 (1973).
16. G. H. Czerlinski and A. Weiss, *Appl. Opt.* **4**, 59 (1965).
17. R. Lumry and R. Legare, *Anal. Chem.* **41**, 551 (1969).

18a. B. C. Bennion, L. K. J. Tong, L. P. Holmes, and E. M. Eyring, *J. Phys. Chem.* **73**, 3288 (1969).

18b. J. D. Owen, B. C. Bennion, L. P. Holmes, E. M. Eyring, M. W. Berg, and J. L. Lords, *Biochim. Biophys. Acta* **209**, 77 (1970).

19. L. DeMaeyer, Messanlagen Studiengesellschaft, Göttingen, Germany, 1971.
20. D. N. Hague, "Fast Reactions," Wiley (Interscience), New York, 1971.

21a. M. Eigen and R. Winkler, *in* "The Neurosciences: Second Study Program" (F. O. Schmitt, ed.), p. 685. Rockefeller Univ. Press, New York, 1970.

21b. R. Winkler, Ph.D. Thesis, Max Planck Inst., Göttingen, and Tech. Univ. of Vienna (1969).

22. C. F. Bernasconi, *J. Amer. Chem. Soc.* **92**, 4682 (1970).
23. C. F. Bernasconi, R. G. Bergstrom, and W. J. Boyle, Jr., *J. Amer. Chem. Soc.* **96**, 4643 (1974).
24. C. F. Bernasconi, *J. Amer. Chem. Soc.* **92**, 129 (1970).

25. C. F. Bernasconi and R. H. de Rossi, *J. Org. Chem.* **38**, 500 (1973).
26. D. Riesner, R. Römer, and G. Maass, *Eur. J. Biochem.* **15**, 85 (1970).
27. G. W. Hoffman, *Rev. Sci. Instrum.* **42**, 1643 (1971).
28. G. Ertl and H. Gerischer, *Z. Electrochem.* **65**, 629 (1961).
29. E. F. Caldin and J. E. Crooks, *J. Sci. Instrum.* **44**, 449 (1967).
30. K. J. Ivin, J. J. McGarvey and E. L. Simmons, *Trans. Faraday Soc.* **67**, 97 (1971).
31. A. F. Yapel, Jr. and R. Lumry, *Methods Biochem. Anal.* **20**, 342 (1971).
32a. P. Brumm, F. P. Kilian, and H. Rüppel, *Ber. Bunsenges.* **72**, 1085 (1968).
32b. H. E. Buchwald and H. Rüppel, *J. Phys. E Sci. Instrum.* **4**, 105 (1971).
33a. W. T. Silvast, J. Asay, and E. M. Eyring, Abstracts, 152nd Meeting, Amer. Chem Soc., Div. Phys, Chem., Paper 136 (Sept. 1966).
33b. E. M. Eyring, Final Tech. Rep., Air Force Office Sci. Res. Grant AFOSR-476-66-A (Dec. 1967).
34. H. Hoffmann, E. Yeager and J. Stuehr, *Rev. Sci. Instrum.* **39**, 649 (1968).
35a. D. H. Turner, G. W. Flynn, N. Sutin, and J. V. Beitz, *J. Amer. Chem. Soc.* **94**, 1554 (1972).
35b. D. H. Turner, Ph.D. Thesis Columbia Univ., New York (1972).
36. E. F. Caldin, J. E. Crooks, and B. H. Robinson, *J. Sci. Instrum.* **4**, 165 (1971).
37. H. Koffer, *Ber. Bunsenges. Phys. Chem.* **75**, 1245 (1971).
38. H. Hirohara, K. J. Ivin, J. J. McGarvey, and J. Wilson, *J. Amer. Chem. Soc.* **96**, 4435 (1974).
39. M. M. Farrow, N. Purdie, A. L. Cummings, W. Herrmann, Jr., and E. M. Eyring, *in* "Chemical and Biological Applications of Relaxation Spectrometry" (E. Wyn-Jones, ed.), p. 69. Reidel, Dordrecht-Holland, 1975.
40a. S. Ameen and L. DeMaeyer, *J. Amer. Chem. Soc.* **97**, 1590 (1975).
40b. F. P. Schäfer, *Angew. Chem. Int. Ed.* **9**, 9 (1970).
41. R. Rigler, A. Jost, and L. DeMaeyer, *Exp. Cell Res.* **62**, 197 (1970).
42. E. F. Caldin, M. W. Grant, B. B. Hasinoff, and P. H. Tregloan, *J. Phys. E Sci. Instrum.* **6**, 349 (1973).
43. H. Strehlow and S. Kalarickal, *Ber. Bunsenges.* **70**, 139 (1966).
44. I. Y. Lee and B. Chance, *Anal. Biochem.* **29**, 331 (1969).
45. F. M. Pohl, *Europ. J. Biochem.* **4**, 373 (1968).
46. G. Hui Bon Hoa and F. Travers, *J. Chim. Phys.* 637 (1972).
47. H. C. Spatz and D. M. Crothers, *J. Mol. Biol.* **42**, 191 (1969).
48. T. Y. Tsong and R. L. Baldwin, *J. Mol. Biol.* **63**, 453 (1972).
49. J. B. Johnson and F. B. Llewellyn, *Bell System Tech. J.* **14**, 85 (1935).
50. D. Skoog and D. M. West, "Fundamentals of Analytical Chemistry," p. 642. Holt, New York, 1966.
51. M. Eigen and L. DeMaeyer, *in* "Rapid Mixing and Sampling Techniques in Biochemistry" (B. Chance, R. H. Eigenhardt, Q. H. Gibson, and K. K. Lonberg-Holm, eds.), p. 175. Academic Press, New York, 1964.
52. G. H. Czerlinski, *in* "Rapid Mixing and Sampling Techniques in Biochemistry" (B. Chance, R. H. Eigenhardt, Q. H. Gibson, and K. K. Lonberg-Holm, eds.), p. 183. Academic Press, New York, 1964.
53. J. E. Erman and G. G. Hammes, *Rev. Sci. Instrum.* **37**, 746 (1966).
54. E. J. Faeder, Ph.D. Thesis, Cornell Univ., Ithaca, New York (1970).
55. L. B. Veil, Ph.D. Thesis, Göttingen (1971).
56. Q. H. Gibson, *Methods Enzymol.* **16**, 187 (1969).
57a. J. E. Erman and G. G. Hammes, *J. Amer. Chem. Soc.* **88**, 5607, 5614 (1966).
57b. G. G. Hammes and J. L. Haslam, *Biochemistry* **8**, 1591 (1969).
57c. E. J. del Rosario and G. G. Hammes, *J. Amer. Chem. Soc.* **92**, 1750 (1970).
58. C. F. Bernasconi, *J. Amer. Chem. Soc.* **93**, 6975 (1971).
59. A. Jost, *Ber. Bunsenges. Phys. Chem.* **78**, 300 (1974).
60. E. F. Caldin, M. W. Grant, B. B. Hasinoff, and P. A. Tregloan, *J. Phys. E Sci. Instrum.* **6**, 349 (1973).

61. A. van Horebeek, Ph.D. Thesis, KUL, Louvain, Belgium (1973).
62. Z. A. Schelly and E. M. Eyring, *J. Chem. Ed.* **48**, A639, A695 (1971).
63. J. E. Stewart and P. Lum, *American Laboratory*, p. 91 (Nov. 1969).
64. G. G. Hammes, *in* "Techniques of Chemistry" (G. G. Hammes, ed.), Vol. VI, part 2, p. 147. Wiley (Interscience), New York, 1973.
65. M. Eigen, *Angew. Chem. Int. Ed.* **3**, 1 (1964).
66. M. Eigen, W. Kruse, G. Maass, and L. DeMaeyer, *Prog. React. Kinet.* **2**, 287 (1964).
67. R. P. Bell, "The Proton in Chemistry," 2nd ed. Cornell Univ. Press, Ithaca, New York, 1973.
68. C. D. Ritchie, *in* "Solute-Solvent Interactions" (J. F. Coetzee and C. D. Ritchie, eds.), p. 219. Dekker, New York, 1969.
69. I. Amdur and G. G. Hammes, "Chemical Kinetics." McGraw-Hill, New York, 1966.
70. M. Eigen and W. Kruse, *Z. Naturforsch.* **18B**, 857 (1963).
71. M. C. Rose and J. Stuehr, *J. Amer. Chem. Soc.* **90**, 7205 (1968); **93**, 4350 (1971); **94**, 5532 (1972).
72. F. Hibbert, *J. Chem. Soc. Perkin II* 1862 (1974).
73. C. F. Bernasconi and F. Terrier, *in* "Chemical and Biological Applications of Relaxation Spectrometry" (E. Wyn-Jones, ed.), p. 379. Reidel, Dordrecht-Holland, 1975.
74. M.-L. Ahrens, M. Eigen, W. Kruse, and G. Maass, *Ber. Bunsenges. Phys. Chem.* **74**, 380 (1970).
75. J. Stuehr, *J. Amer. Chem. Soc.* **89**, 2826 (1967).
76. P. J. Dynes, G. S. Chapman, E. Kebede, and F. W. Schneider, *J. Amer. Chem. Soc.* **94**, 6356 (1972).
77. E. F. Caldin and J. E. Crooks, *J. Chem. Soc.* (*B*) 959 (1967).
78a. E. F. Caldin, J. E. Crooks, and D. O'Donnell, *J. Chem. Soc. Faraday Trans. I* **69**, 993 (1973).
78b. E. F. Caldin, J. E. Crooks, and D. O'Donnell, *J. Chem. Soc. Faraday Trans. I* **69**, 1000 (1973).
79. G. D. Burfoot, E. F. Caldin, and H. Goodman, *J. Chem. Soc. Faraday Trans. I* **70**, 105 (1974).
80. K. J. Ivin, J. J. McGarvey, E. L. Simmons, and R. Small, *J. Chem. Soc. Faraday Trans. I* **67**, 104 (1971); **69**, 1016 (1973).
81. J. E. Crooks and B. H. Robinson, *J. Chem. Soc. Chem. Commun.* 979 (1970).
82. J. E. Crooks and B. H. Robinson, *Trans. Faraday Soc.* **66**, 1436 (1970); **67**, 1707 (1971).
83. M. Eigen, *Discuss. Faraday Soc.* **24**, 25 (1957).
84. M. Eigen, *Ber. Bunsenges. Phys. Chem.* **67**, 753 (1963).
85. H. Brintzinger and G. G. Hammes, *Inorg. Chem.* **5**, 1286 (1966).
86. M. Eigen and R. G. Wilkins, *Advan. Chem. Ser.* **49**, 55 (1965).
87. M. Eigen, *Pure Appl. Chem.* **6**, 97 (1963).
88. A. McAuley and J. Hill, *Quart. Rev.* (*London*) **23**, 18 (1963).
89. K. Kustin and J. Swinehart, *Progr. Inorg. Chem.* **13**, 107 (1970).
90. J. E. Crooks, *MTP* (*Med. Tech. Publ.*) *Int. Rev. Sci. Phys. Chem. Ser. One* **9**, 299 (1972).
91. H. Diebler, *Proc. 3rd Symp. Coord. Chem.*, Vol. 2, p. 53. Akademiai Kiado, Budapest, 1971.
92. H. Diebler, M. Eigen, G. Ilgenfritz, G. Maass, and R. Winkler, *Pure Appl. Chem.* **20**, 93 (1969).
93. L. J. Kirschenbaum and K. Kustin, *J. Chem. Soc.*(*A*) 684 (1970).
94. C. M. Frey, J. L. Banyasz, and J. E. Stuehr, *J. Amer. Chem. Soc.* **94**, 9198 (1972).
95. G. G. Hammes and J. I. Steinfeld, *J. Amer. Chem. Soc.* **84**, 4639 (1962).
96. R. F. Pasternak, L. Gibb, and H. Sigel, *J. Amer. Chem. Soc.* **94**, 8031 (1972).
97. L. Michaelis and M. L. Menten, *Biochem. Z.* **49**, 333 (1913).
98. M. Eigen, *Quart. Rev. Biophys.* **1**, 3 (1968).
99. M. Dixon and E. C. Webb, "Enzymes," 2nd ed. Longmans, Green, London, 1964.
100. H. Gutfreund, "An Introduction to the Study of Enzymes," Oxford Univ. Press (Blackwell), London and New York, 1965.
101. G. G. Hammes, *Advan. Protein Chem.* **23**, 1 (1968).

102. G. G. Hammes and P. R. Schimmel, *J. Phys. Chem.* **71**, 917 (1967).
103. G. G. Hammes and P. R. Schimmel, *Enzymes* **2**, 67 (1970).
104. M. Eigen and G. G. Hammes, *Advan. Enzymol.* **25**, 1 (1963).
105. G. G. Hammes, *Accounts Chem. Res.* **1**, 321 (1968).
106. E. J. del Rosario and G. G. Hammes, *J. Amer. Chem. Soc.* **92**, 1750 (1970).
107. G. G. Hammes and J. L. Haslam, *Biochemistry* **8**, 1591 (1969).
108. G. G. Hammes and F. G. Walz, Jr., *J. Amer. Chem. Soc.* **91**, 7179 (1969).
109. F. Guillain and D. Thusius, *J. Amer. Chem. Soc.* **92**, 5534 (1970).
110. A. H. Colen, *J. Biol. Chem.* **245**, 738 (1970).
111. G. Czerlinski, *Science* **132**, 1490 (1960).
112. G. Czerlinski and G. Schreck, *Biochemistry* **3**, 89 (1964).
113. E. J. del Rosario and G. G. Hammes, *Biochemistry* **10**, 716 (1971).
114. M. Eigen, G. Ilgenfritz, and W. Kruse, *Chem. Ber.* **98**, 1623 (1965).
115. M. Eigen, K. Kustin, and H. Strehlow, *Z. Phys. Chem.* (*Frankfurt*) **31**, 140 (1961).
116. M.-L. Ahrens, *Ber. Bunsenges. Phys. Chem.* **72**, 691 (1968).
117. H. G. Schecker and G. Schulz, *Z. Phys. Chem.* (*Frankfurt*) **65**, 221 (1969).
118. M.-L. Ahrens and G. Maass, *Angew. Chem. Int. Ed.* **10**, 72 (1971).
119. L. R. Green and J. Hine, *J. Org. Chem.* **38**, 2801 (1973).
120. D. R. Trentham, C. H. McMurray, and C. I. Pogson, *Biochem. J.* **114**, 19 (1969).
121. M.-L. Ahrens, G. Maass, P. Schuster, and H. Winkler, *J. Amer. Chem. Soc.* **92**, 6134 (1970).
122. H. Heck, A. Iwata, and F. Mah, *J. Chem. Ed.* **50**, 141 (1973).
123. H. Diebler and R. N. F. Thorneley, *J. Amer. Chem. Soc.* **95**, 896 (1973).
124. R. N. F. Thorneley and H. Diebler, *J. Amer. Chem. Soc.* **96**, 1072 (1974).
125. P. Schuster and H. Winkler, *Tetrahedron* 2249 (1970).
126. M. R. Crampton, *Advan. Phys. Org. Chem.* **7**, 211 (1969).
127. M. J. Strauss, *Chem. Rev.* **70**, 667 (1970).
128. C. F. Bernasconi, *MTP* (*Med. Techn. Publ.*) *Int. Rev. Sci. Org. Chem. Ser. One* **3**, 33 (1973).
129. C. F. Bernasconi, *in* "Chemical and Biological Applications of Relaxation Spectrometry" (E. Wyn-Jones, ed.), p. 343. Reidel, Dordrecht-Holland, 1975.
130. C. F. Bernasconi and C. L. Gehriger, *J. Amer. Chem. Soc.* **96**, 1092 (1974).
131. C. F. Bernasconi and F. Terrier, *J. Amer. Chem. Soc.* **97**, 7458 (1975).
132. G. G. Hammes and D. C. Hubbard, *J. Phys. Chem.* **70**, 2889 (1966).
133. D. E. V. Schmeckel and D. M. Crothers, *Biopolymers* **10**, 465 (1971).
134. R. Bittman, *J. Mol. Biol.* **46**, 251 (1969).
135. T. R. Tritton and S. C. Mohr, *Biochem. Biophys. Res. Commun.* **45**, 1240 (1971).
136. D. Riesner and R. Römer, *in* "Physical Properties of Nucleic Acids" (J. Duchesne, ed.), Vol. 2, p. 237. Academic Press, New York, 1973.
137. D. Pörschke and M. Eigen, *J. Mol. Biol.* **62**, 361 (1971).
138. D. Pörschke, O. C. Uhlenbeck, and F. H. Martin, *Biopolymers* **12**, 1313 (1973).
139a. M. Eigen, *in Nobel Symp., 5th* (S. Cleasson, ed.), p. 333. Wiley (Interscience), New York, 1967.
139b. M. Eigen, *in Nobel Symp., 5th* (S. Cleasson, ed.), p. 355. Wiley (Interscience), New York, 1967.
140. M. Eigen, *in* "The Neurosciences" (G. C. Quarton, T. Melnechuck, and F. O. Schmitt, eds.), p. 130. Rockefeller Univ. Press, New York, 1967.
141. D. Pörschke, Ph.D. Thesis, Univ. of Braunschweig (1968).
142. G. W. Hoffmann and D. Pörschke, *Biopolymers* **12**, 1625 (1973).
143. R. J. H. Davies, *J. Mol. Biol.* **63**, 117 (1972).
144. B. C. Bennion, L. K. Tong, L. P. Holmes, and E. M. Eyring, *J. Phys. Chem.* **73**, 3288 (1969).
145. B. C. Bennion and E. M. Eyring, *J. Colloid. Interface Sci.* **32**, 286 (1970).
146. J. Lang and E. M. Eyring, *J. Polym. Sci. A-2* **10**, 89 (1972).
147. H. Hoffmann, *in* "Chemical and Biological Applications of Relaxation Spectrometry" (E. Wyn-Jones, ed.), p. 181. Reidel, Dordrecht-Holland, 1975.

148. H. Diebler, M. Eigen, and P. Matthies, *Z. Naturforsch.* **16B**, 629 (1961).
149. P. Hurwitz and K. Kustin, *Inorg. Chem.* **3**, 823 (1964).
150. K. G. Brandt, P. C. Parks, G. H. Czerlinski, and G. P. Hess, *J. Biol. Chem.* **241**, 4180 (1966).
151. J. H. Swinehart, *J. Amer. Chem. Soc.* **88**, 1056 (1966).
152. M. W. Cheung and J. H. Swinehart, *J. Phys. Chem.* **76**, 1875 (1972).
153. J. H. Fuhrhop, P. Wasser, D. Riesner, and D. Mauzerall, *J. Amer. Chem. Soc.* **94**, 7996 (1972).
154. A. Yamagishi, Y. Iida, and M. Fujimoto, *Bull. Chem. Soc. Japan* **45**, 3482 (1972).
155. R. D. Farina, D. J. Halko, J. H. Swinehart, *J. Phys. Chem.* **76**, 2343 (1972).
156. T. Yasunaga and S. Nishikawa, *Bull. Chem. Soc. Japan* **45**, 1262 (1972).
157. R. F. Pasternack *et al.*, *J. Amer. Chem. Soc.* **94**, 4512 (1973).
158. D. H. Turner, G. W. Flynn, S. K. Lundberg, L. D. Faller, and N. Sutin, *Nature* (*London*) **239**, 215 (1972).
159. F. Cramer, W. Saenger, and H.-Ch. Spatz, *J. Amer. Chem. Soc.* **89**, 14 (1967).
160. G. G. Hammes and D. E. Tallman, *J. Amer. Chem. Soc.* **92**, 6042 (1970).
161. H. Träuble, *Naturwissenschaften* **58**, 277 (1971).
162. J. D. Owen, P. Hemmes, and E. M. Eyring, *Biochim. Biophys. Acta* **219**, 276 (1970).
163. D. Thusius, P. Dessen, and J.-M. Jallon, *J. Mol. Biol.* **92**, 413 (1975).

Chapter 12 | Pressure-Jump Methods

12.1 Principles and Apparatus

The pressure-jump methods which have been authoritatively reviewed (*1–4*), are based on the fact that chemical equilibria display a more or less marked pressure dependence. This dependence is given by the well-known thermodynamic relationship

$$\left(\frac{\partial \ln K}{\partial p}\right)_T = -\frac{\Delta V}{RT} \tag{12.1}$$

where ΔV is the standard molar volume change of the reaction, conveniently expressed in milliliters, p is the pressure in atmospheres, and R is the gas constant (82 ml·atm mole^{-1} deg^{-1}).

Analogous to Eq. 11.2 we can write

$$\frac{\Delta K}{K} = -\frac{\Delta V}{RT}\Delta p \tag{12.2}$$

for small perturbations. In practice the number of reactions for which ΔV is significant enough to make the pressure-jump method attractive is limited. Reactions most likely to be accompanied by relatively large volume changes are ion-forming or ion-consuming equilibria because of solvation and electrostriction effects. For example, for $CH_3COOH \rightleftharpoons CH_3COO^- + H^+$, $\Delta V = -10.9$ ml (*2*).

In a typical experiment $\Delta p = 60$ atm. Assuming $\Delta V = 10$ ml, one obtains $\Delta K/K = 2.45 \times 10^{-2}$ at room temperature, that is, a 2.45% change in K. This calculation shows that the effects to be expected are relatively small even under favorable conditions and make the use of the very sensitive conductometric method (see Section 12.1.2) advisable. In fact this is the most frequently used detection technique.

In some situations the equilibrium displacement may be larger than that calculated on the basis of Eq. 12.2. This is because strictly speaking the pressure jump occurs adiabatically, thus also causing a concomitant temperature change in the solution. The adiabatic pressure dependence of K is given by

$$\left(\frac{d \ln K}{\partial p}\right)_S = -\frac{\Delta V}{RT} + \frac{\alpha}{\rho c_P} \frac{\Delta H}{RT} + \beta_S \Delta \nu \tag{12.3}$$

where ΔV and ΔH are the molar volume and enthalpy changes, respectively, c_P is the specific heat at constant pressure, α is the thermal expansion coefficient, β_S is the isentropic compressibility, and $\Delta \nu$ is the change in number of moles for the reaction. In aqueous solution at 25°C a pressure jump of 50 atm is accompanied by a temperature jump of about 0.1°, with $\Delta V = 30$ ml/mole, $\Delta H = 10$ kcal/mole, and $\Delta \nu = 1$ the relative magnitude of the three terms in Eq. 12.3 would be about 20:2:1 (*5*). However, in nonaqueous solvents the last two terms may become much more important; in a favorable case, where all terms have the same sign, the equilibrium displacement may therefore be considerably augmented by these terms.

12.1.1 The Pressure Jump

The pressure jump can be produced either by a sudden application of pressure to the system or by a sudden release of pressure. However, the time in which the pressure change can be achieved is considerably longer than the heating time in a typical temperature-jump apparatus. Ljunggren and Lamm (*6*) described the first experimental realization of a pressure-jump apparatus. Their sample cell was connected to a nitrogen tank. By rapidly opening the valve a pressure increase to 150 atm could be obtained in about 50 msec; chemical relaxation was monitored conductometrically.

In 1959 Strehlow *et al.* (*5, 7*) described a greatly superior pressure-jump apparatus. Their conductivity cell, which contains the reaction solution, and a reference cell (Section 12.1.2), are enclosed under an inert liquid such as xylene in an autoclave. Using compressed air the reaction and reference solutions can be pressurized to about 60 atm. By the blow of a steel needle a thin metal disk used to close the autoclave is punctured whereby the pressure is released within about 60 μsec. Apart from the much shorter rise time a further advantage of this design is that kinetics can be studied at the conventional pressure of 1 atm rather than at high pressures.

Somewhat modified versions of Strehlow and Becker's apparatus have been described by several workers (*2–4, 8, 9*). In the newest design by Knoche and Wiese (*9*) special care has been given to the problem of reducing mechanical disturbances following the pressure jump, which is often the primary cause for large dead times. Thus relaxation times as short as 30 μsec can now be measured. Figure 12.1 shows the basic features of the apparatus by Knoche and Wiese (*9*) which is described in the next section.

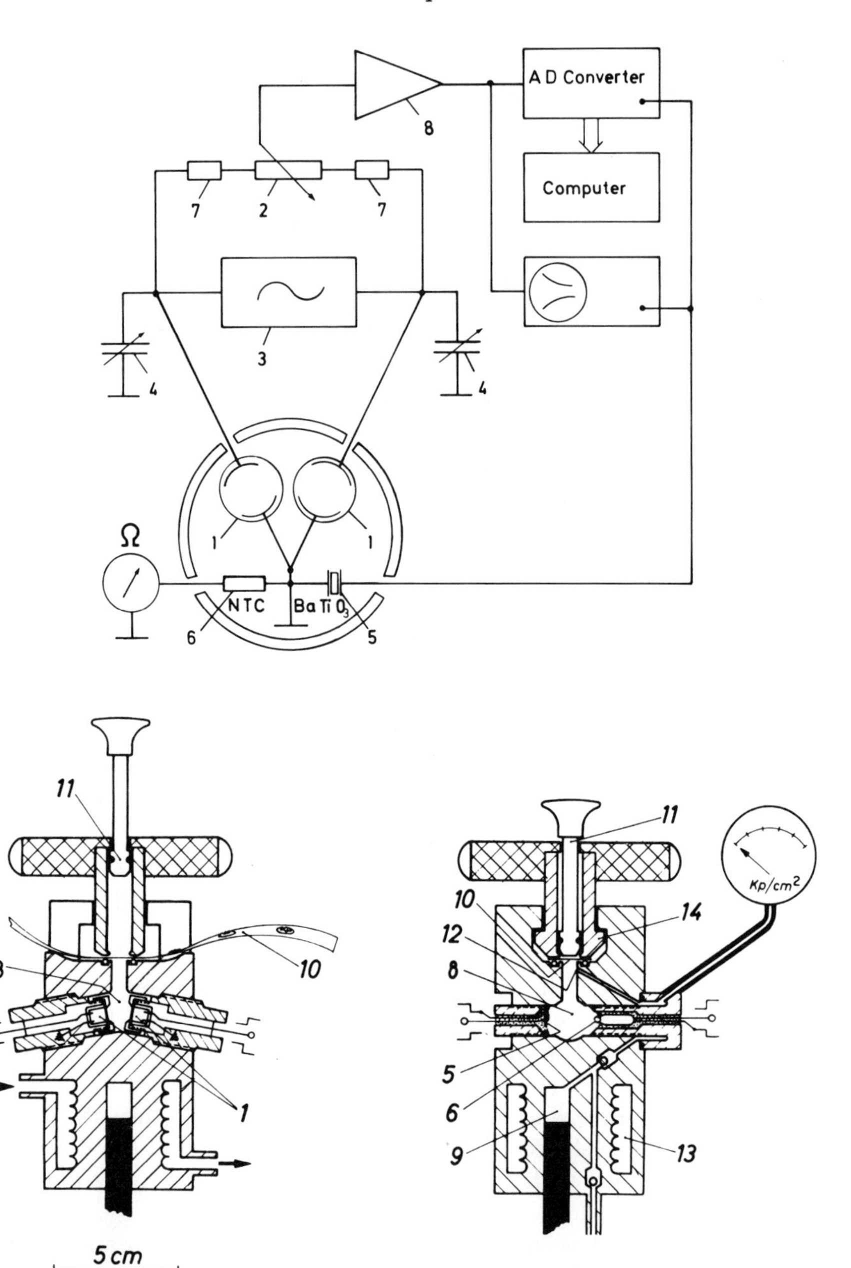

FIGURE 12.1. Schematic diagram and sectional views of the autoclave of the pressure-jump apparatus by Knoche and Wiese (*9*): 1, conductivity cells; 2, potentiometer; 3, 40 kHz generator for Wheatstone bridge; 4, tunable capacitors; 5, piezoelectric capacitor; 6, thermistor; 7, 10-turn helipot for tuning bridge; 8, experimental chamber; 9, pressure pump; 10, rupture diaphragm; 11, vacuum pump; 12, pressure inlet; 13, heat exchanger; 14, bayonet socket. (Reproduced by permission of Marcel Dekker, Inc.)

12.1.2 Conductometric Detection

Since the equilibrium displacement following a pressure jump is usually quite small, the very sensitive conductometric detection method is commonly used. The specific conductivity, σ (in Ω^{-1} cm^{-1}), of an electrolyte solution is given by

$$\sigma = \frac{\mathbf{F}}{1000} \sum c_j |z_j| u_j = \frac{\mathbf{F}}{1000} \rho \sum m_j |z_j| u_j \tag{12.4}$$

where **F** is the Faraday constant (96,493 C/equiv), z_j the valence of ion j, c_j the molar and m_j the molal concentration of ion j, u_j its electrical mobility (in cm^2 V^{-1} sec^{-1}), and ρ the density of the solution. It is advantageous to express the concentrations in terms of molality in order to separate the concentration changes caused by chemical relaxation from those caused by volume or density changes. For small perturbations we can write (*1*)

$$\Delta\sigma = \frac{\mathbf{F}}{1000} \left(\rho \sum |z_j| u_j \, \Delta m_j + \rho \sum |z_j| m_j \, \Delta u_j + \sum |z_j| m_j u_j \, \Delta\rho \right) \tag{12.5}$$

The first term on the right-hand side of the equation corresponds to chemical relaxation. The other terms are "physical effects," i.e., the change of the ionic mobility and the density as a consequence of the pressure and temperature changes; they follow the pressure jump essentially instantaneously. The effect of the temperature change can be particularly disturbing because the temperature starts equilibrating with the surroundings, thus leading to a drift during the measurement of the relaxation effect. However, this problem can be eliminated by using a reference cell filled with a nonrelaxing solution but having the same temperature dependence of the conductivity as the sample cell. If the reference cell is subjected to the same pressure jump and the conductivity changes of both the sample and the reference cell are measured in a suitable bridge arrangement as a difference signal, the temperature effect can be fully compensated for.

A typical modern setup is that of Knoche and Wiese (*9*) shown in Fig. 12.1. The sample and reference cell are both inside the pressure autoclave and form two arms of an alternating current Wheatstone bridge whose off-balance voltage is read with an oscilloscope. The bridge must of course be operated at a frequency which is greater than the reciprocal value of the relaxation time to be measured, typically 40 kHz or higher.

It is evident that conductometric detection is at its best when the sample solution contains only those ions involved in the reaction under study. Addition of other electrolytes, for example an acid–base buffer, decreases the sensitivity of the method. Nevertheless Strehlow and Wendt (*5*) were able to obtain a reasonable precision even in a situation where 98% of the total conductivity was due to inert electrolytes and only 2% due to the reactants.

When the addition of salts or buffer is necessary one may be able to minimize the undesired extra conductivity by using salts with ions of low mobility, for example tetraalkylammonium ions instead of sodium or potassium ions. In some cases amphoteric reagents such as amino acids near their isolectric point might prove suitable as buffers and eliminate the problem altogether.

12.1.3 Data Evaluation and Other Detection Methods

The evaluation of the relaxation time(s) from the oscilloscope traces is the same as for the temperature-jump method. The vertical axis on the oscilloscope measures the relative change in resistance in the two conductivity cells. Since the bridge is driven by an alternating current, a typical oscilloscope picture looks like a ringing pattern whose frequency is the oscillating frequency of the bridge and whose envelopes correspond to the exponential relaxational decay, as shown in Fig. 12.2.

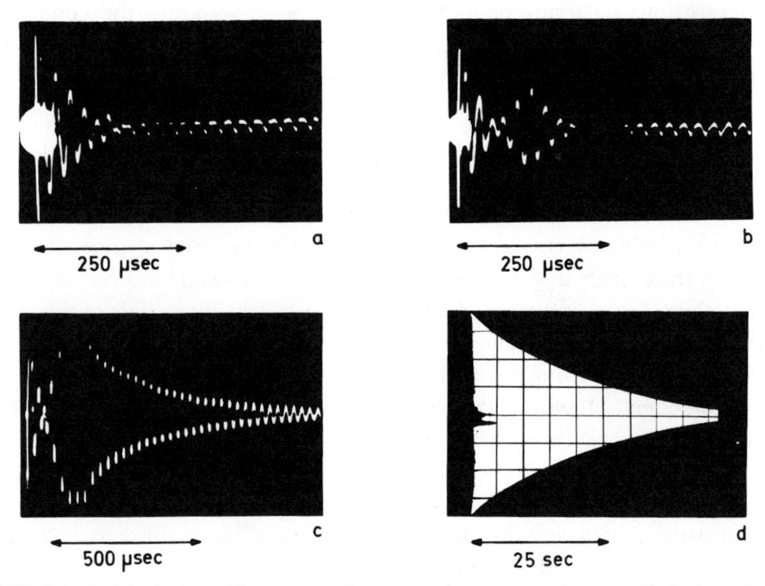

FIGURE 12.2. Typical oscillograms of pressure-jump experiments. Relative change in conductivity for pressure-jumps of 130 atm in solutions of 0.05 *M* $InCl_3$, pH 3.25. (a) At 60°C showing only pressure decay. (b) At 27.5°C, $\tau = 50 \pm 15$ μsec. (c) At 0.7°C, $\tau = 215 \pm 10$ μsec. (d) Solution of 0.1 *M* α-ketoglutaric acid, pH 1.69, at 1°C, $\tau = 25.8$ sec. [From Knoche and Wiese (*9*), by permission of Marcel Dekker, Inc.]

Instead of evaluating the data from the oscillograms, interfacing with a computer system is possible. A sampling technique that allows the evaluation of a large number of repeated experiments (thus increasing the S/N ratio) coupled with an on-line computer processing has been described recently (*10*).

A pressure-jump apparatus based on Strehlow's design but using spectrophotometric detection has also been described (*11*). Helisch and Knoche (*12*) have also developed a technique where chemical relaxation is monitored with a rapidly responding thermistor (NTC resistor) which measures increases or decreases in temperature depending on the thermicity of the reaction. Relaxation times in the range of 0.01–100 sec can be measured with this device.

12.2 Shock Wave Apparatus

A technique closely related to the pressure-jump method makes use of a shock wave in a shock tube. In a design by Jost (*13*) the sample cell is mounted to the lower end of a 3 m high-pressure tube (Fig. 12.3). The tube is divided into a high-

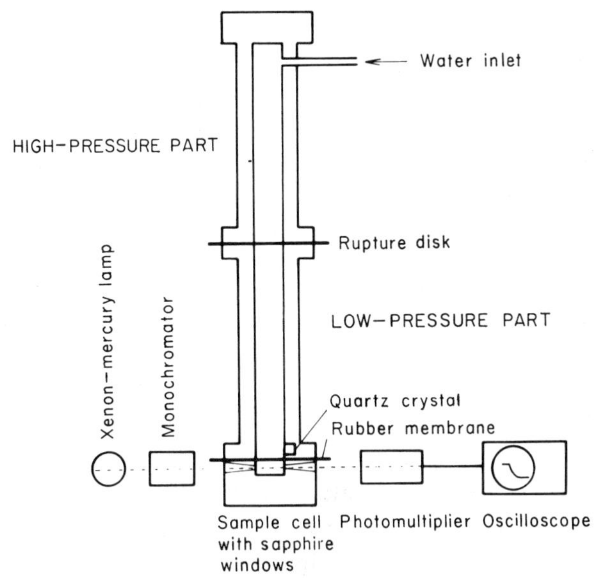

FIGURE 12.3. Schematic diagram of shock-wave apparatus, after Jost (*13*). [From Knoche (*3*), by permission of John Wiley & Sons.]

and a low-pressure compartment of equal size by a rupture disk. The tube is filled with water whereby the high-pressure compartment is pressurized to about 1000 atm until the rupture disk breaks. This produces a shock wave that propagates down into the low-pressure region at the speed of sound; at the same time a "dilution wave" expands into the high-pressure compartment. Since the velocity of sound in a liquid increases with pressure, the leading front of the pressure pulse expanding into the low-pressure region steepens as it travels through the liquid. By the time the wave reaches the sample cell, which is covered by a rubber membrane, it produces a 1000 atm pressure pulse with a rise time of about 1 μsec and a duration of about 2 msec; the duration of the pulse is determined by the time it takes for the dilution wave to be reflected at the upper end of the tube and reach the sample cell. Thus relaxation times in the range of 1 msec to 2 μsec can be measured by this technique which is coupled with an optical detection system.

The main disadvantage of the method is that rates are measured at very high pressures, the cumbersome nature of the experiments, and the inherent problems in designing an apparatus able to withstand pressures of 1000 atm. A somewhat simpler design has been described by Hoffmann *et al.* (*14, 15*). They use the more sensitive conductometric readout which allows the use of smaller pressures (about

150 atm). Instead of using water in both compartments of the pressure tube, alcohol or hexane is used in the low-pressure, gas in the high-pressure chamber. A still more refined shock-wave apparatus able to measure relaxation times as short as 10^{-7} sec is under development (*16*).

12.3 Applications

Despite the simplicity of the pressure-jump method (excluding the shock-wave technique), it has so far not found as wide an application as the temperature-jump method. Apart from the nonavailability of commercial equipment there are several more fundamental reasons for this. One is the short time limit, currently at 50–30 μsec as compared to 1 μsec in most commercial temperature-jump methods. Another constraint is the restriction to reactions with relatively large ΔV; these are much rarer than the reactions having an appreciable ΔH. However, the possibility of coupling a reaction whose ΔV is small to a process with a large ΔV should be kept in mind, in the same way as temperature-jump amplitudes can be enhanced as described in Section 7.1.2.

12.3.1 Metal Complexes

There are a number of metal complex formations for which the processes of Eq. 11.39 or 11.40 are slow enough for study by the pressure-jump method; in fact the study of such reactions has been the principal application of the method. The most widely quoted example is the intensive and very careful study of the formation of $BeSO_4$ (*17, 18*). The reciprocal relaxation time at 25°C is in the 300–400 sec^{-1} range and shows the typical curvilinear dependence according to Eq. 11.42, indicating that the rapidly formed outer sphere complex ③ is present at detectable concentrations.

$$\underset{①}{Be^{2+} + SO_4^{2-}} \underset{}{\overset{\text{fast}}{\rightleftharpoons}} \underset{③}{Be^{2+}(H_2O)SO_4^{2-}} \overset{\text{slow}}{\rightleftharpoons} \underset{④}{BeSO_4}$$

In conjunction with ultrasonic measurements, pressure-jump relaxation *amplitudes* were also exploited to determine reaction volumes (*18*).

Further examples include complex formation of other metal ions with sulfate ions (*19*), complexation of nickel ion with malonate (*20–23*), tartrate (*20*), succinate (*22*), glycolate (*24*), lactate (*25*), or phthalate (*26*), the complexation of Mg^{2+}, Mn^{2+}, Ni^{2+}, Co^{2+}, Cu^{2+}, and Zn^{2+} with *m*-benzenedisulfonate ion in methanol (*27*), the complexation of lanthanides with oxalate (*28*), and the complexation of Co^{2+}, Fe^{2+}, Mn^{2+}, Ni^{2+} with chloride or thiocyanate ion in methanol (*29*) and in dimethylsulfoxide (*30*).

Examples illustrating the application of the shock-wave method include complexation of Mg^{2+} with malonate or tartrate (*15*), and of Ni^{2+} with malonate (*31*), with chloride and SCN^- in dimethylsulfoxide (*32*), and with various ligands in acetonitrile (*33*).

For reviews and discussions of some of the results see Hoffmann *et al.* (*4*), Knoche (*3*), and Crooks (*34*).

12.3.2 Hydration of Carbonyl Compounds

The hydration kinetics of several carbonyl compounds such as pyruvic acid (*35*), ninhydrin (*36*), ketoglutaric acid (*37*), and propionaldehyde (*12*) have been reported. Since in neutral or acidic solution no ions are produced or consumed in the hydration reaction

$$\mathrm{>C{=}O + H_2O \rightleftharpoons >C(OH)_2}$$

the conductometric detection might appear unsuited in such investigations.

However, in the cases of ketocarboxylic acids advantage can be taken of the difference in the acidity of the hydrated and nonhydrated forms. Let us illustrate this for pyruvic acid (*35*). The reaction scheme is

$$\begin{array}{ccc}
① \ \mathrm{CH_3C({=}O)COOH} & \underset{\text{slow}}{\overset{\mathrm{H_2O}}{\rightleftharpoons}} & \mathrm{CH_3C(OH)_2COOH} \ ② \\
K_a^1 \updownarrow \text{fast} & & \text{fast} \updownarrow K_a^2 \\
③ \ \mathrm{CH_3C({=}O)COO^- + H^+} & \underset{\text{slow}}{\overset{\mathrm{H_2O}}{\rightleftharpoons}} & \mathrm{CH_3C(OH)_2COO^- + H^+} \ ④
\end{array}$$

Since $K_a^1 > K_a^2$, a shift in the hydration equilibrium ① ⇌ ② say to the left will produce a net increase in the conductivity of the system because of a net gain in the concentration of ionic species through the dissociation of $CH_3COCOOH$ which is not compensated for by an equal loss of ions through process ④ → ②. Note that this is similar to the use of an acid–base indicator in conjunction with optical detection methods, only that here the change in c_{H^+} is detected directly and no special indicator is needed.

In the case of the hydration of propionaldehyde the relaxation was detected thermometrically (*12*).

12.3.3 Other Applications

The first application of the pressure-jump technique involved the study of the hydration of CO_2 to form H_2CO_3 (*6*). In the last few years processes in micellar systems have also been investigated by the pressure-jump (*38–42*) or the shock-wave method (*42*). For more on micelle systems see Sections 10.4 and 11.5.6. A few measurements of the conformational changes in proteins have been reported (*2*).

Problems

1. Discuss the relative merits of spectrophotometric, fluorimetric and conductometric detection in the pressure-jump and temperature-jump apparatus.

2. What is the percent change in K brought about in a shock-tube experiment in aqueous solution with a pressure pulse of 1000 atm when $\Delta V = -3$ cm^3/mole, at 25°C?

References

1. M. Eigen and L. DeMaeyer, *in* "Technique of Organic Chemistry" (S. L. Friess, E. S. Lewis, and A. Weissberger, eds.), Vol. VIII, part 2, p. 895. Wiley (Interscience), 1963.
2. M. T. Takahashi and R. A. Alberty, *Methods Enzymol.* **16**, 21 (1969).
3. W. Knoche, *in* "Techniques of Chemistry" (G. G. Hammes, ed.), Vol. VI, part 2, p. 187. Wiley (Interscience), New York, 1973.
4. H. Hoffmann, J. Stuehr, and E. Yeager, *in* "Chemical Physics of Ionic Solutions" (B. E. Conway and R. G. Barradas, eds.), p. 255. Wiley, New York, 1966.
5. H. Strehlow and H. Wendt, *Inorg. Chem.* **2**, 6 (1963).
6. S. Ljunggren and O. Lamm, *Acta Chem. Scand.* **12**, 1834 (1958).
7. H. Strehlow and M. Becker, *Z. Elektrochem.* **63**, 457 (1959).
8. G. Macri and S. Petrucci, *Inorg. Chem.* **9**, 1009 (1970).
9. W. Knoche and G. Wiese, *Chem. Instrum.* **5**, 91 (1973–1974).
10. M. Križan and H. Strehlow, *Chem. Instrum.* **5**, 99 (1973–1974).
11. D. E. Goldsack, R. E. Hurst, and J. Love, *Anal. Biochem.* **28**, 273 (1969).
12. J. Helisch and W. Knoche, *Ber. Bunsenges. Phys. Chem.* **75**, 951 (1971).
13. A. Jost, *Ber. Bunsenges. Phys. Chem.* **70**, 1057 (1966).
14. H. Hoffmann and E. Yeager, *Rev. Sci. Instrum.* **39**, 1151 (1968).
15. G. Platz and H. Hoffmann, *Ber. Bunsenges. Phys. Chem.* **76**, 491 (1972).
16. H. Hoffmann, cited in Knoche (*3*).
17. H. Strehlow and W. Knoche, *Ber. Bungenges Phys. Chem.* **73**, 427 (1969).
18. W. Knoche, C. A. Firth, and D. Hess, *Advan. Mol. Relax. Proc.* **6**, 1 (1974).
19. C. Kalidas, W. Knoche, and D. Papadopoulos, *Ber. Bunsenges. Phys. Chem.* **75**, 106 (1971).
20. H. Hoffmann, *Ber. Bunsenges. Phys. Chem.* **69**, 916 (1965).
21. D. Saar, G. Macri, and S. Petrucci, *J. Inorg. Nucl. Chem.* **33**, 4227 (1971).
22. J. L. Bear and C. T. Lin, *J. Phys. Chem.* **72**, 2026 (1968).
23. H. Hoffmann and J. Stuehr, *J. Phys. Chem.* **70**, 955 (1966).
24. S. Harada, Y. Okuue, H. Kan, and T. Yasunaga, *Bull. Chem. Soc. Japan* **47**, 769 (1974).
25. S. Harada, H. Tanabe, and T. Yasunaga, *Bull. Chem. Soc. Japan* **46**, 3125 (1973).
26. S. Harada, H. Tanabe, and T. Yasunaga, *Bull. Chem. Soc. Japan* **46**, 2450 (1973).
27. G. Macri and S. Petrucci, *Inorg. Chem.* **9**, 1009 (1970).
28. A. J. Graffeo and J. L. Bear, *J. Inorg. Nucl. Chem.* **30**, 1577 (1968).
29. F. Dickert, P. Fischer, H. Hoffmann, and G. Platz, *J. Chem. Soc. Chem. Commun.* 106 (1972).
30. F. Dickert and H. Hoffman, *Ber. Bunsenges. Phys. Chem.* **74**, 641 (1970).
31. H. Hoffmann and E. Yeager, *Ber. Bunsenges. Phys. Chem.* **74**, 642 (1970).
32. F. Dickert and H. Hoffmann, *Ber. Bunsenges. Phys. Chem.* **75**, 1320 (1971).
33. H. Hoffmann, T. Janjic, and R. Sperati, *Ber. Bunsenges. Phys. Chem.* **78**, 223 (1974).
34. J. E. Crooks, *MTP* (*Med. Tech. Publ.*) *Int. Rev. Sci. Phys. Chem. Ser. One* **9**, 229 (1972).
35. H. Strehlow, *Z. Elektrochem.* **66**, 392 (1962).
36. W. Knoche, H. Wendt, M.-L. Ahrens, and H. Strehlow, *Collect. Czech. Chem. Commun.* **31**, 388 (1966).
37. J. Jen and W. Knoche, *Ber. Bunsenges. Phys. Chem.* **73**, 539 (1969).

38. K. Takeda and T. Yasunaga, *J. Colloid Interface Sci.* **45**, 406 (1973).
39. T. Yasunaga, K. Takeda, N. Tatsumoto, and H. Uehara, *in* "Chemical and Biological Applications of Relaxation Spectrometry" (E. Wyn-Jones, ed.), p. 143. Reidel, Dordrecht-Holland, 1975.
40. U. Herrmann and M. Kahlweit, *Ber. Bunsenges. Phys. Chem.* **77**, 1119 (1973).
41. T. Janjic and H. Hoffmann, *Z. Phys. Chem.* (*NF*) **86**, 322 (1973).
42. H. Hoffmann, *in* "Chemical and Biological Applications of Relaxation Spectrometry" (E. Wyn-Jones, ed.), p. 181. Reidel, Dordrecht-Holland, 1975.

Chapter 13 | The Electric Field-Jump Method

13.1 Principles

The application of an electric field to a solution has several effects. If the solution contains dipolar species, they will tend to orient their dipole axes with the direction of the applied field. Polarizable molecules will undergo a deformation known as dielectric polarization. If the solution contains free ions, the electric field induces a migration of these ions, thereby producing an electric current. Furthermore, and more relevant to our present discussion, the electric field can have an effect on chemical equilibria.

For example an equilibrium such as

$$A \rightleftharpoons B$$

where B has a large, A a small, dipole moment, is shifted toward B in an electric field. Typical cases are the conversion of the neutral form of an amino acid into its zwitterionic form, the helix–coil transitions in polypeptides and polynucleotides, or other conformational changes in biopolymers. Reactions of higher molecularity where reactants and product(s) have different dipole moments are subject to the same effects, e.g., the association of carboxylic acids to form hydrogen-bonded dimers, or the association of nucleotide bases.

Equilibria involving ions

$$A^+ + B^- \rightleftharpoons AB \tag{13.1}$$

are often even more sensitive to the application of an electric field; the field induces a shift toward producing more ions. This is known as the dissociation field effect (DFE) or the "second Wien effect" (*1*).

In principle the effect of an electric field on chemical equilibria can be described by the thermodynamic relation

$$\left(\frac{\partial \ln K}{\partial |E|}\right)_{p,T} = \frac{\Delta M}{RT} \tag{13.2}$$

where E is the field strength (in volts per centimeter), and ΔM is the molar change in the macroscopic electric moment or the "molar polarization." For nonionic systems the relaxation theory is in fact based on this thermodynamic approach just as for the temperature-jump and pressure-jump methods. In the case of ionic equilibria it must be realized that one can never reach a true thermodynamic equilibrium in the presence of an electric field because of the field-induced flow of the ions. However, the DFE can still be used as a means of perturbing ionic equilibria as long as one restricts oneself to very short electric pulses.

13.1.1 Ionic Equilibria

The theory of the DFE has been developed by Onsager (*2*). For fields of moderate strength one can write

$$\left(\frac{\partial \ln K}{\partial |E|}\right)_{p,T} = \frac{z_A u_A - z_B u_B}{u_A + u_B} \frac{|z_A z_B| e_0^3}{8\pi\epsilon\epsilon_0 (kT)^2} \qquad (13.3)$$

where z_A and z_B are the valences of the ions, u_A and u_B are the mobilities of the ions (defined as the mechanical mobility which is related to the diffusion coefficient D by $u_j = D_j/kT$), $e_0 = 1.60 \times 10^{-19}$ A sec is the charge of the electron, ϵ is the dielectric constant of the medium, $\epsilon_0 = 8.85 \times 10^{-14}$ F/cm is the absolute permittivity in vacuo, and $k = 1.38 \times 10^{-23}$ W sec deg^{-1} is the Boltzmann constant.

For a univalent electrolyte Eq. 13.3 simplifies to

$$\left(\frac{\partial \ln K}{\partial |E|}\right)_{p,T} = \frac{e_0^3}{8\pi\epsilon\epsilon_0 (kT)^2} = \frac{9.64}{\epsilon T^2} \qquad (13.4)$$

For small perturbations we can write

$$\frac{\Delta K}{K} = \frac{9.64}{\epsilon T^2} |\Delta E| \qquad (13.5)$$

In an aqueous solution ($\epsilon = 78$) at 25°C, K changes by about 14% if a field of 100 kV/cm is applied. Smaller fields are required for achieving comparable shifts in less polar solvents.

Nonionic equilibria can also be perturbed by the DFE if they are coupled to a rapid ionic equilibrium. A possible scheme is

$$A^- + H^+ \underset{}{\overset{\text{fast}}{\rightleftharpoons}} AH \rightleftharpoons BH \qquad (13.6)$$

where the (slow) equilibrium $AH \rightleftharpoons BH$ is coupled to an acid–base equilibrium. This is the same principle as coupling a temperature-independent equilibrium to a strongly temperature-dependent one. An interesting example where this possibility was exploited is the study of the helix-coil transition of poly-α,L-glutamic acid (*3–5*); dissociation of protons from the side chains increases the electric charge on the polypeptide which in turn induces a transition from the helix to the coil form.

13.1.2 Dipole Equilibria

To the best of our knowledge the electric field pulse method has not been applied to nonionic equilibria except when coupled to an ionic equilibrium. The principal problem is that the equilibrium displacements by the electric field are very small unless very high field intensities are used. However, stationary relaxation methods based on the same principle have been developed and are discussed in Chapter 16.

13.2 Experimental Techniques

As mentioned in the introduction a side effect of the electric field when applied to ionic solutions is to induce an electric current. This leads to the dissipation of energy as heat through frictional losses by the moving ions. At the field strengths necessary to cause an appreciable DFE (50–100 kV/cm) this heating effect is a very serious problem if the field is applied for a long time. In fact, even pure water would start to boil if subjected to a field of 100 kV/cm for more than 1 sec (*6*). We recognize of course that this heating effect is the principle of the Joule heating temperature-jump method. From the above it is clear that only short pulses can be used and the reaction solution should contain as low an electrolyte concentration as possible.

13.2.1 Rectangular Pulse Method

In this method, also known as the square pulse method, a rectangular electric pulse (see also Section 8.3.4) is applied to the electrodes of the sample cell. The rise time, i.e., the time to establish a constant value of the electric field, is typically on the order of 20 nsec or less (see Section 13.2.3).

Two types of experiments can be visualized. In the first chemical relaxation is studied while the field is on. In this mode, only relaxation times shorter than the pulse duration can be measured and the field must be carefully controlled so as to have a constant value throughout the duration of the pulse. In view of the discussed heating effect, pulse durations of more than about 30 μsec are hardly acceptable. Even for such short pulses some heating cannot be avoided altogether. This means that a linear temperature increase is superimposed on the rectangular field pulse which may slightly distort the relaxation signal.

A more desirable mode of operation is to monitor the relaxation after the field is switched off again. Here the problems of keeping the field constant and of Joule heating do not exist, and much longer relaxation times can be measured. However, the earlier experimental realizations of the rectangular pulse technique did not allow this second mode of operation (see Section 13.2.3).

13.2.2 Damped Harmonic Impulse Method

In this method, which is also known as the amplitude dispersion method, a damped harmonic oscillation is used as the impulse. This can be achieved by special adjustments of the inductance and capacity of the circuit. One makes the damping so large that in fact only one "oscillation" occurs. For a very rapid reaction the adjustment of the chemical equilibrium to the "oscillating" field is practically instantaneous, but for less rapid reactions there is a time lag. This is shown in Fig. 13.1. The relaxation time can be determined from the dependence

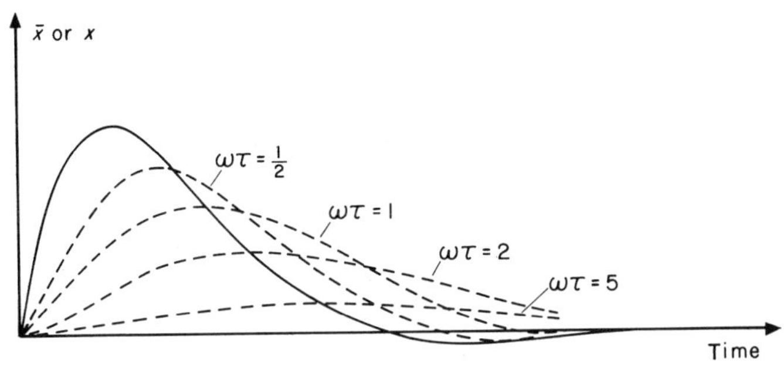

FIGURE 13.1. Relaxational response to a strongly damped harmonic oscillation; (———) = $\bar{x}(t)$, (- - -) = $x(t)$. [From Eigen and DeMaeyer (*8*), by permission of John Wiley & Sons.]

of the time lag on the oscillating frequency; thus this is quite similar to the situation in stationary relaxation methods (see Section 8.4) only that there the perturbation is an *un*damped harmonic oscillation. For a mathematical treatment and more details see Eigen *et al.* (*7*, *8*).

The main disadvantage of this method is that several experiments (time lag determined as a function of the oscillating frequency) must be performed in order to evaluate one relaxation time. In contrast to this, the rectangular pulse methods allow the relaxation time to be determined in a single experiment; currently the amplitude dispersion method is no longer used much and our further discussion will be restricted to the rectangular pulse methods.

13.2.3 Electric Pulse and Detection Systems

In the design by Eigen and DeMaeyer (*8*, *9*) the discharge of a high-voltage capacitor was used to generate the field pulse. The sample cell was built into a special high-field Wheatstone bridge arrangement for conductometric detection of the concentration changes. In this arrangement the voltage of the capacitor could be switched to the sample cell within about 0.1 μsec by firing a spark gap. After a few microseconds the capacitor could be shortened to ground by a second spark gap which brought about the sudden collapse of the field. Since the Wheatstone bridge was driven by the perturbing pulse itself, the measuring voltage in the bridge dropped to zero the moment the field collapsed, making it impossible to measure relaxation in the switched-off mode. However, optical detection systems have now been developed where this problem does not arise (*10–12*).

More recently, by discharging a coaxial delay cable rather than a capacitor, considerably shorter relaxation times have been achieved (*10*, *11*, *13*, *14*). The most advanced equipment appears to be the one by Ilgenfritz (*10*, *11*); it has a rise time of a few nanoseconds and a variable pulse duration of 1.5–20 μsec. However, there are some disadvantages associated with coaxial cables; although the *rise* time of the pulse can be made very short, the *fall* time is on the order of several hundred nanoseconds (*12*), which makes it impossible to measure relaxation times shorter than about 300 nsec at zero field. In response to this problem, Eyring's group (*12*) has now developed an instrument in which a low-inductance 0.005 μF plastic capacitor charged to 60 kV produces 30–52 kV high-voltage pulses of a duration up to 29 μsec with rise and fall times of less than 22 nsec.

As far as optical detection methods are concerned, it was shown in Section 11.2.1 that the sensitivity becomes worse for very short relaxation times because of an unfavorable S/N ratio. Since the electric field-jump technique is primarily designed for very rapid processes, the S/N ratio can become a serious problem. According to Eq. 11.17 S/N is proportional to the square root of the intensity of the sampling light beam. Considerably enhanced light intensities and thus enhanced S/N ratios have recently been achieved by using a pulsed light source which yields an intense square wave flash of light with a duration of 0.1–10 msec (*15*).

13.3 Applications

Relaxation times as short as about 10^{-8} sec, and as long as about 10^{-4} sec (measured at zero field) can be determined by the electric field-jump technique. This is an important time range in fast reaction kinetics and would seem to make this method a very useful and popular one. However, as a participant at a recent summer school on relaxation techniques remarked, "The reviews written since 1967 that consider this technique and the descriptions of equipment modifications approach in number those papers actually devoted to new chemical results obtained by this method" (*16*).

There are several reasons for this.

1. Construction of an electric field jump apparatus is more complicated and expensive than that of the more versatile temperature-jump apparatus.
2. The application is restricted to ion-producing and ion-consuming reactions, or reactions that can be coupled to an ionic equilibrium such as in 13.6.
3. Only solutions of very low conductivity can be used. This is a serious handicap because in many systems like 13.1 the equilibrium constant is small, and thus AB is only produced when A^+ and B^- are present at relatively high concentrations which leads to a high conductivity.
4. For the very fast reactions the older and better established ultrasonic techniques are usually used instead.

Early applications of the electric field jump method involved the determination of rates of proton transfers between weak bases and H^+, or weak acids and OH^-, e.g., $CH_3COO^- + H^+ \rightleftharpoons CH_3COOH$. These measurements, which have been reviewed (*8*, *17*) numerous times (see also Sections 11.5.1 and 15.4.1), were carried out mostly with the amplitude dispersion method.

One of the earliest applications of the rectangular pulse method was the measurement of the "prototype" of all proton transfer reactions (*9*):

$$H^+ + OH^- \xrightleftharpoons{1.4 \times 10^{11} M^{-1} \sec^{-1}} H_2O \tag{13.7}$$

The method continues to be used for the determination of proton transfer rates (*18–23*); from our discussion in Section 11.5.1 we see that the electric field-jump method complements the temperature-jump method for acids and bases whose equilibration rates are beyond the reach of this latter, i.e., systems with p*K*-values lower than 4.7 or higher than 9.3.

The binding of ligands to metal ions (*24–27*) including metal ion hydrolyses (*28–30*) has also been the subject of several studies. The kinetics of helix–coil transitions of polypeptides have recently aroused a great deal of interest (see also Section 10.2) and several relaxation techniques, among them the electric field-jump method (*3–5*), have been applied. The mechanism by which the conformational transition can be induced by the dissociation field effect has been briefly discussed at the end of Section 13.1.1; the same mechanism can also be exploited for detecting the relaxation. That is, the electric field jump not only induces a perturbation of the helix–coil equilibrium via the perturbation of the acid–base equilibria in the side chains, but the relaxation of the helix–coil equilibrium in turn induces an adjustment of the acid–base equilibria which can be monitored conductometrically (*3, 4*) or spectrophotometrically (*5*) with an acid–base indicator.

Other applications of biological interest include studies of the cooperative binding of toluidine blue to poly-α,L-glutamic acid (*31*), the mechanism of oxygen binding to hemoglobin (*32*), electric field-induced transitions in hemoglobin (*33*), and the ionization of the iron-bound water of methemoglobin and metmyoglobin in basic solution (*34*).

An interesting phenomenon in this latter study is that the relaxation after the field jump is toward a state of smaller OD in the case of metmyoglobin, but of increased OD for methemoglobin (both experiments with field on). This was taken to indicate that the electric field shifts the equilibrium toward the acidic form for metmyoglobin, but toward the basic form for methemoglobin (*34*). A possible explanation (*34*) is that in metmyoglobin the acidic form is positively charged so that the reaction with OH^- forms a neutral species whereas in methemoglobin the acidic form is negatively charged so that the reaction with OH^- produces a

$$\left(\gtrless Fe^{III}\cdot H_2O\right)^+ + OH^- \rightleftharpoons\ \gtrless Fe^{III}OH + H_2O \tag{13.8}$$

doubly negative species; the DFE would shift the first rection to the left, the second to the right.

$$\left(\gtrless Fe^{III}\cdot H_2O\right)^- + OH^- \rightleftharpoons \left(\gtrless Fe^{III}OH\right)^{2-} + H_2O \tag{13.9}$$

Although there may be other reasons (*34*) for the opposite signs in the relaxation effects, this example illustrates how the electric field-jump method can provide structural information which is otherwise hard to obtain. Note that the temperature-jump method, which in this example would in fact have been fast enough to determine the relaxation times, would not reveal this information; the reaction enthalpies for both 13.8 and 13.9 have the same sign (*34*), leading to the same sign in the temperature-jump relaxation effect.

Problem

1. What is the percent change in the concentration of A^{2+} in an equilibrium

$$A^{2+} + B^{2-} \rightleftharpoons AB$$

if an electric field of 85 kV/cm is applied at 25 °C for $[A^{2+}]_0 = 10^{-5}\ M$, $[B^{2-}]_0 = 3 \times 10^{-4}\ M$, $K = 3 \times 10^3\ M^{-1}$, (a) in aqueous solution, (b) in ethanol.

References

1. M. Wien, *Phys. Z.* **32**, 545 (1931).
2. L. Onsager, *J. Chem. Phys.* **2**, 599 (1934).
3. T. Yasunaga, T. Sano, K. Takahashi, H. Takenaka, and S. Ito, *Chem. Lett.* 405 (1973).
4. T. Yasunaga, Y. Tsuji, T. Sano, and H. Takenaka, *in* "Chemical and Biological Applications of Relaxation Spectrometry" (E. Wyn-Jones, ed.), p. 493. Reidel, Dordrecht-Holland, 1975.
5. A. L. Cummings and E. M. Eyring, *in* "Chemical and Biological Applications of Relaxation Spectrometry" (E. Wyn-Jones, ed.), p. 505. Reidel, Dordrecht-Holland, 1975.
6. L. DeMaeyer, *Methods Enzymol.* **16**, 80 (1969).
7. M. Eigen and J. Schoen, *Z. Elektrochem.* **59**, 483 (1955).
8. M. Eigen and L. DeMaeyer, *in* "Techniques of Organic Chemistry" (S. L. Friess, E. S. Lewis, and A. Weissberger, eds.), Vol. VIII, part 2, p. 895. Wiley (Interscience), New York, 1963.
9. M. Eigen and L. DeMaeyer, *Z. Elektrochem.* **59**, 986 (1955).
10. G. Ilgenfritz, Ph.D. Thesis, Georg August Univ., Göttingen (1966).
11. G. Ilgenfritz, *in* "Probes of Structure and Function of Macromolecules and Membranes" (B. Chance, ed.), Vol. 1, p. 505. Academic Press, New York, 1971.
12. S. L. Olsen, R. L. Silver, L. P. Holmes, J. J. Auborn, P. Warrick, Jr., and E. M. Eyring, *Rev. Sci. Instrum.* **42**, 1247 (1971).
13. D. T. Rampton, L. P. Holmes, D. L. Cole, R. P. Jensen, and E. M. Eyring, *Rev. Sci. Instrum.* **38**, 1637 (1967).
14. B. R. Staples, D. J. Turner, and G. Atkinson, *Chem. Instrum.* **2**, 127 (1969).
15. S. L. Olsen, L. P. Holmes, and E. M. Eyring, *Rev. Sci. Instrum.* **45**, 859 (1974).
16. E. M. Eyring, *in* "Chemical and Biological Applications of Relaxation Spectrometry" (E. Wyn-Jones, ed.), p. 85. Reidel, Dordrecht-Holland, 1975.
17. M. Eigen, *Angew. Chem. Int. Ed.* **3**, 1 (1964).
18. L. P. Holmes, A. Silzars, D. L. Cole, L. D. Rich, and E. M. Eyring, *J. Phys. Chem.* **73**, 737, 738 (1969).
19. J. J. Auborn, P. Warrick, Jr., and E. M. Eyring, *J. Phys. Chem.* **75**, 2488, 3026 (1971).
20. P. Warrick, Jr., J. J. Auborn, and E. M. Eyring, *J. Phys. Chem.* **76**, 1184 (1972).
21. R. G. Sandberg, G. H. Henderson, R. D. White, and E. M. Eyring, *J. Phys. Chem.* **76**, 4023 (1972).
22. D. J. Lentz, J. E. C. Hutchins, and E. M. Eyring, *J. Phys. Chem.* **78**, 1021 (1974).

23. M. W. Massey, Jr. and Z. A. Schelly, *J. Phys. Chem.* **78**, 2450 (1974).
24. M. Eigen and E. M. Eyring, *Inorg. Chem.* **2**, 636 (1963).
25. H. Diebler, M. Eigen, G. Ilgenfritz, G. Maass, and R. Winkler, *Pure Appl. Chem.* **20**, 93 (1969).
26. M. M. Farrow, N. Purdie, and E. M. Eyring, *Inorg. Chem.* **13**, 2024 (1974).
27. H. Hirohara, K. J. Ivin, J. J. McGarvey, and J. Wilson, *J. Amer. Chem. Soc.* **96**, 4435 (1974).
28. D. L. Cole, L. D. Rich, J. D. Owen, and E. M. Eyring, *Inorg. Chem.* **8**, 682 (1969).
29. L. D. Rich, D. L. Cole, and E. M. Eyring, *J. Phys. Chem.* **73**, 713 (1969).
30. P. Hemmes, L. D. Rich, D. L. Cole, and E. M. Eyring, *J. Phys. Chem.* **74**, 2859 (1970); **75**, 929 (1971).
31. T. Yasunaga, H. Takenaka, T. Sano, and Y. Tsuji, *in* "Chemical and Biological Applications of Relaxation Spectrometry" (E. Wyn-Jones, ed.), p. 467. Reidel, Dordrecht-Holland, 1975.
32. T. M. Schuster and G. Ilgenfritz, *in* "Symmetry and Function of Biological Systems at the Macromolecular Level" (*Nobel Symp. II*) (A. Engstrom and B. Strandberg, eds.), p. 181. Wiley, New York, 1969.
33. G. Ilgenfritz and T. M. Schuster, *in* "Probes of Structure and Function of Macromolecules and Membranes" (B. Chance, T. Yonetani, and A. S. Mildvan, eds.), Vol. II, p. 399. Academic Press, New York, 1971.
34. G. Ilgenfritz and T. M. Schuster, *in* "Probes of Structure and Function of Macromolecules and Membranes" (B. Chance, T. Yonetani, and A. S. Mildvan, eds.), Vol. II, p. 299. Academic Press, New York, 1971.

Chapter 14 | *The Concentration-Jump Method*

14.1 Principles

The success of the relaxation methods in measuring very fast reactions may induce one to forget that the principles of relaxation kinetics apply just as well to slow reactions, particularly the attractive possibility of linearizing the rate equations when dealing with small perturbations. There are various ways to generate small concentration jumps. Some of the more obvious ones are

1. Dilution of the reaction mixture with a small amount of solvent.
2. Adding a solution which is more (or less) concentrated in the reactants.
3. Adding a different solvent in order to change the solvent composition.
4. Adding base or acid to induce a pH jump.
5. Changing the ionic strength by a salt jump.

As long as the equilibrium is dependent on the nature of the solvent, the pH, or the ionic strength, methods 3, 4, and 5 are applicable to equilibria of any molecularity. On the other hand, methods 1 and 2 are only applicable to equilibria that can be shifted by the change of the total concentration of the solutes; for example, this is the case for $A + B \rightleftharpoons C$ but not for $A \rightleftharpoons B$.

All concentration-jump experiments involve a mixing process. Consequently the shortest relaxation time which is measurable depends on the mixing time. For very slow reactions (τ in the order of minutes or more) ordinary mixing will be satisfactory. With relatively cheap devices, mixing times can easily be reduced to about 2 sec (*1*, *2*); in fact Swinehart (*1*) has suggested the use of such devices in a simple concentration-jump experiment for a junior level physical chemistry laboratory.

14.2 The Stopped-Flow Technique

If shorter mixing times are required, flow techniques must be used, most typically the stopped-flow method. A stopped-flow apparatus is basically a fast mixing device that is connected to a fast detection system similar to the ones used with the temperature-jump method. The method is a well-established one and is routinely used for measuring rates of "relatively" fast reactions, the bulk of them essentially irreversible. Its use in concentration-jump relaxation experiments thus constitutes only a very small fraction of all the applications. A detailed discussion of this technique and of the general applications is therefore quite beyond the scope of this book and we shall restrict ourselves to a few very brief remarks; for more detailed accounts, including other flow techniques, the reviews by Roughton and Chance (*3*), Gibson (*4*), Chance (*5*), or Caldin (*6*) (elementary treatment) should be consulted.

The instruments that are currently used in a majority of laboratories, including a commercial product (Durrum–Gibson), are modifications of a design by Gibson and Milnes (*4*, *7*). Figure 14.1 shows the basic features of it. The pistons of two

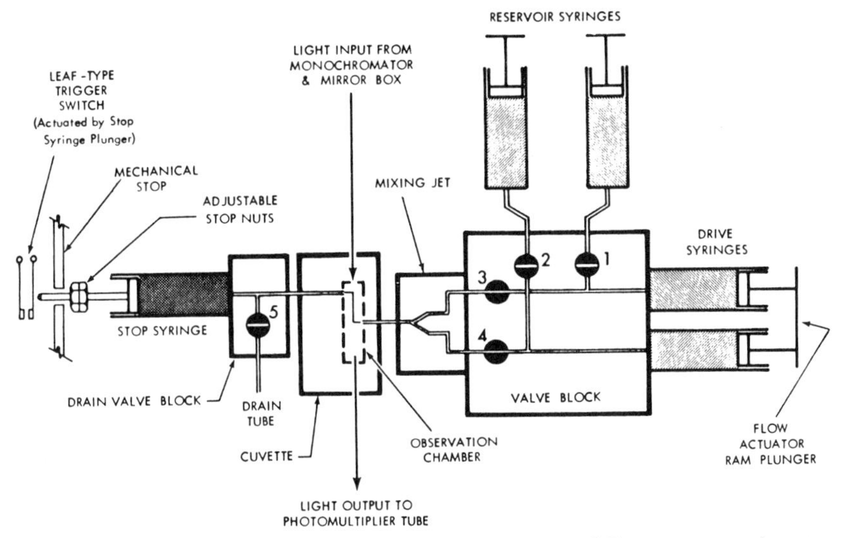

FIGURE 14.1. Schematic diagram of Durrum–Gibson stopped-flow apparatus (courtesy of Durrum Instrument Corp.).

"drive syringes," which contain the two solutions to be mixed, are pushed simultaneously by a pneumatic actuator or ram plunger. This drives the solutions into a specially designed mixing chamber or "mixing jet" where they are thoroughly mixed within about 1–2 msec. From there the mixture flows into a spectrophotometric or fluorimetric observation cell and finally into the "stop syringe"; when the plunger of this latter syringe reaches a mechanical stop, the flow is instantaneously stopped and a trigger switch activates the oscilloscope sweep.

14.3 Applications

In the context of this chapter we shall deal mainly with small concentration jumps produced by the methods enumerated in Section 14.1. Note, however, that in principle every kinetic experiment that is initiated by the mixing of two or more reagents could be called a "concentration jump"; in the conventional expriment, say for the reaction $A + B \rightleftharpoons C$, the "jump" is typically from zero up to the initial reactant concentrations and thus constitutes a large perturbation. If the reaction can be conducted under pseudo-first-order conditions, the formalism of relaxation kinetics nevertheless applies (see Section 2.1) and performing the experiments in the "conventional way" is often superior to the small jump technique because the concentration changes during relaxation (relaxation amplitude) are larger and thus more easily measured with high precision. Examples where the stopped-flow technique was applied in this way have been discussed in Sections 4.5.2 and 7.3.

When pseudo-first-order conditions are not feasible, for example in systems such as

$$2A \rightleftharpoons B \tag{14.1}$$

$$2A \rightleftharpoons B + C \tag{14.2}$$

small jumps are necessary for the linearization of the rate equations. Examples illustrating the solvent-jump technique on equilibria such as 14.1 are the study of the dimerization of the tetrasodium salt of cobalt(II) 4,4′,4″,4‴-tetrasulfophthalocyanine (*8*), and the dimerization of rhodamine-type dyes (*9, 10*). The experiments were carried out as dilution jumps in the stopped-flow apparatus (*8, 9*) and some in a continuous-flow apparatus (*10*).

Another example of solvent-jump experiments that are particularly appropriate to the problems at hand has been reported by Robinson *et al.* (*11*) in a system in which micellization occurs. By diluting a solution containing detergent at a concentration above the cmc to one where the concentration is slightly below the cmc the process of disintegration of the micelles could be measured.

A different type of solvent-jump experiment for studying the helix I $\rightleftharpoons$ helix II interconversion of poly-L-proline was reported (*12*). Here the greater stability of the helix I in aliphatic alcohols but greater stability of the helix II in trifluoroethanol, water, or benzyl alcohol was exploited by inducing the concentration jump with a sudden change in solvent composition.

Method 2, i.e., adding a solution with a slightly different reactant concentration, was used in the study of the chromate–dichromate reaction (*1, 2*).

$$2HCrO_4^- \rightleftharpoons Cr_2O_7^{2-} + H_2O$$

A pH-jump study on a system which in principle corresponds to Eq. 14.2 but is more complex owing to various acid–base equilibria, namely the electron transfer reactions of an azaviolene (*13*), scheme 4.107, has already been discussed in Section 4.5.1. A problem of considerable practical importance in such investigations is that of selecting the appropriate size of the pH jump so that the rate equations can still be linearized but the relaxation effect is large enough to be easily measured. This problem has been treated in some detail in Section 5.2 for a simplified version of scheme 4.107, namely reaction 5.23.

Another case where pH jumps were found to be useful involved the study of the stepwise addition of methanol to an olefin (*14*).

Examples involving salt jumps have been reported in the study of the conformational change of a synthetic DNA from one type of double helix to another (*15*), the cooperative binding of ethidium bromide to the same DNA (*16*), and the salt-induced micelle formation of sodium dodecyl sulfate (*11*).

References

1. J. H. Swinehart, *J. Chem. Ed.* **44**, 524 (1967).
2. J. H. Swinehart and G. W. Castellan, *Inorg. Chem.* **3**, 278 (1964).
3. F. J. W. Roughton and B. Chance, *in* "Technique of Organic Chemistry" (S. L. Friess, E. S. Lewis, and A. Weissberger, eds.), Vol. VIII, part 2, p. 703. Wiley (Interscience), New York, 1963.
4. Q. H. Gibson, *Methods Enzymol.* **16**, 187 (1969).
5. B. Chance, *in* "Techniques of Chemistry" (G. G. Hammes, ed.), Vol. VI, part 2, p. 5. Wiley (Interscience), New York, 1973.
6. E. F. Caldin, "Fast Reactions in Solution." Wiley, New York, 1964.
7. Q. H. Gibson and L. Milnes, *Biochem. J.* **91**, 161 (1964).
8. Z. A. Schelly, R. D. Farina, and E. M. Eyring, *J. Phys. Chem.* **74**, 617 (1970).
9. M. M. Wong and Z. A. Schelly, *J. Phys. Chem.* **78**, 1891 (1974).
10. Z. A. Schelly and M. M. Wong, *Int. J. Chem. Kinet.* **6**, 687 (1974).
11. B. H. Robinson, N. C. White, C. Mateo, K. J. Timmins, and A. James, *in* "Chemical and Biological Applications of Relaxation Spectrometry" (E. Wyn-Jones, ed.), p. 201. Reidel, Dordrecht-Holland, 1975.
12. D. Winklmair, J. Engel, and V. Ganser, *Biopolymers* **10**, 721 (1971).
13. C. F. Bernasconi, R. G. Bergstrom, and W. J. Boyle, Jr., *J. Amer. Chem. Soc.* **96**, 4643 (1974).
14. C. F. Bernasconi and W. J. Boyle, Jr., *J. Amer. Chem. Soc.* **96**, 6070 (1974).
15. F. M. Pohl and T. M. Jovin, *J. Mol. Biol.* **67**, 375 (1972).
16. F. M. Pohl, T. M. Jovin, W. Baehr, and J. J. Holbrook, *Proc. Nat. Acad. Sci. U.S.* **69**, 3805 (1972).

Chapter 15 | Ultrasonic Techniques

15.1 Principles

The ultrasonic or acoustic techniques belong to the category of stationary relaxation techniques; they are the oldest of all commonly used relaxation methods. As the plural implies there are a number of such techniques which differ in many important details. The specific technique applicable to a given problem primarily depends on the frequency range of interest. Taken as a group the ultrasonic techniques span a frequency range of about 10^4–10^{10} Hz or even higher; this allows the measurement of relaxation times in the range of 10^{-5}–10^{-11} sec. This is an important time range; the very short times are not covered by any of the transient techniques and thus the ultrasonic methods nicely complement the transient techniques. The acoustic techniques have been, after the temperature-jump method, the second most frequently used type of relaxation technique; however, it should be pointed out at the outset that they are much more difficult to use than the temperature-jump method.

The theory of the varied phenomena associated with the propagation of sound waves in liquids is rather complicated and beyond the scope of this book; for more in-depth treatments the reader is referred to Herzfeld and Litovitz (*1*), Stuehr and Yeager (*2*), or Blandamer (*3*) (elementary). We shall concentrate on the formalism relevant to the attenuation of a sound wave by chemical relaxation phenomena and emphasize qualitative understanding; some equations will be introduced without complete derivation. For a more elaborate mathematical treatment Eigen and DeMaeyer (*4*) should be consulted.

15.1.1 Attenuation of a Sound Wave

Soundwaves are propagated adiabatically through a medium, giving rise to small periodic pressure, temperature, and density fluctuations; the pressure variations are on the order of 0.03 atm (*5*), the temperature variations on the order of 0.002° (*5*), and are thus much smaller than in a typical pressure-jump or temperature-jump experiment, respectively.

The fundamental equation describing the propagation of a planar sound wave in a liquid is the so-called wave equation. For pressure fluctuations it takes the form

$$\frac{\partial^2 p}{\partial t^2} = v^2 \frac{\partial p^2}{\partial x^2} \tag{15.1}$$

where p is the alternating sound pressure in the liquid, v is the propagation velocity or "phase velocity" of the wave, x is the distance traveled by the wave, and t is the time. The solution of Eq. 15.1 for a nonattenuated sinusoidal wave is given by

$$p(x, t) = p_0 \exp[i\omega(t - x/v)] \tag{15.2}$$

where p_0 is the amplitude and $\omega = 2\pi f$ is the angular frequency, where f is the frequency in hertz.

In practice there is always a drop in the amplitude of the sound pressure with increasing distance. Like many other attenuation processes (e.g., Beer's law), the attenuation of a sound wave obeys an exponential law

$$p(x, t) = p_0 \exp(-\alpha x) \exp[i\omega(t - x/v)] \tag{15.3}$$

where α is known as the attenuation or absorption coefficient; it has the dimensions of $(\text{length})^{-1}$ and is usually expressed in nepers per centimeter (Np/cm).

Equations analogous to 15.1–15.3 can be formulated, e.g., for changes in temperature or density.

Frequently a dimensionless absorption coefficient is introduced which is defined by

$$\mu = \alpha\lambda = \alpha v/f = 2\pi\alpha v/\omega \tag{15.4}$$

where $\lambda = v/f$ is the wavelength.

Let us now turn to the three main factors which contribute to the attenuation of the sound wave.

1. Absorption of energy by the solvent molecules as a consequence of shearing motions, the so-called viscous energy losses.
2. Heat losses because all liquids conduct heat to some extent.

These two sources of sound absorption are often referred to as "classical" absorption (α_{class}); note that the first is much more important than the second.

3. The third factor is energy absorption due to chemical or physical relaxation processes. This occurs whenever the equilibrium position can be affected by changes in pressure ($\partial \ln K/\partial p = -\Delta V/RT$) and/or changes in temperature ($\partial \ln K/\partial T = \Delta H/RT^2$). This so-called nonclassical or relaxational absorption is of course our main concern. However, in order to be able to interpret sound absorption data correctly it is important to know how classical absorption depends on frequency and to have an idea of the magnitude of the effect; α_{class} is given by

$$\alpha_{\text{class}} = \frac{2\pi^2 f^2}{v^3 \rho} \left(\frac{4}{3}\eta_s + \frac{\gamma - 1}{c_p}\chi \right) \tag{15.5}$$

where ρ is the density, η_s is the shear viscosity, χ is the thermal conductivity, c_P is the specific heat at constant pressure, $\gamma = c_P/c_V$ with c_V being the specific heat at constant volume, while the other symbols have the same meaning as before.

An important feature of Eq. 15.5 is that α_{class} depends on the *square* of the sound frequency, making $\alpha_{class}/f^2 = \text{const}$. As we shall see, it is therefore common practice to report sound absorption data in terms of α/f^2. The magnitude of α_{class}/f^2 in pure water at 25°C is 8.5×10^{-17} sec²/cm (*6*) while the measured total α/f^2 of the same solvent is 22.5×10^{-17} sec²/cm (*7, 8*). The difference between the two numbers is due to the relaxation of the equilibrium between the various water structures (*7, 8*). For more on solvent absorption see Section 15.1.4.

15.1.2 Sound Absorption Due to Chemical Relaxation

Assume that the equilibrium

$$A + B \rightleftharpoons C + D \tag{15.6}$$

is associated with a positive ΔV and consider what would happen in a pressure-jump experiment. If the pressure is increased, the equilibrium constant is decreased according to $\partial \ln K/\partial p = -\Delta V/RT$, which induces the equilibrium to shift toward the left. To put it into the terminology of Le Chatelier's principle, the pressure increase drives the system toward a state of smaller volume (A + B), a situation quite comparable to the compression of a gas, an analogy we will use below.

If the pressure rise time is very short compared to the chemical relaxation time, and the pressure is maintained at its high value for a time long compared to the relaxation time, the equilibrium will shift by its maximum possible amount, at a rate given by

$$\tau^{-1} = k_1(\bar{c}_A + \bar{c}_B) + k_{-1}(\bar{c}_C + \bar{c}_D)$$

This corresponds to the rectangular step discussed in Section 8.3.1. If, on the other hand, the pressure rises very much slower than relaxation, the equilibrium adjusts itself at a rate given by the rate of pressure increase, but still shifts by the maximum possible amount.

Let us now consider the energy balance in these two processes. In both cases the "compression" of the system (driving the equilibrium to the left) requires pressure–volume work ($\int p\, dv$) to be performed. For the rapid pressure jump followed by slow relaxation (slow volume contraction), the amount of work is a maximum and is given by the shaded area of Fig. 15.1a. For the slow jump we have a situation approaching the ideal of a "reversible compression" in the thermodynamic sense of the word (volume contraction keeps pace with pressure increase); thus the work needed is at a minimum and is given by the shaded area of Fig. 15.1c.

Next assume the pressure is released again and restored to its original value. If the pressure release is very rapid, the situation shown in Fig. 15.1b prevails, i.e., little work is returned by the system; if the pressure drops slowly, the situation corresponds to a "reversible" volume expansion with a maximum amount of work

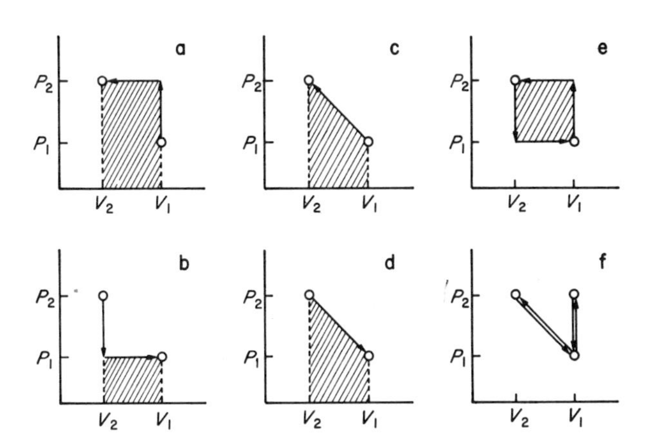

FIGURE 15.1. Pressure–volume work balance in various types of pressure-jump experiments. For explanations see text.

returned by the system (Fig. 15.1d). The overall result of such a cycle is that for the rapid pressure jumps there is a maximum *net* loss of work energy which is dissipated as heat (Fig. 15.1e); for the slow jumps there is no change in the work balance (Fig. 15.1f, diagonal arrows).

It is easily visualized that for intermediate situations where the rise and fall time of the pressure is comparable to the relaxation time, some net work energy loss occurs which amounts to anything between zero and the maximum amount. Also, if the pressure falls again before the equilibrium has time to fully adjust to the high pressure, the equilibrium shift will be less than the maximum possible amount, with a correspondingly smaller loss and gain in work.

A further type of experiment would be to have a very fast rise in pressure which is immediately followed by a very fast return to the original conditions. Here the equilibrium has no time to react to the pressure changes at all, no shifts and no volume changes occur, and thus no work is involved, as shown in Fig. 15.1f (vertical arrows). Similar considerations apply to the energy balance in analogous cycles of temperature jumps.

The situations described above reflect roughly what happens in a sound absorption experiment except that the pressure and temperature fluctuations in an ultrasonic study are sinusoidal and take place in small volume elements rather than in the bulk solution. Thus at very low frequencies ($\omega \ll \tau^{-1}$) the equilibrium adjusts itself instantaneously to the fluctuations (Fig. 8.5); the energy absorbed in the first half of the cycle is returned during the second half and no net sound absorption occurs. At very high frequencies ($\omega \gg \tau^{-1}$) the system cannot follow the fluctuations at all, and again the result is no absorption of energy. It is only at intermediate frequencies ($\omega \sim \tau$) that absorption occurs; the mathematical analysis will show that this absorption, as measured by μ (Eq. 15.4), has the same frequency dependence as σ_{im} (Eq. 8.58).

15.1.3 Mathematical Analysis

The relevant physical property for the formal description of the interaction of a sound wave with a liquid is the adiabatic compressibility. It is defined as

$$\kappa_S = -\frac{1}{V}\left(\frac{\partial V}{\partial p}\right)_S = \frac{1}{\rho}\left(\frac{\partial \rho}{\partial p}\right)_S \tag{15.7}$$

where S is the entropy. The sound velocity in the fluid is directly related to κ_S by

$$v^2 = 1/\rho\kappa_S \tag{15.8}$$

The adiabatic compressibility of any fluid can be written as the sum of two terms

$$\kappa_S = \kappa_\infty + \kappa' \tag{15.9}$$

where κ' is a frequency-dependent term that vanishes for very high frequencies ($\omega \gg \tau^{-1}$) since at very high frequencies the volume changes induced by the shift of the chemical equilibria cannot keep pace with the pressure fluctuations; thus κ_∞ is the limiting compressibility at very high frequencies. Note that usually $\kappa' \ll \kappa_\infty$.

Since the chemical contribution to the compressibility is directly related to the position of the chemical equilibrium as expressed by the concentration variable Δc_j or x (Eq. 1.9 or 1.12), the mathematical problem of finding the dependence of κ' on the frequency of the pressure fluctuations is the same as that of finding the frequency dependence of x. The forcing function to which κ' has to adjust can be defined, in analogy to $\bar{x}$, as the chemical contribution to the compressibility which prevails when the system does have time to adjust to the pressure changes ($\omega\tau \ll 1$); it is usually symbolized by κ^{ch}. Thus, in analogy to Eq. 8.52, 8.54, and 8.55, we have

$$\kappa' = \sigma\kappa^{\text{ch}} = \kappa^{\text{ch}}\left(\frac{1}{1 + \omega^2\tau^2} - \frac{i\omega\tau}{1 + \omega^2\tau^2}\right) \tag{15.10}$$

It can be shown *(4, 6, 7)* that the relaxational absorption per wavelength (Eq. 15.4) is proportional to the imaginary part of κ' and is given by

$$(\alpha\lambda)^{\text{ch}} = \mu^{\text{ch}} = \frac{\pi\kappa^{\text{ch}}}{\kappa_0}\,\frac{\omega\tau}{1 + \omega^2\tau^2} \tag{15.11}$$

with

$$\kappa_0 = \kappa_\infty + \kappa^{\text{ch}} \approx \kappa_\infty \tag{15.12}$$

Thus μ^{ch} has the same frequency dependence as σ_{im} (Fig. 8.6), with a maximum at $\omega\tau = 1$ given by

$$(\mu^{\text{ch}})_{\max} = \pi\kappa^{\text{ch}}/2\kappa_0 \tag{15.13}$$

In systems characterized by several relaxation times, Eqs. 15.9 and 15.10 become

$$\kappa_S = \kappa_\infty + \sum_j \kappa_j' = \kappa_\infty + \sum \sigma_j\kappa_j^{\text{ch}} \tag{15.14}$$

and

$$\mu^{\mathrm{ch}} \approx \sum_j \frac{\pi \kappa_j^{\mathrm{ch}}}{\kappa_\infty} \frac{\omega \tau_j}{1 + \omega^2 \tau_j^2} \tag{15.15}$$

There are two alternative ways to express sound absorption data. One that is particularly popular with U.S. scientists is to take the quantity α^{ch}/f^2, which according to Eq. 15.4 is equivalent to μ^{ch}/vf. Thus, with Eq. 15.11 and $f = \omega/2\pi$ we obtain

$$\frac{\alpha^{\mathrm{ch}}}{f^2} = \frac{2\pi^2 \tau \kappa^{\mathrm{ch}}}{\kappa_0 v} \frac{1}{1 + \omega^2 \tau^2} = A \frac{1}{1 + \omega^2 \tau^2} \tag{15.16}$$

where A is a constant under a given set of experimental conditions; A is the relaxation amplitude or "relaxation strength." The *total* absorption can then be expressed by

$$\frac{\alpha}{f^2} = A \frac{1}{1 + \omega^2 \tau^2} + B \tag{15.17}$$

or, if there are several relaxation effects, by

$$\frac{\alpha}{f^2} = \sum_j \left(A_j \frac{1}{1 + \omega^2 \tau_j^2} \right) + B \tag{15.18}$$

where B is the background absorption ($B = \alpha/f^2$ when all $\omega^2 \tau_j^2 \gg 1$); B includes classical absorption (Eq. 15.5) and usually other effects associated with relaxation in the solvent such as associations, rotational isomerizations, and dipole orienta-

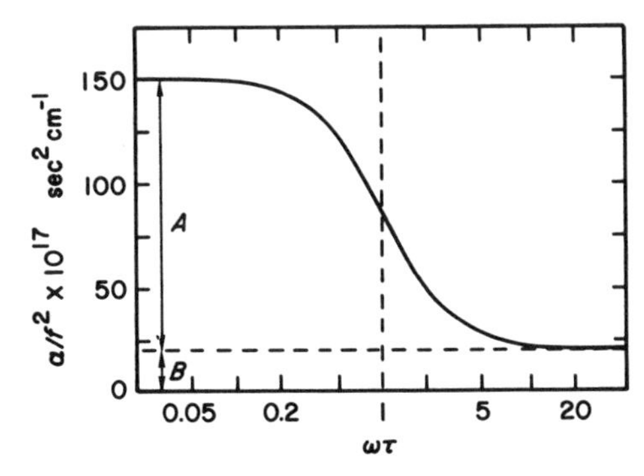

FIGURE 15.2. Plot of α/f^2 versus $\omega\tau$; curve calculated for $A = 128 \times 10^{-17}$ sec^2/cm and $B = 22 \times 10^{-17}$ sec^2/cm. [From Stuehr (*6*), by permission of John Wiley & Sons.]

tions, as mentioned in Section 15.1.1. Figure 15.2 shows α/f^2 as function of $\omega\tau$ for the case of one relaxation time only.

German scientists frequently represent sound absorption data with a quantity called the "absorption volume per molecule"; it is defined (*9*) as

$$Q\lambda = 2\mu^{\mathrm{ch}}/N_A c_0 \quad (\mathrm{cm}^3/\mathrm{molecule}) \tag{15.19}$$

where N_{A} is Avogadro's number and c_0 is the total stoichiometric solute concentration (in moles per cubic centimeter); Q is called the "absorption cross section."

15.1.4 Relaxation Amplitudes

Whether relaxational sound absorption is significant and measurable depends on the relative magnitude of the A_j values compared to B (Eq. 15.17 or 15.18). In aqueous solution B is particularly small ($\sim 22.5 \times 10^{-17}$ sec²/cm at room temperature) which makes water a favorable solvent; approximate B values for some other common solvents are 30×10^{-17} for methanol, 55×10^{-17} for acetonitrile, 74×10^{-17} for *n*-hexane, 77×10^{-17} for *n*-butanol, 490×10^{-17} for carbon tetrachloride, and 780×10^{-17} for benzene. For tabulations of B values for various liquids see Stuehr (*6*), Heasell and Lamb (*10*), Andreae and Joyce (*11*), and Edmonds *et al.* (*12*); a good number of common solvents have B values of less than 100×10^{-17} sec²/cm.

From Eq. 15.16 we see that the relaxation amplitude A is proportional to κ^{ch}; in analogy to the overall adiabatic compressibility (Eq. 15.7), κ^{ch} is defined by

$$\kappa^{\mathrm{ch}} = -\left[\frac{1}{V}\left(\frac{\partial V}{\partial p}\right)_S\right]^{\mathrm{ch}} \tag{15.20}$$

which for a single-step system can also be written as

$$\kappa^{\mathrm{ch}} = -\frac{1}{V}\left(\frac{\partial V}{\partial c_j}\right)_S\left(\frac{\partial c_j}{\partial \ln K}\right)_S\left(\frac{\partial \ln K}{\partial p}\right)_S \tag{15.21}$$

It can be shown (*4*) that this is equivalent to

$$\kappa^{\mathrm{ch}} = \frac{1}{RT}\left[\Delta V - \frac{\alpha_p}{\rho c_P}\Delta H\right]^2 \Gamma \tag{15.22}$$

where ΔV and ΔH are the molar volume and molar enthalpy of reaction, respectively, α_p is the thermal expansion coefficient at constant pressure $[-V^{-1}(\partial V/\partial T)_p]$, c_P is the specific heat at constant pressure, ρ is the density, and Γ is given by

$$\Gamma = \left(\sum_j \nu_j^2/\bar{c}_j\right)^{-1} \tag{15.23}$$

(which is also Eq. 6.36; see Section 6.1). Thus combining Eq. 15.16 and 15.22 and making use of the relationship of Eq. 15.8 (with $\kappa_S = \kappa_0$) we obtain

$$A = \frac{2\pi^2 \rho v \tau}{RT}\left(\Delta V - \frac{\alpha_p}{\rho c_P}\Delta H\right)^2 \Gamma \tag{15.24}$$

The most interesting feature about Eq. 15.24 is that A depends on τ; no such dependence of the relaxation amplitude occurs with transient relaxation techniques. For very short relaxation times a measurable amplitude can only be achieved by using high reagent concentrations which increase Γ. To illustrate this let us consider a simple example, namely the reaction

$$\mathrm{A} \rightleftharpoons \mathrm{B}$$

If the experiment is conducted in aqueous solution the $\alpha_p\, \Delta H/\rho c_P$ term is usually very small and negligible (note that in organic solvents this is usually not the case).

Since at 25°C the sound velocity in water is $v = 1498$ m/sec (*13*), Eq. 15.24 becomes

$$A = 1.2 \times 10^{-7} (\Delta V)^2 \tau \Gamma \tag{15.25}$$

Let us assume a typical value of 10 cm^3/mole for ΔV, and assume $K = 1$ so that $\Gamma = 0.25 c_0$ ($c_0 = \bar{c}_A + \bar{c}_B$). For $c_0 = 4 \times 10^{-3}$ *M*, we obtain

$$A = 1.2 \times 10^{-8} \tau$$

If $\tau = 10^{-7}$, 10^{-8}, or 10^{-9} sec, respectively, we have $A = 120 \times 10^{-17}$, 12×10^{-17}, and 1.2×10^{-17} sec^2/cm. Since $B = 22.5 \times 10^{-17}$ sec^2/cm, we see that for $\tau = 10^{-7}$ or 10^{-8} sec the difference between A and B is quite appreciable, but for $\tau = 10^{-9}$ sec it is very small. In the latter case a higher c_0 would have to be chosen for obtaining good data.

15.1.5 Relaxation Times from Sound Velocity Dispersion

It should be mentioned that relaxation times can also be determined from the frequency dependence of the sound velocity. That the sound velocity must be frequency dependent can be seen from its relation to the frequency-dependent κ_S, Eq. 15.8. It can be shown (*14*) that

$$v^2 = v_0^2 + \frac{\kappa^{ch}}{\kappa_0} v_0 v_\infty \frac{\omega^2 \tau^2}{1 + \omega^2 \tau^2} \tag{15.26}$$

where v_0 is the sound velocity at low frequencies ($\omega\tau \ll 1$), and v_∞ is the sound velocity at high frequencies ($\omega\tau \gg 1$). Expressing κ^{ch}/κ_0 as $(2\mu^{ch})_{max}/\pi$ (Eq. 15.13) affords

$$v^2 = v_0^2 + \frac{2}{\pi} (\mu^{ch})_{max} v_0 v_\infty \frac{\omega^2 \tau^2}{1 + \omega^2 \tau^2} \tag{15.27}$$

We now define $v = v_0 + \Delta v$ and call Δv the velocity dispersion; for $\Delta v \ll v_0$ we obtain ($v_\infty \approx v_0$)

$$\Delta v = \frac{(\mu^{ch})_{max}}{\pi} v \frac{\omega^2 \tau^2}{1 + \omega^2 \tau^2} \tag{15.28}$$

Thus τ can in principle be evaluated from the frequency dependence of Δv. In practice, Δv is usually very small and difficult to measure with high precision. Hence most investigators prefer to determine τ from absorption data.

15.2 General Experimental Considerations and Treatment of Data

15.2.1 Evaluation of Relaxation Times

Assume we suspect a relaxation in a certain time range. This range determines the choice of the particular ultrasonic technique to be used. Typically one would then proceed as follows. First α/f^2 is determined at the highest and lowest frequencies possible with the technique at hand, with one of the following results.

1. α/f^2 is the same at the high- and low-frequency ends. This can mean there is no relaxation occurring in the frequency range under consideration or the relaxation amplitude is too small. If an increase of the reagent concentration still does not produce any effect, there may still be a relaxation outside the frequency range of the method at hand. Comparison of α/f^2 with B for the pure solvent allows us to decide whether we should investigate a lower or higher frequency range. If $\alpha/f^2 = B$, this means we may be located on the right-hand side of the plot in Fig. 15.2 and a lower frequency range should be tried; if $\alpha/f^2 > B$, we obviously are on the left-hand side of Fig. 15.2. Note, however, that although in this situation we are bound to find a relaxation effect at higher frequencies, there is still the possibility of an additional relaxation effect at lower frequencies also.

2. α/f^2 is lower at the high-frequency end. This indicates that there is a relaxation effect and we may proceed to measure α/f^2 over the whole frequency range in order to map out a curve such as that in Fig. 15.2. Since the ultrasonic frequencies are measured in hertz, it is common practice to replace ω by $2\pi f$ and to define f_r as the frequency where $2\pi f_r = \tau^{-1}$ which transforms Eq. 15.17 into

$$\frac{\alpha}{f^2} = \frac{A}{1 + (f/f_r)^2} + B \tag{15.29}$$

There are several ways to recast Eq. 15.29 to yield a straight line (*6*) thereby allowing an easy evaluation of A and f_r; computer fitting of the experimental α/f^2 values to Eq. 15.29 is, however, an increasingly popular method of data analysis. It should be pointed out that a frequency range spanning at least a decade but preferably more is necessary to obtain highly accurate values for f_r.

If the data do not fit Eq. 15.29 well, there is probably more than one relaxation time in the frequency range under study, and

$$\frac{\alpha}{f^2} = \sum_j \frac{A_j}{1 + (f/f_{r_j})^2} + B \tag{15.30}$$

holds. Needless to say, when there are two or more relaxation times a much wider frequency range must be investigated which usually requires the combination of at least two different experimental techniques.

The problems in calculating the different f_{r_j} values are similar to those in calculating the τ_j values from transient relaxation experiments (Section 9.2), i.e., when two f_r values are very close and the amplitudes are similar there will be several sets of f_{r_j} and A_j values which fit the data equally well. Rassing and Lassen (*15*) have analyzed the situation for two relaxation processes by simulating various relaxation curves for strongly overlapping relaxation times in a way similar to that used for transient methods in Section 9.2. They conclude that unless $\tau_2/\tau_1 \geq 4$ and the amplitude ratio is between 0.25 and 4 it is hardly possible to recognize the presence of two relaxation effects.

An alternative method of evaluating ultrasonic data (for one relaxation time) is based on

$$\mu^{ch} = \frac{\pi\kappa^{ch}}{\kappa_0} \frac{f/f_r}{1 + (f/f_r)^2} \tag{15.31}$$

which is equivalent to Eq. 15.11. Taking logarithms affords

$$\log \mu^{ch} = \log \frac{\pi\kappa^{ch}}{\kappa_0} + \log \frac{f/f_r}{1 + (f/f_r)^2} \tag{15.32}$$

A plot of μ^{ch} versus f on a logarithmic scale leads to a curve whose shape is solely determined by the frequency term but whose vertical position depends on $\pi\kappa^{ch}/\kappa_0$ and whose horizontal position depends on f_r. By using a template with the appropriate theoretical shape one can immediately check whether the experimental data obey Eq. 15.32, i.e., whether there is indeed only one relaxation time; f_r and μ^{ch} are easily determined from the position of the maximum.

For a more detailed discussion of the advantages and fallacies of the various data treatment procedures see Stuehr (*6*).

15.2.2 Interpretation of Relaxation Times in Terms of Mechanisms

In transient relaxation methods a special detection device is necessary for monitoring the relaxation. This is not the case in ultrasonic techniques, which is both an advantage and a disadvantage. The advantage is that no restrictions on the types of processes that can be studied are imposed by the specific detection technique, such as different extinction coefficients in reactant and product.

On the other hand, well-defined changes in a physical property, e.g., an increase in OD following a temperature jump, can often give a clue as to what reaction the observed relaxation effect should be attributed. No such diagnosis is possible from sound absorption data and one is essentially restricted to evaluating the concentration dependence of τ and possibly the relaxation amplitude. A survey of the literature shows that in fact it is not an uncommon situation that different scientists disagree as to what processes are responsible for certain relaxation effects; some examples are discussed in Section 15.4.

15.3 Experimental Methods

The basic components of acoustical techniques are devices which generate ultrasound by converting electric energy into sound energy and which measure sound intensities; they are called transducers. Quartz crystals are the most commonly used transducers. When subjected to an alternating electric field, the quartz crystal oscillates, thereby radiating sound into the adjacent medium. This effect is reversible so that quartz crystals can also be used as receivers. This property of quartz crystals and of some other materials is known as the piezoelectric effect.

As mentioned earlier, several different experimental techniques have to be used, depending on the frequency range of interest. The principal reason for this is the fact that the attenuation of ultrasound increases with the square of the frequency. As a consequence, at say 1 MHz a sound wave suffers the same attenuation by traveling a distance of 1 mm as it does by traveling 1 km at 1 kHz; this clearly suggests that a single instrument will hardly be suitable over a wide frequency range. In fact any individual technique covers only slightly more than one order of magnitude in frequency range.

Table 15.1 gives a summary of the most common techniques. The basic principles of a selection of these methods are briefly mentioned in the following sections;

Table 15.1. Ultrasonic Absorption Techniques

Frequency range	Method	Remarks
10–100 kHz	Resonance (decay rate)	Very large volumes
100 kHz–1 MHz	Resonance (line width); reverberation	Large volumes
1–10 MHz	Resonance (line width)	Volumes < 40 ml
	Interferometer; differential techniques	High accuracy
10–100 MHz	Pulse; optical	High accuracy, most convenient range
100 MHz–1 GHz	Pulse	Difficult
1–10 GHz	Brillouin scattering	

for more details and numerous references see Eigen and DeMaeyer (*4*), Eggers and Kustin (*16*), and Stuehr (*6*). It should be pointed out that no ultrasonic spectrometers can be purchased commercially; although the electronic components are easily available the mechanical parts must be built by a good machine shop. Good practical and technical advice can be found in the review by Eggers and Kustin (*16*).

15.3.1 Frequency Range 10 kHz to 1 MHz

At low frequencies α is very small and very long pathways (huge vessels) would in principle be necessary in order to notice a significant attenuation. This difficulty is circumvented by using so-called *resonance* or *reverberation techniques* whereby the sound wave is reflected numerous times from the walls of the vessel and thus travels through the liquid many times.

In a design by Kurtze and Tamm (*9*) a spherical vessel (*resonant sphere method*) is used and a train of ultrasonic pulses, generated by a piezoelectric transducer, is applied radially, producing symmetrical radial vibration modes. One way to determine the attenuation is to measure the half-power bandwidth Δf for a resonance mode of vibration with continuous wave excitation; Δf is related to μ by (*16*)

$$\mu = \pi \Delta f / f \tag{15.33}$$

An alternative way is to stop the excitation after a short time and to measure the rate of decay in the sound intensity by a second transducer.

The resonant sphere method works well at frequencies up to ~100 kHz. Above 100 kHz subsequent resonances (harmonics) of a spherical vessel become too close and thus difficult to excite separately. Here it becomes advantageous to excite a narrow band of frequencies by a source that generates "random noise." Again, one may measure the decay rate after the pulse is turned off. This method is known as the *reverberation technique*.

Despite making use of multiple reflections, these resonance and reverberation techniques require relatively large sample volumes (up to 50 liters), particularly at the low-frequency end, making them impractical when expensive chemicals are involved.

15.3.2 Frequency Range 1–10 MHz

Eggers and co-workers (*16*, *17*) have recently developed a resonance technique which allows measurements on volumes as low as 40 ml in the range of 0.2–10 MHz, or on volumes of 5 ml in the range of 1–20 MHz. The main feature of this design is a cylindrical resonator consisting of two disk-shaped quartz transducers which enclose the liquid. The first transducer is driven by a sinusoidal voltage and produces standing waves in the solution at particular resonance frequencies. The second transducer (receiver) generates a voltage peak whenever there is a resonance condition. The half-power frequency bandwidth Δf in the receiver is again the measured quantity.

Other methods used in the range of 1–10 MHz include the *acoustic interferometer* (*4*, *6*), which is similar to the cylindrical resonator, and the *differential methods* (*4*, *6*), which directly yield the attenuation *difference* between solution and solvent and thus allow highly precise measurements of even small effects.

15.3.3 Frequency Range 10–100 MHz

The range of 10–100 MHz is the easiest to work in. The attenuation of sound waves is of a convenient magnitude in volumes of "reasonable" size and highly accurate measurements (errors $\pm 2\%$ or better) can usually be made; also the electronic components necessary to build an apparatus are most easily available.

The most popular technique in this range is the *pulse method*. In this technique an ultrasonic pulse is generated by subjecting a piezoelectric quartz to a short train of radiofrequency pulses (of a few microseconds' duration) generated by an oscillator. After traveling a variable distance through the solution, the sound pulses are received by another quartz transducer and the attenuation is measured. The contribution to the sound attenuation that arises from the solution is found by comparing the amplitude of the received signals with that of the transmitted pulses which have passed through a calibrated attenuator. Note that the sound velocity is also quite easily measured with this technique by simply determining the delay time between transmitted and received pulse.

At the higher frequency end the distance between emitting and receiving transducer has to be very small because of the high attenuation. This can lead to interference between the electric signals of transmitter and receiver (electrical cross-talking). The problem is avoided by increasing the distance with fused quartz rods which act as acoustical delay lines.

A variation of the pulse method is to use only one transducer which acts both as transmitter and receiver. The sound waves sent into the solution are reflected back from plane-parallel reflectors which can be moved with respect to the transducer. This procedure is known as the *pulse–echo method.*

Another method, the *optical technique*, makes use of the Debye–Sears (*18*) effect. The periodic variations in the density which are produced when a plane sound wave passes through a liquid act like a diffraction grating toward a light beam crossing the ultrasonic beam at right angles. The intensity of the diffracted light is proportional to the sound intensity which can be used to measure the attenuation of the sound wave; the method works well in the range of 10–100 MHz.

15.3.4 Frequency Range above 100 MHz

The pulse technique can be used up to about 1 GHz but the difficulties in designing a good sample cell are substantial at the high-frequency end. One of the principal technical problems is to have a very accurate parallel alignment of the two transducers and to maintain it while changing the distance between them; for example, at 500 MHz the tolerance is about 2.5×10^{-4} cm (*6*).

For even higher frequencies (1–10 GHz) a method called *Brillouin scattering* can be used (*6*). Here the sound waves are generated by thermal excitation. In fact these sound waves do not have to be generated by any special source; they are always present and cover a wavelength range from the size of the container down to interatomic distances. However, their intensities are extremely small and thus a very sensitive detection device is needed. Brillouin (*19*) has shown that, since the sound waves generate a diffraction grating, light waves can be used for the detection in a way similar to that discussed in connection with the optical technique (Section 15.3.3).

15.4 Applications of the Ultrasonic Methods

The ultrasonic techniques were the first relaxation techniques to be applied to chemical problems; similar to the temperature-jump method the applications have been too numerous for a comprehensive discussion and we shall again be quite selective. Among the most frequently investigated processes we find inter- and intramolecular proton transfer reactions, ion association phenomena, dimerization through hydrogen-bond formation (e.g., carboxylic acids), aggregation phenomena such as the formation of micelles, and conformational and rotational isomerizations in large and small molecules. Note that some of these latter are often no longer regarded as "chemical reactions" in the usual sense of the word.

15.4.1 Proton Transfer Reactions

As pointed out in Section 11.5.1, many proton transfer processes are too rapid for the temperature-jump technique or even the electric field-jump method and thus have to be studied by ultrasonics. In fact the majority of proton transfer relaxation studies have been done by ultrasonic methods. The general features of proton transfer reactions have been discussed in Sections 4.4 and 11.5.1 and references to numerous reviews of the early work have been given in Section 11.5.1.

A typical example of a rather straightforward study from Eigen's laboratory (*20*) refers to proton transfer processes in aqueous solutions of aliphatic amines. A typical set of data with absorption curves of ethylamine solutions of increasing concentration is shown in Fig. 15.3. The shape of the curves matches the theoretical curves

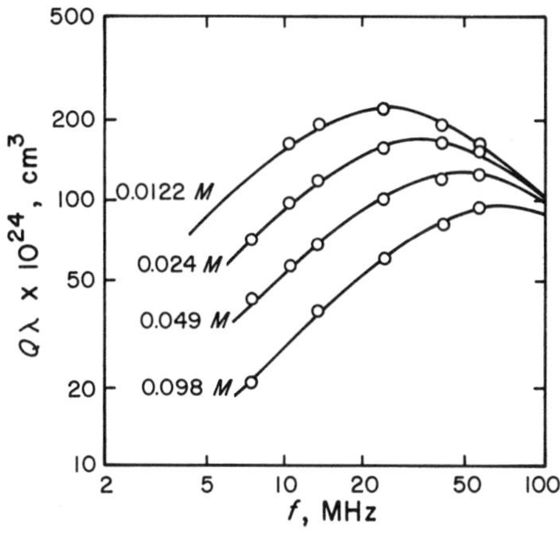

FIGURE 15.3. Concentration dependence of ultrasonic absorption in aqueous solutions of ethylamine at 20°C. [From Eigen *et al.* (*20*), by permission of Akademische Verlagsgesellschaft.]

calculated for one relaxation time only; the concentration dependence of the relaxation time is consistent with that expected for the reaction

$$R_3NH^+ + OH^- \underset{k_{32}}{\overset{k_{23}}{\rightleftharpoons}} R_3N + H_2O \tag{15.34}$$

$$\tau_2^{-1} = k_{23}(\bar{c}_{OH} + \bar{c}_{R_3NH}) + k_{32} \tag{15.35}$$

(which is the same as Eq. 4.71).

In view of our discussion in Section 4.4 this is reasonable; although the full reaction scheme should be represented by Eq. 4.64 and give rise to two relaxation processes, it was shown that in an alkaline solution of an aliphatic amine only the second relaxation time (Eq. 4.71) has a measurable amplitude.

The list of papers dealing with ultrasonic proton transfer studies is still growing steadily. Apart from continuing to investigate simple organic acids (*21–23*), amines (*24–29*), or amino acids (*30–32*), studies that have essentially confirmed Eigen's (*31a*) theory, some recent and current efforts have also been directed to more subtle questions and to systems involving molecules that are more complex.

An interesting recent study concerns the question of whether the transformation between the zwitterionic (Z) and the neutral (N) forms of cysteine and related compounds occurs via *intra*- or *inter*molecular proton transfer; the two possibilities are

$$\underset{\text{Z}}{R\langle^{\overset{+}{N}H_3}_{S^-}} \underset{k_{-1}}{\overset{k_1}{\rightleftharpoons}} \underset{\text{N}}{R\langle^{NH_2}_{SH}} \tag{15.36}$$

$$\underset{2\text{Z}}{R\langle^{\overset{+}{N}H_3 \dashrightarrow {}^-S}_{S^- \dashleftarrow H_3\overset{+}{N}}\rangle R} \underset{k_{-2}}{\overset{k_2}{\rightleftharpoons}} \underset{2\text{N}}{R\langle^{NH_2 \dashleftarrow HS}_{SH \dashrightarrow H_2N}\rangle R} \tag{15.37}$$

The relaxation time expressions are given by

$$1/\tau = k_1 + k_{-1} \tag{15.38}$$

$$1/\tau = 2k_2\bar{c}_Z + 2k_{-2}\bar{c}_N \tag{15.39}$$

At pH 9.5 two relaxations were observed (*34*). In the case of cysteine one was in the frequency range of 10^5–10^6 Hz, the other around 10^8 Hz. The former was concentration dependent and was attributed to the reaction

$$R\langle^{NH_3^+}_{S^-} + OH^- \rightleftharpoons R\langle^{NH_2}_{S^-} + H_2O$$

The second was concentration *in*dependent and attributed to reaction 15.36 with $\tau^{-1} = 3.6 \times 10^8\ \text{sec}^{-1}$. For glutathione and cysteamine the high-frequency process showed a concentration dependence that is consistent with reactions 15.36 and 15.37 operating concurrently; i.e., τ^{-1} is given by the sum of Eqs. 15.38 and 15.39.

Another study, which involved *o*-, *m*-, and *p*-aminobenzoic acids, showed that the intramolecular proton transfer between the zwitterionic and the neutral forms is only significant for the ortho derivative and only in methanol and acetone, but not in aqueous solution (*35*).

Current interest centers around inter- and intramolecular proton transfers in nucleosides and nucleotides (*36a,b*), and in proteins and polypeptides (*37, 38*). There have been disagreements as to whether the observed relaxation effects are indeed to be attributed to proton transfer processes (*37–43*) or to conformational changes (*31, 44–48*), highlighting one of the basic difficulties in interpreting sound absorption data (see Section 15.2.2).

15.4.2 Metal Complex Formation

As far as the number of studies are concerned metal complex formation reactions have been the most fertile ground for the application of ultrasonic techniques. Numerous reviews summarizing and discussing the general features of these reactions have already been mentioned in Section 11.5.2; some additional reviews which focus on ultrasonic studies are those by Stuehr and Yeager (*2*), Blandamer (*3*), Stuehr (*6*), Tamm (*49*), and Petrucci (*50*).

As pointed out in Section 11.5.2 the first two steps in a metal complex formation reaction

$$\underset{①}{M^{m+}(\text{sol}) + L^{l-}(\text{sol})} \underset{k_{21}}{\overset{k_{12}}{\rightleftharpoons}} \underset{②}{M^{m+}(\text{sol, sol})L^{l-}} \underset{k_{32}}{\overset{k_{23}}{\rightleftharpoons}} \underset{③}{M^{m+}(\text{sol})L^{l-}} \underset{k_{43}}{\overset{k_{34}}{\rightleftharpoons}} \underset{④}{ML^{(m-l)+}} \tag{15.40}$$

(which is also reaction 11.38) are usually too fast for any of the transient relaxation techniques and thus must be studied by ultrasonics. In fact even numerous inner sphere complex formations (③ ⇌ ④) have rates in a time range outside that of the transient methods, particularly the reactions of alkali or alkaline earth metal ions with chelating agents such as EDTA and NTA (*51*); these complexes are rather weak, thus making k_{43} very high ($\sim 10^8$–10^9 sec^{-1}).

Many studies have been reported on the reactions of divalent cations with divalent anions. According to the mechanism of Eq. 15.40 a maximum of three relaxation times should be observed. In the early studies only two relaxation effects were usually observed. Typically the low-frequency absorption was strongly dependent on the nature of the cation, whereas the high-frequency absorption was independent of it and usually around 200 MHz. This is shown in Fig. 15.4 for a

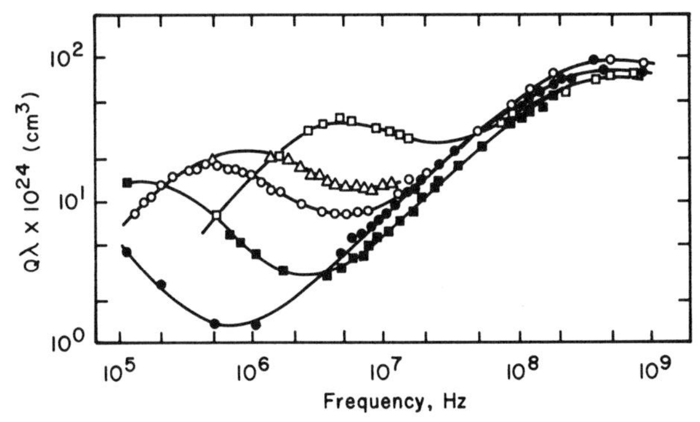

FIGURE 15.4. Ultrasonic relaxation spectra for some divalent sulfates: □ Mn^{2+}; △ Fe^{2+}; ○ Co^{2+}; ■ Mg^{2+}; ● Ni^{2+}. [From Stuehr (*6*), by permission of John Wiley & Sons.]

few representative examples. This is consistent with the high-frequency absorption being due to outer sphere complex formation, ① ⇌ ③, where ② is assumed to be present at very low equilibrium concentrations (see discussion in Section 11.5.2) and the low-frequency absorption being due to inner sphere complex formation, ③ ⇌ ④ (*52*); another possibility is that the short relaxation time refers to ② ⇌ ③ with the reaction ① ⇌ ② remaining invisible because ΔV is approximately zero and thus has a negligible amplitude (*52*).

In subsequent work Atkinson and Kor (*53*) claimed discovery of a third relaxation time with a small amplitude in the frequency range of about 35 MHz, a phenomenon that had already been reported earlier (*54*). This now led to the assignment of the fastest process to the reaction ① ⇌ ② and the middle relaxation time to the anion desolvation process ② ⇌ ③. However, Jackopin and Yeager (*55, 56*) disputed this interpretation and attributed the 35 MHz absorption to an equipment artifact or impurities.

Still different data have come from Tamm's (*57*) laboratory; though no distinct maximum around 35 MHz was observed the data were nevertheless interpreted in terms of three relaxation times as indicated in Fig. 15.5. A possible criticism

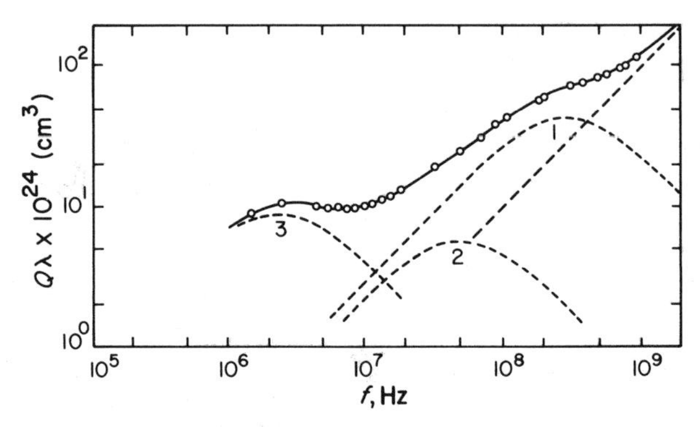

FIGURE 15.5. Analysis of ultrasonic absorption of 2 *M* $MnSO_4$ solution at 5°C in terms of three relaxation effects. [From Bechtler *et al.* (*57*), by permission of the American Institute of Physics.]

of these last data is that they were obtained at very high concentrations (0.5–3 *M*) which may have a serious effect on the "solvent absorption" or background absorption because of possible structural effects of the ions on the solvent (*58*); this could be part of the reason why now the high-frequency absorption is at 400–600 MHz instead of 200 MHz. The problem of knowing what the true background absorption might be in such studies is a serious one at these very high frequencies because at low concentration where structural effects of the ions on the solvent would be minimal the relaxation amplitude is too small (Eq. 15.24). The problem is compounded by the inherently smaller precision attainable at very high frequencies. Measurements by the Brillouin scattering technique are somewhat helpful in establishing the solvent contribution. For example, Fritsch *et al.* (*59, 60*) report that at 6–7 GHz solutions of $CuSO_4$ and $ZnSO_4$ at concentrations of up to 0.2 *M* have, within $\pm 5\%$, the same absorption as the pure solvent; on this basis they interpret the absorption in the 30–900 MHz range as being the result of two relaxation processes (① ⇌ ② and ② ⇌ ③) in the case of $CoSO_4$, $NiSO_4$, $CuSO_4$, and $ZnSO_4$.

However, it seems that the precision of the Brillouin measurements is still not high enough to settle the problem; using the example of $MnSO_4$, Jackopin and Yeager (*56*) have shown that with the precision currently available it is very difficult to distinguish between one or two relaxation effects in the high-frequency region and the data can equally well be fitted assuming only one effect, just as had been done by Hemmes *et al.* (*61*).

This discussion illustrates well the inherent difficulties in interpreting sound absorption data; for more details on these problems see Petrucci's review (*50*).

Other areas of recent interest include the complexation of lanthanide ions with sulfate and nitrate ions (*62–67*), of lanthanide ions with murexide (*62, 68*), and the association kinetics of electrolytes (in nonaqueous solvents) such as alkylammonium salts (*69–71*) and divalent sulfates (*72, 73*), among others (*74, 75*). In studying the complexation of Li^+ with NO_3^- in tetrahydrofuran, Wang and Hemmes (*75*) came to the interesting conclusion that the conversion of the solvent-separated ion pair (or outer sphere complex ③) into the contact ion pair (④) corresponds to a desolvation of the anion, in contrast to the usual conclusion that this step is equivalent to solvent loss from the cation.

15.4.3 Conformational Changes

A. Small Molecules

Conformational changes in small molecules as a consequence of the rotation about single bonds have been studied for a number of systems. Examples include the isomerization of α,β-unsaturated aldehydes (*76*), e.g.,

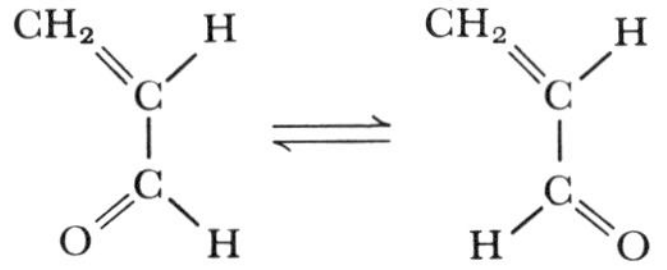

$$CH_2{=}CH{-}CHO \;(s\text{-}trans) \rightleftharpoons CH_2{=}CH{-}CHO \;(s\text{-}cis) \qquad (15.41)$$

substituted vinylic compounds (*77*), and carboxylic esters (*78*); rotational isomerism of *n*-alkanes (*79*),

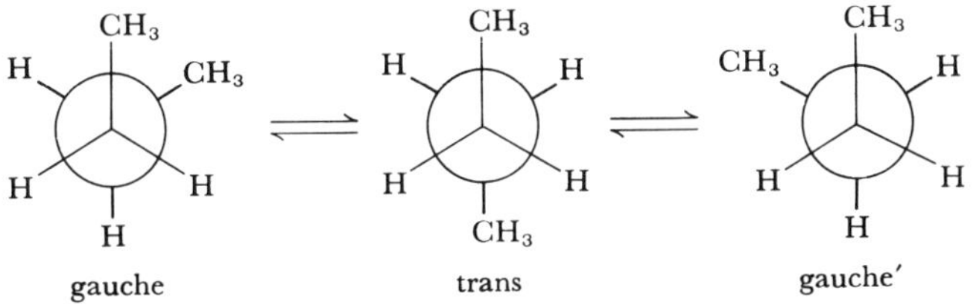

gauche ⇌ trans ⇌ gauche′ (15.42)

substituted alkanes (*78, 80*), and triethylamine (*81*); and interconversions of isomeric chair forms of cyclohexane derivatives (*79, 82, 83*), e.g.,

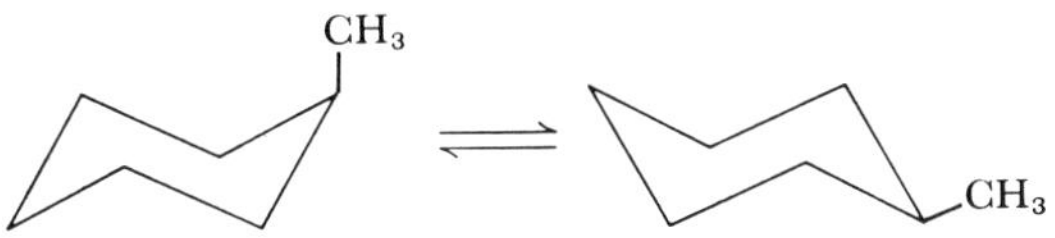

(15.43)

and heterocyclic systems such as 2-alkyl-1,3-dioxanes (*84*). More examples can be found in some reviews (*3, 85–90*).

Note that these systems can generally be represented as A ⇌ B reactions, which means that the measurements of the relaxation time ($\tau^{-1} = k_1 + k_{-1}$) have to be supplemented by an equilibrium determination if the individual rate constants are desired. The case of the rotational isomerism in alkanes (Eq. 15.42) is interesting. Here there are in principle two relaxation times. The process can be written as

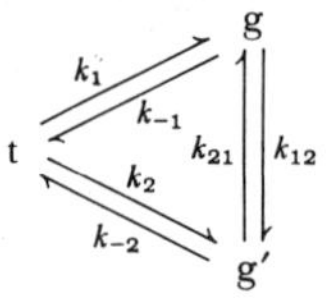

$$\mathrm{t} \underset{k_{-1}}{\overset{k_1}{\rightleftharpoons}} \mathrm{g}, \qquad \mathrm{t} \underset{k_{-2}}{\overset{k_2}{\rightleftharpoons}} \mathrm{g}', \qquad \mathrm{g} \underset{k_{21}}{\overset{k_{12}}{\rightleftharpoons}} \mathrm{g}' \tag{15.44}$$

Since g and g' are enantiomers and thus equivalent from a reactivity point of view, we have $k_2 = k_1$, $k_{-2} = k_{-1}$, and $k_{12} = k_{21}$. The system is thus similar to Eq. 7.1 (with $k_{-2} = k_{-1}$) only that there is an additional direct pathway ($\mathrm{g} \rightleftharpoons \mathrm{g}'$) and that there is even more degeneracy ($k_2 = k_1$). The relaxation times are easily shown to be

$$1/\tau_1 = 2k_1 + k_{-1} \tag{15.45}$$

$$1/\tau_2 = k_{-1} + 2k_{12} \tag{15.46}$$

The two normal reactions (see Section 7.3.1) are

$$\mathrm{t} \underset{k_{-1}}{\overset{2k_1}{\rightleftharpoons}} (\mathrm{g}, \mathrm{g}') \tag{15.47}$$

$$\mathrm{g} \underset{\overleftarrow{k} + k_{12}}{\overset{\overrightarrow{k} + k_{12}}{\rightleftharpoons}} \mathrm{g}' \tag{15.48}$$

with $\overrightarrow{k} = \overleftarrow{k} = 0.5\, k_{-1}$.

Because of the equivalence of g and g′ the reaction volume and reaction enthalpy for the second normal reaction are zero and with it the relaxation amplitude in an ultrasonic experiment.

A recent example of a conformational change of biological interest is the syn–anti isomerization of purine nucleosides such as adenosine (15.49) (*91*), for which $\tau^{-1} = k_1 + k_{-1} = 2.5 \times 10^8\ \mathrm{sec}^{-1}$ at room temperature in aqueous solution.

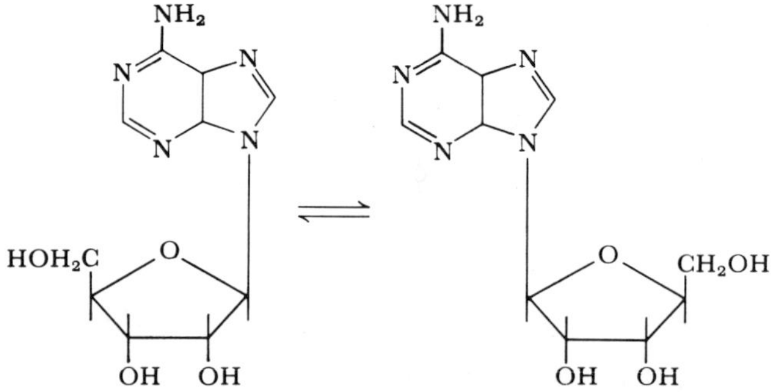

$$\text{syn} \rightleftharpoons \text{anti} \tag{15.49}$$

NH₂ N N N N HOH₂C O OH OH syn ⇌ NH₂ N N N N O CH₂OH OH OH anti

In a more sophisticated approach Hemmes *et al.* (*92*) exploited measurements of the relaxation amplitudes at different temperatures for evaluating the equilibrium constant, which allowed k_1 and k_{-1} to be calculated separately ($k_1/k_{-1} = 3.6$ at

25°C). However, it should be pointed out that the interpretation of the relaxation effect as being due to reaction 15.49 has recently been questioned (*36a*).

B. Large Molecules

As mentioned earlier (Section 10.2) helix–coil transitions of polypeptides have served as models for similar transitions in proteins; we have given a brief account of Schwarz's theory of these processes. Ultrasonic studies with polypeptides have in fact generally been interpreted in terms of Schwarz's theory which predicts that for long chains the mean relaxation time is given by

$$1/\tau^* = k_F\{(s - 1)^2 + 4\sigma\} \tag{15.50}$$

with k_F, s, and σ defined in Section 10.2; Eq. 15.50 predicts that at the transition midpoint ($s = 1$) τ^* goes through a maximum given by

$$\tau^* = (4\sigma k_F)^{-1} \tag{15.51}$$

In a transient relaxation experiment the evaluation of τ^* (from the slope of a tangent to the relaxation curve at time $t = 0$) is quite straightforward (see Section 9.3.1) regardless of whether the relaxation spectrum is wide or narrow. In an ultrasonic experiment the relaxation curve becomes quite complicated for a broad relaxation spectrum and τ^* is not easily determined; for a narrow relaxation spectrum the situation is simple. Here the relaxational contribution to $\mu = \alpha\lambda$, for the case of a helix–coil transition, is given by (*93*)

$$\mu^{ch} = \pi \frac{\rho v^2(\Delta V)^2}{RT} s \frac{\partial\theta}{\partial s} c_0 \frac{2\omega\tau^*}{1 + \omega^2\tau^{*2}} \tag{15.52}$$

where θ is the degree of helicity (Section 10.2); i.e., the relaxation curve looks the same as that for a single relaxation process with $\tau = \tau^*$.

Sound absorption studies have in fact demonstrated that the relaxation spectrum must be narrow; an example is shown in Fig. 15.6 for poly-L-glutamic acid (*48*) where the shape of the relaxation curve comes close to the one predicted for one discrete relaxation time. The predictions of Eq. 15.50 could also be confirmed as

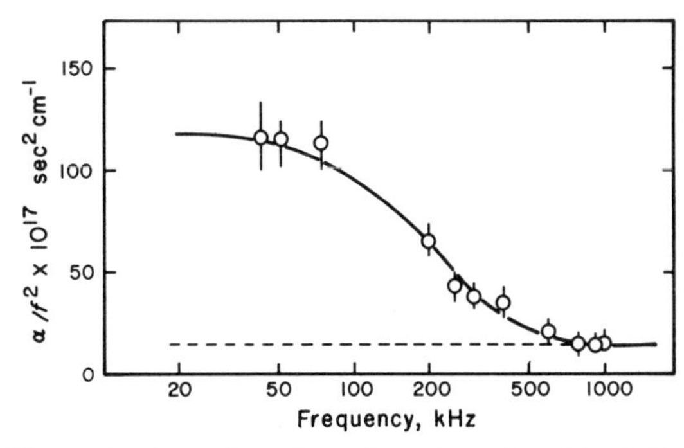

FIGURE 15.6. Ultrasonic absorption of a poly-L-glutamic acid solution at pH 5.11, 37°C. Dashed line is the high-frequency extrapolation. [From Barksdale and Stuehr (*48*), by permission of the American Chemical Society.]

follows. The helicity of a polypeptide is easily controlled by the pH because it is mainly through hydrogen bonding of protonated carboxylic groups to the nitrogens that the helical structure comes about. Thus increasing the pH decreases the helicity. The kinetic studies show that τ^* indeed goes through a maximum at the pH value which corresponds to the transition midpoint. Examples are poly-L-glutamic acid (*48*, *94*, *95*), poly-L-ornithine (*96*), and poly-L-lysine (*47*).

By means of Eq. 15.51 k_F can usually be calculated since the nucleation parameter σ is available from equilibrium studies. Some authors (*47*, *96*) report k_F values around 10^9 to $>10^{10}\ M^{-1}\ \text{sec}^{-1}$, indicating that growth (Section 10.2) is diffusion controlled or nearly so; others (*48*) find k_F to be $<10^8\ M^{-1}\ \text{sec}^{-1}$. Studies by other relaxation techniques have also led to reports of both high (*97*) and low (*98*) k_F values; for a discussion of possible reasons see Section 10.2.

It should be mentioned that Zana *et al.* (*37*, *40*, *42*, *43*) have challenged the interpretation of the relaxation effects in terms of a helix–coil transition in some cases; they have pointed out that proton transfers or counterion binding which can lead to a similar pH dependence of the relaxation time may be responsible for the observed relaxation. Hussey and Edmonds (*99*) have suggested that measurements of ultrasonic relaxation amplitudes over a wide pH range should allow the contribution of helix–coil transitions to be distinguished from that of other processes.

15.4.4 Molecular Aggregation Phenomena

Molecular associations through hydrogen bonding have generally been found to be very fast and to approach diffusion-controlled rates ($k > 10^9$) in the thermodynamically favored direction. The best known examples refer to the dimerization of carboxylic acids (*100–103*) which is usually represented by

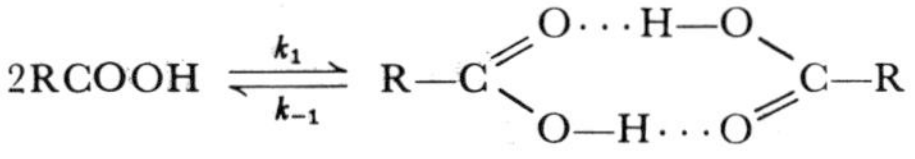

$$2\text{RCOOH} \underset{k_{-1}}{\overset{k_1}{\rightleftharpoons}} \text{R—C}\begin{matrix}\nearrow\text{O}\cdots\text{H—O}\searrow \\ \searrow\text{O—H}\cdots\text{O}\nearrow\end{matrix}\text{C—R} \qquad (15.53)$$

More recently it was suggested that the ultrasonic absorption data are more consistent with an interconversion between an open and the cyclic dimer (*104*, *105*).

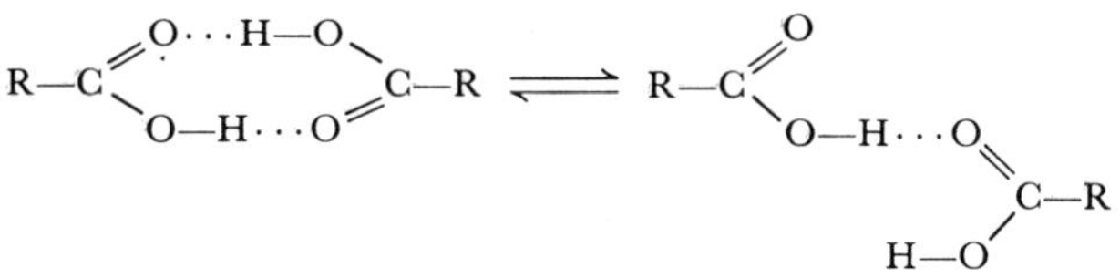

$$\text{R—C}\begin{matrix}\nearrow\text{O}\cdots\text{H—O}\searrow \\ \searrow\text{O—H}\cdots\text{O}\nearrow\end{matrix}\text{C—R} \rightleftharpoons \text{R—C}\begin{matrix}\nearrow\text{O} \\ \searrow\text{O—H}\cdots\text{O=C(—R)—O—H}\end{matrix} \qquad (15.54)$$

Other dimerization studies include that of α-pyridone in hydrogen-bonding solvents (*106*) and ϵ-caprolactam in cyclohexane (*107*); an example for an *intra*molecular hydrogen bond-forming process is the reaction

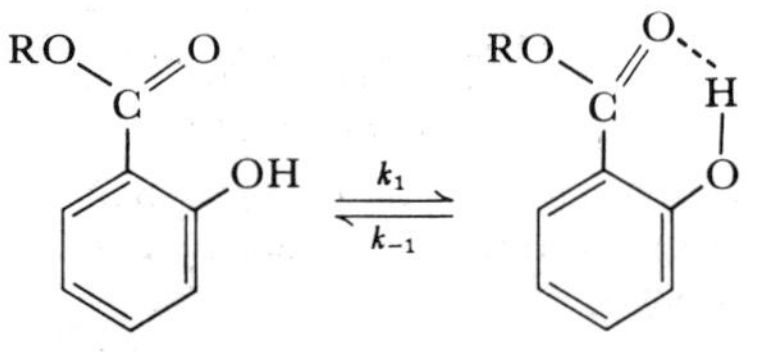

$$\text{(o-ROOC—C}_6\text{H}_4\text{—OH)} \underset{k_{-1}}{\overset{k_1}{\rightleftharpoons}} \text{(o-RO—C(=O}\cdots\text{H—O)—C}_6\text{H}_4\text{, intramolecular H-bond)} \qquad (15.55)$$

where for R = CH_3, $k_1 = 9.5 \times 10^5 \text{ sec}^{-1}$ and $k_{-1} = 2.6 \times 10^7 \text{ sec}^{-1}$ at 25°C (*108*).

Aggregates higher than dimers have also been investigated, for example in the case of *N*-methylacetamide (*103*), aliphatic amines (*109a*), and alcohols (*109b*). For more examples on aggregation through hydrogen bonding see the reviews by Rassing (*103*) and DeMaeyer (*110*).

Self-association of purine bases (base stacking) has also been investigated recently (*111*); the stacking of the bases in nucleic acids is believed to be important in contributing to the helix stability but the nature of the forces involved is still somewhat unclear. Kinetic studies on model systems may help provide some answers. Thus one popular view according to which stacking is principally due to hydrophobic interactions has been challenged on the basis of thermodynamic and kinetic data obtained for the self-associaton of N^6,N^9-dimethyladenine (*111*).

As mentioned in Section 10.4 the kinetics and mechanisms of micelle formation comprise a subject of great current interest but also of controversy as far as the interpretation of the results and even the facts is concerned. Most ultrasonic studies have been carried out on ionic detergents such as alkali alkyl sulfates (*112–114*) or the salts of long-chain carboxylic acids (*115–118*). While Yasunaga *et al.* (*112, 118*) have interpreted the ultrasonic relaxation effect as being due to an interaction between micelles and counterions, Zana *et al.* (*115, 116, 117*) demonstrated that different counterions had little effect on the data and thus proposed that relaxation is due to an equilibrium between monomers and micelles, a view shared by Rassing *et al.* (*113*) and Adair *et al.* (*114*). Some authors have found a pronounced dependence of the relaxation time on surfactant concentration as soon as the cmc is passed (*113, 114, 116*), others report a very weak or no dependence for dilute micellar solutions but a strong dependence at high concentrations (*114*). Some of the interpretations, along with those from studies involving other relaxation techniques, have been discussed in Section 10.4, where additional references have also been given.

15.4.5 Miscellaneous Processes

Ultrasonic studies on solute–solvent interactions of glycine, diglycine, and triglycine (*119*), on the effect of urea or polyethylene glycol on solvent structure of aqueous solutions (*120*), on conformational changes in the polymer chain of polystyrene (*121*) and *N*-polyvinylpyrrolidone (*122*), and on molten salt micellar systems (*123*) have been reported. A recent report suggests that the two relaxation processes found in 95% (v/v) aqueous acetone solutions of *N*-methylpyridinium iodide are due to the formation of solvent separated and contaction pairs (*124*).

Problems

1. Fittipaldi *et al.* (*125*) determined the sound absorption coefficient of a 0.2 *M* aqueous $CuSO_4$ solution at 20°C in the frequency range of 15–225 MHz whereas Plass and Kehl (*126*) made similar measurements on the same solution at frequencies >300 MHz. Their results are summarized in the following table.

f (MHz)	α(Np/cm)	f (MHz)	α(Np/cm)
15	0.165	225	24.9
25	0.456	310	43.0
55	2.13	525	101
75	3.81	690	162
105	7.25	935	276
125	9.70	1470	609
175	16.1		

(a) Plot α/f^2 versus f and determine A and B (choose a logarithmic scale for the frequency).

(b) Plot μ^{ch} versus f. Use B as determined under (a) but also try $B = 25 \times 10^{-17}$, 26×10^{-17}, and 27.3×10^{-17} sec²/cm, respectively. Comment on what you find by slightly changing B. (Assume $v = 1500$ m/sec).

2. Hemmes *et al.* (*61*) report the following α values for a 0.2 M aqueous solution of $MnSO_4$ at 25°C:

f (MHz)	α(Np/cm)	f (MHz)	α(Np/cm)
3	0.061	110	4.23
5	0.144	150	7.60
9	0.181	190	11.5
15	0.249	250	19.3
30	0.530	270	22.6
45	0.915	310	28.6
70	1.97	330	31.9
90	3.31		

How many relaxation times do you detect? Determine their values by plotting μ^{ch} versus f. Use $\alpha_0/f^2 = 22 \times 10^{-17}$ sec²/cm for the solvent, and $v = 1500$ m/sec. To what processes do you attribute the relaxations?

3. Grimshaw and Wyn-Jones (*29*) report the following sound absorption data for the reaction

$$\text{(piperidine, NH)} + H_2O \rightleftharpoons \text{(piperidinium, } \overset{+}{N}H_2\text{)} + OH^-$$

$10^3 \times c_0$ (*M*)	$10^6 \times (\mu^{ch})_{max}$	f_r (MHz)
1.15	160	13.5
1.60	245	15.1
1.92	275	16.0
2.40	350	18.0
2.56	405	18.3
3.00	460	20.3
3.20	460	21.3

The reaction solutions were made up by dissolving piperidine in water where c_0 is the stoichiometric piperidine concentration. Evaluate the rate constants and the molar volume ΔV of the reaction. You may assume that the ΔH term in Eq. 15.22 is negligible. Use Eq. 15.8 to find κ_0 ($v = 1500$ m/sec).

References

1. K. F. Herzfeld and T. A. Litovitz, "Absorption and Dispersion of Ultrasonic Waves." Academic Press, New York, 1959.
2. J. Stuehr and E. Yeager, *Phys. Acoust.* **IIA**, 351 (1965).
3. M. J. Blandamer, "Introduction to Chemical Ultrasonics." Academic Press, New York, 1973.
4. M. Eigen and L. DeMaeyer, *in* "Technique of Organic Chemistry" (S. L. Friess, E. S. Lewis, and A. Weissberger, eds.), Vol. VIII, part 2, p. 895. Wiley (Interscience), New York, 1963.
5. J. Rassing, *in* "Chemical and Biological Applications of Relaxation Spectrometry" (E. Wyn-Jones, ed.), p. 1. Reidel, Dordrecht-Holland, 1975.
6. J. Stuehr, *in* "Techniques of Chemistry" (G. G. Hammes, ed.), Vol. VI, part 2, p. 237. Wiley (Interscience), New York, 1973.
7. L. Hall, *Phys. Rev.* **73**, 775 (1948).
8. T. A. Litovitz and E. H. Carnevale, *J. Appl. Phys.* **26**, 816 (1955).
9. G. Kurtze and K. Tamm, *Acustica* **3**, 33 (1953).
10. E. L. Heasell and J. Lamb, *Proc. Phys. Soc. B.* **69**, 869 (1956).
11. J. H. Andreae and P. L. Joyce, *Brit. J. Appl. Phys.* **13**, 462 (1962).
12. P. D. Edmonds, V. F. Pearce, and J. H. Andreae, *Brit. J. Appl. Phys.* **13**, 551 (1962).
13. Handbook of Chemistry and Physics. Chem. Rubber Co.
14. J. Markham, R. T. Beyer, and R. Lindsay, *Rev. Mod. Phys.* **23**, 353 (1951).
15. J. Rassing and H. Lassen, *Acta Chem. Scand.* **23**, 1007 (1969).
16. F. Eggers and K. Kustin, *Methods Enzymol.* **16**, 55 (1969).
17. F. Eggers and T. Funk, *Rev. Sci. Instrum.* **44**, 969 (1973).
18. P. Debye and F. W. Sears, *Proc. Nat. Acad. Sci. U.S.* **18**, 410 (1932).
19. L. Brillouin, *Ann. Phys.* (*Paris*) **17**, 88 (1922).
20. M. Eigen, G. Maass, and G. Schwarz, *Z. Phys. Chem.* (*NF*) **74**, 319 (1971).
21. T. Yasunaga, M. Tanoura, and M. Miura, *J. Chem. Phys.* **43**, 2735, 3512 (1965).
22. T. Sano, T. Miyazaki, N. Tatsumoto, and T. Yasunaga, *Bull. Chem. Soc. Japan* **46**, 43 (1973).
23. L. Jackopin and E. B. Yeager, *J. Acoust. Soc. Amer.* **52**, 831 (1972).
24. M. M. Emara, G. Atkinson, and E. Baumgartner, *J. Phys. Chem.* **76**, 334 (1972).
25. M. J. Blandamer, D. E. Clarke, N. J. Hidden, and M. C. R. Symons, *Trans. Faraday Soc.* **63**, 66 (1967).
26. S. Nishikawa, T. Yasunaga, and N. Tatsumoto, *Bull. Chem. Soc. Japan* **46**, 1657 (1973).

27. R. S. Brundage and K. Kustin, *J. Phys. Chem.* **74**, 672 (1970).
28. K. Applegate, L. J. Slutsky, and R. C. Parker, *J. Amer. Chem. Soc.* **90**, 6909 (1968).
29. D. Grimshaw and E. Wyn-Jones, *J. Chem. Soc. Faraday Trans. II* **69**, 168 (1973).
30. M. Hussey and P. D. Edmonds, *J. Acoust. Soc. Amer.* **49**, 1907 (1971).
31. R. D. White, L. J. Slutsky, and S. Pattison, *J. Phys. Chem.* **75**, 161 (1971).
31a. M. Eigen, *Angew. Chem. Int. Ed.* **3**, 1 (1964).
32. S. Brun, J. E. Rassing, and E. Wyn-Jones, *Advan. Mol. Relax. Proc.* **5**, 313 (1973).
33. D. Grimshaw, P. J. Heywood and E. Wyn-Jones, *J. Chem. Soc. Faraday Trans. II* **69**, 756 (1973).
34. G. Maass and F. Peters, *Angew. Chem. Int. Ed.* **11**, 428 (1972).
35. R. D. White and L. J. Slutsky, *J. Phys. Chem.* **76**, 1327 (1972).
36a. J. Lang, J. Sturm, and R. Zana, *J. Phys. Chem.* **77**, 2329 (1973).
36b. J. Lang, J. Sturm, and R. Zana, *J. Phys. Chem.* **78**, 80 (1974).
37. R. Zana and J. Lang, *J. Phys. Chem.* **74**, 2734 (1970).
38. R. D. White and L. J. Slutsky, *Biopolymers* **11**, 1973 (1972).
39. L. J. Slutsky and R. D. White, *in* "Chemical and Biological Applications of Relaxation Spectrometry" (E. Wyn-Jones, ed.), p. 407. Reidel, Dordrecht-Holland, 1975.
40. J. Lang, C. Tondre, and R. Zana, *J. Phys. Chem.* **75**, 374 (1971).
41. C. Tondre and R. Zana, *Biopolymers* **10**, 2635 (1971).
42. R. Zana and C. Tondre, *J. Phys. Chem.* **76**, 1737 (1972).
43. R. Zana, *in* "Chemical and Biological Application of Relaxation Spectrometry" (E. Wyn-Jones, ed.), p. 487. Reidel, Dordrecht-Holland, 1975.
44. F. Dunn and L. W. Kessler, *J. Phys. Chem.* **74**, 2736 (1970).
45. L. W. Kessler and F. Dunn, *J. Phys. Chem.* **73**, 4256 (1969).
46. W. D. O'Brien, Jr. and F. Dunn, *J. Phys. Chem.* **76**, 528 (1972).
47. R. Parker, L. J. Slutsky, and K. Applegate, *J. Phys. Chem.* **72**, 3177 (1968).
48. A. D. Barksdale and J. E. Stuehr, *J. Amer. Chem. Soc.* **94**, 3334 (1972).
49. K. Tamm, *in* "Encyclopedia of Physics," Vol. 11, part I. Springer, New York, 1961.
50. S. Petrucci (ed.), *in* "Ionic Interactions," Vol. II, p. 40. Academic Press, New York, 1971.
51. M. Eigen and G. Maass, *Z. Phys. Chem.* (*NF*) **49**, 163 (1966).
52. M. Eigen and K. Tamm, *Z. Elektrochem.* **66**, 93, 107 (1962).
53. G. Atkinson and S. K. Kor, *J. Phys. Chem.* **69**, 128 (1965); **71**, 673 (1967).
54. R. J. Smithson and T. A. Litovitz, *J. Acoust. Soc. Amer.* **28**, 462 (1956).
55. L. G. Jackopin and E. Yeager, *J. Phys. Chem.* **70**, 313 (1966).
56. L. G. Jackopin and E. Yeager, *J. Phys. Chem.* **74**, 3766 (1970).
57. A. Bechtler, K. G. Breitschwerdt, and K. Tamm, *J. Chem. Phys.* **52**, 2975 (1970).
58. S. Petrucci (ed.), *in* "Ionic Interactions," Vol. II, p. 95. Academic Press, New York, 1971.
59. K. Fritsch, J. L. Hunter, J. F. Dill, C. J. Montrose, and T. A. Litovitz, *Meeting Acoust. Soc. Amer., 76th, Cleveland, Ohio* (1968).
60. K. Fritsch, C. J. Montrose, J. L. Hunter, and J. F. Dill, *J. Chem. Phys.* **52**, 2242 (1970).
61. P. Hemmes, F. Fittipaldi, and S. Petrucci, *Acustica* **21**, 228 (1969).
62. N. Purdie and C. A. Vincent, *Trans. Faraday Soc.* **63**, 2745 (1967).
63. R. Garnsey and D. W. Edbon, *J. Acoust. Soc. Amer.* **91**, 50 (1969).
64. D. P. Fay and N. Purdie, *J. Phys. Chem.* **74**, 1160 (1970).
65. G. S. Darbari, F. Fittipaldi, and S. Petrucci, *Acustica* **25**, 125 (1971).
66. H.-C. Wang and P. Hemmes, *J. Phys. Chem.* **78**, 261 (1974).
67. H. B. Silver, N. Scheinin, G. Atkinson, and J. J. Grecsek, *J. Chem. Soc. Faraday Trans. II* **68**, 1200 (1972).
68. D. P. Fay, D. Litchinsky, and N. Purdie, *J. Phys. Chem.*, **73**, 544 (1969).
69. S. Petrucci and M. Battistini, *J. Phys. Chem.* **71**, 1181 (1967).
70. M. J. Blandamer, M. J. Foster, N. J. Hidden, and M. C. R. Symons, *Trans. Faraday Soc.* **64**, 3247 (1968).
71. T. Noveske, J. E. Stuehr, and D. F. Evans, *J. Sol. Chem.* **1**, 93 (1972).

72. S. Petrucci, *J. Phys. Chem.* **71**, 1174 (1967).
73. P. Hemmes, F. Fittipaldi, and S. Petrucci, *Acustica* **21**, 228 (1969).
74. A. Diamond, A. Fanelli, and S. Petrucci, *Inorg. Chem.* **12**, 611 (1973).
75. H.-C. Wang and P. Hemmes, *J. Amer. Chem. Soc.* **95**, 5115 (1973).
76. M. S. de Groot and J. Lamb, *Proc. Roy. Soc. London* **243A**, 84 (1957).
77. E. Wyn-Jones, K. R. Crook, and W. J. Orville-Thomas, *Advan. Mol. Relax. Proc.* **4**, 193 (1972).
78. E. Wyn-Jones and W. J. Orville-Thomas, *Trans. Faraday Soc.* **64**, 2907 (1968).
79. J. E. Piercy and M. G. S. Rao, *J. Chem. Phys.* **46**, 3957 (1967).
80. R. A. Pethrick and E. Wyn-Jones, *J. Chem. Phys.* **49**, 5349 (1968).
81. E. L. Heasell and J. Lamb, *Proc. Roy. Soc. London* **237A**, 233 (1956).
82. J. Lamb and J. Sherwood, *Trans. Faraday Soc.* **51**, 1674 (1955).
83. J. E. Piercy, *J. Acoust. Soc. Amer.* **33**, 198 (1961).
84. G. Eccleston and E. Wyn-Jones, *J. Chem. Soc.* (*C*) 2469 (1971).
85. E. Wyn-Jones and W. J. Orville-Thomas, *in* "Molecular Relaxation Processes," Chem. Soc. Spec. Publ. 20, p. 209. Academic Press, New York, 1966.
86. J. E. Piercy, *Proc. IEEE* **53**, 1346 (1965).
87. J. Lamb, *Phys. Acoust.* **2**, part A, 203 (1965).
88. R. A. Pethrick and E. Wyn-Jones, *Quart. Rev.* **23**, 301 (1969).
89. E. Wyn-Jones and W. J. Orville-Thomas, *Advan. Mol. Relax. Proc.* **2**, 201 (1972).
90. E. Wyn-Jones, *J. Mol. Struct.* **6**, 65 (1970).
91. L. M. Rhodes and P. R. Schimmel, *Biochemistry* **10**, 4426 (1971).
92. P. R. Hemmes, L. Oppenheimer, and F. Jordan, *J. Amer. Chem. Soc.* **96**, 6023 (1974).
93. G. Schwarz, *J. Mol. Biol.* **11**, 64 (1965).
94. T. K. Saksena, B. Michels, and R. Zana, *J. Chim. Phys. Physiochem. Biol.* **65**, 597 (1968).
95. R. Zana, *J. Amer. Chem. Soc.* **94**, 3646 (1972).
96. G. G. Hammes and P. B. Roberts, *J. Amer. Chem. Soc.* **91**, 1812 (1969).
97. G. Schwarz and J. Seelig, *Biopolymers* **6**, 1263 (1968).
98. T. Yasunaga, Y. Tsuji, T. Sano, and H. Takenaka, *in* "Chemical and Biological Applications of Relaxation Spectrometry" (E. Wyn-Jones, ed.), p. 493. Reidel, Dordrecht-Holland 1975.
99. M. Hussey and P. D. Edmonds, *J. Phys. Chem.* **75**, 4012 (1971).
100. J. E. Piercy and J. Lamb, *Trans. Faraday Soc.* **52**, 930 (1956).
101. W. Maier, *Z. Elektrochem.* **64**, 132 (1960).
102. J. Rassing, O. Osterberg, and T. A. Bar, *Acta Chem. Scand.* **21**, 1443 (1967).
103. J. Rassing, *Advan. Mol. Relax. Proc.* **4**, 55 (1972).
104. T. Sano, N. Tatsumoto, T. Niwa, and T. Yasunaga, *Bull. Chem. Soc. Japan* **45**, 2669 (1972).
105. N. Tatsumoto, T. Sano, and T. Yasunaga, *Bull. Chem. Soc. Japan* **45**, 3096 (1972).
106. G. G. Hammes and P. J. Lillford, *J. Amer. Chem. Soc.* **92**, 7578 (1970).
107. M. M. Emara and G. Atkinson, *Advan. Mol. Relax. Proc.* **4**, 203 (1972).
108. T. Yasunaga, N. Tatsumoto, H. Inoue, and M. Miura, *J. Phys. Chem.* **73**, 477 (1969).
109a. S. Nishikawa, T. Yasunaga, and K. Takahashi, *Bull. Chem. Soc. Japan* **46**, 2992 (1973).
109b. J. Rassing and B. N. Jensen, *Acta Chem. Scand.* **25**, 3663 (1971).
110. L. C. M. DeMaeyer, *Israel J. Chem.* **9**, 351 (1971).
111. D. Pörschke and F. Eggers, *Eur. J. Biochem.* **26**, 490 (1972).
112. T. Yasunaga, H. Oguri, and M. Miura, *J. Colloid. Interface Sci.* **23**, 352 (1967).
113. J. Rassing, P. J. Sams, and E. Wyn-Jones, *J. Chem. Soc. Faraday Trans. II* **69**, 180 (1969); **70**, 1247 (1974).
114. D. A. W. Adair, V. C. Reinsborough, N. Plavac, and J. P. Valleau, *Can. J. Chem.* **52**, 429 (1974).
115. E. Graber, J. Lang, and R. Zana, *Kolloid Z. Z. Polym.* **238**, 470 (1970).
116. E. Graber and R. Zana, *Kolloid Z. Z. Polym.* **238**, 479 (1970).

117. R. Zana, *in* "Chemical and Biological Applications of Relaxation Spectrometry" (E. Wyn-Jones, ed.), p. 139. Reidel, Dordrecht-Holland, 1975.
118. T. Yasunaga, S. Fujii, and M. Miura, *J. Colloid Interface. Sci.* **30**, 399 (1969).
119. G. G. Hammes and C. N. Pace, *J. Phys. Chem.* **72**, 2227 (1968).
120. G. G. Hammes and P. R. Schimmel, *J. Amer. Chem. Soc.* **89**, 442 (1967).
121. W. Ludlow, E. Wyn-Jones, and J. Rassing, *Chem. Phys. Lett.* **13**, 477 (1972).
122. J. Rassing, *Acta Chem. Scand.* **25**, 1506 (1971).
123. V. C. Reinsborough and J. P. Valleau, *Aust. J. Chem.* **21**, 2905 (1968).
124. P. Hemmes, J. Costanzo, and F. Jordan, *J. Chem. Soc. Chem. Commun.* 696 (1973).
125. F. Fittipaldi, P. Hemmes, and S. Petrucci, *Acustica* **23**, 322 (1970).
126. K. Plass and A. Kehl, *Acustica* **20**, 360 (1968).

Chapter 16 | Stationary Electric Field Methods

16.1 Principles and Experimental Techniques

The detailed theory on which these methods are based is somewhat complex; we shall stress the basic features necessary for a qualitative understanding of the phenomena involved. For more details the articles by DeMaeyer (*1*), DeMaeyer and Persoons (*2*), Schwarz (*3*), and Persoons (*4*) should be consulted.

16.1.1 Dielectric Relaxation and Chemical Field Effect in Dipole Equilibria

In Section 13.1 we discussed the effects of applying an electric field to a solution of ionic or dipole equilibria. Since, in contrast to the electric field-jump methods, the stationary electric field techniques have been mainly applied to dipole equilibria we shall concentrate our discussion on these. Thus for equilibrium

$$A \underset{k_{-1}}{\overset{k_1}{\rightleftharpoons}} B \tag{16.1}$$

where A and B have different dipole moments, the application of an electric field tends to align the dipoles with the field and shifts the equilibrium toward the species with the larger dipole moment. The reorientation of the dipoles is called *dielectric relaxation* while the shift of the equilibrium is commonly referred to as the *chemical field effect*; for a quantitative formulation of this latter,

$$\left(\frac{\partial \ln K}{\partial |E|}\right)_{p,T} = \frac{\Delta M}{RT} \tag{16.2}$$

applies (which is the same as Eq. 13.2), where the symbols have the same meaning as in Eq. 13.2.

Macroscopically the dielectric relaxation and the chemical field effect manifest themselves by a change in the dielectric constant of the system. In a rapidly alternating field these effects may or may not be able to keep pace with the changing field, depending on the relation between the frequency of the alternating field and the orientational and chemical relaxation times. This is completely analogous to the effect of an ultrasound wave on the compressibility of the solution (Chapter 15), and therefore the changes in the dielectric constant must also be frequency dependent. Thus we can write

$$\epsilon_{tot} = \epsilon_{\infty} + \epsilon_{or} + \epsilon_{ch} \tag{16.3}$$

where ϵ_{or} and ϵ_{ch} are the frequency-dependent contributions to the dielectric constant arising from dipole orientation and chemical relaxation, respectively, and ϵ_{∞} is the limiting dielectric constant at very high frequencies (when $\epsilon_{or} = \epsilon_{ch} = 0$).

Since typically the alternating electric field is a harmonic oscillation [$\delta E = \delta E^0 \exp(i\omega t)$] we expect the frequency dependence of ϵ_{or} and ϵ_{ch} to be that of the "transfer function" σ (Eqs. 8.55–8.58, Fig. 8.6), and thus

$$\epsilon_{or} = \epsilon_{or}^0 \left(\frac{1}{1 + \omega^2\tau_{or}^2} - \frac{i\omega\tau_{or}}{1 + \omega^2\tau_{or}^2} \right) \tag{16.4}$$

$$\epsilon_{ch} = \epsilon_{ch}^0 \left(\frac{1}{1 + \omega^2\tau_{ch}^2} - \frac{i\omega\tau_{ch}}{1 + \omega^2\tau_{ch}^2} \right) \tag{16.5}$$

This is usually true although Eq. 16.4 holds only in the special though common case where $\tau_{or}^{-1} \gg \tau_{ch}^{-1}$ as we shall see in Section 16.1.2; τ_{ch} has its usual meaning, i.e., for equilibrium 16.1 we have $\tau_{ch}^{-1} = k_1 + k_{-1}$, while the orientational relaxation time is given by

$$1/\tau_{or} = 2D_r \tag{16.6}$$

where D_r is the rotational diffusion coefficient, assumed to be equal for A and B.

Again, in analogy to the ultrasonic experiments, the loss of energy from the alternating field is proportional to the imaginary parts of ϵ_{or} and ϵ_{ch}, respectively, and τ_{ch} (and τ_{or}) can be determined from the frequency dependence of this energy loss. (The *dispersion* of the dielectric constant is proportional to the real parts of ϵ_{or} and ϵ_{ch}, respectively). In practice (see Section 16.1.3) significant equilibrium shifts which produce a measurable ϵ_{ch}^0 occur only at very high field intensities which often makes it rather difficult to obtain meaningful data. However, under certain circumstances τ_{ch} can also be determined from the chemically induced dielectric relaxation effect which is measurable at much lower field strengths (Section 16.1.2).

16.1.2 Chemically Induced Dielectric Relaxation

For simplicity let us assume that in equilibrium 16.1 only B has a dipole moment significantly different from zero. Thus upon the application of the electric field only B tends to orient its axis with the field. One can visualize two competing mechanisms accomplishing this orientation: (1) rotation of a B molecule from an unfavorable into a favorable orientation; (2) chemical transformation of an A molecule, which just happens to have the right orientation, into a B molecule, compensated by a transformation $B \rightarrow A$ of an unfavorably oriented B molecule so that no net chemical change occurs.

Hence, as shown by Schwarz (*3*), dielectric relaxation is characterized by two rather than one relaxation time, given by

$$1/\tau_{\mathrm{I}} = 1/\tau_{\mathrm{or}} = 2D_{\mathrm{r}} \tag{16.7}$$

$$1/\tau_{\mathrm{II}} = 1/\tau_{\mathrm{ch}} + 1/\tau_{\mathrm{or}} \tag{16.8}$$

whereas Eq. 16.4 must be extended to

$$\epsilon_{\mathrm{or}} = \epsilon_{\mathrm{or}}^{0\mathrm{I}}\left(\frac{1}{1+\omega^2\tau_{\mathrm{I}}^2} - \frac{i\omega\tau_{\mathrm{I}}}{1+\omega^2\tau_{\mathrm{I}}^2}\right) + \epsilon_{\mathrm{or}}^{0\mathrm{II}}\left(\frac{1}{1+\omega^2\tau_{\mathrm{II}}^2} - \frac{i\omega\tau_{\mathrm{II}}}{1+\omega^2\tau_{\mathrm{II}}^2}\right) \tag{16.9}$$

For small molecules rotational diffusion is very fast so that usually $\tau_{\mathrm{or}}^{-1} \gg \tau_{\mathrm{ch}}^{-1}$ and $\tau_{\mathrm{II}}^{-1} \approx \tau_{\mathrm{I}}^{-1} = \tau_{\mathrm{or}}^{-1}$ whereby Eq. 16.9 simplifies to Eq. 16.4 with $\epsilon_{\mathrm{or}}^{0} = \epsilon_{\mathrm{or}}^{0\mathrm{I}} + \epsilon_{\mathrm{or}}^{0\mathrm{II}}$. If the rotational diffusion can be slowed down sufficiently, for example by using a solvent of high viscosity, or if A and B are macromolecules, it is possible to have $\tau_{\mathrm{ch}}^{-1} \gg \tau_{\mathrm{or}}^{-1}$ so that

$$1/\tau_{\mathrm{II}} \approx 1/\tau_{\mathrm{ch}} \tag{16.10}$$

In this case both τ_{I} and τ_{II} can be observed as shown schematically in Fig. 16.1, and τ_{ch} can be determined. This is remarkable because this can occur without any

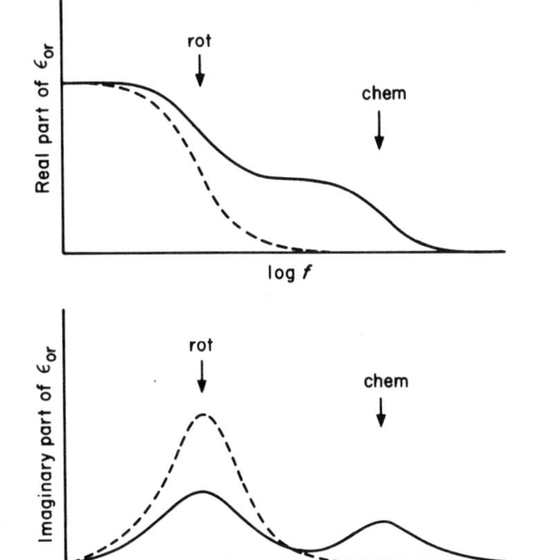

FIGURE 16.1. Chemically induced dielectric relaxation. Solid curve: $\tau_{\mathrm{ch}}^{-1} \gg \tau_{\mathrm{or}}^{-1}$; dashed curve: $\tau_{\mathrm{ch}}^{-1} \ll \tau_{\mathrm{or}}^{-1}$. [Adapted from Schwarz and Seelig (*5*), by permission of John Wiley & Sons.]

displacement of the chemical equilibrium, a phenomenon reminiscent of nmr exchange experiments. It provides a method of measuring τ_{ch} in cases where the applied field is not strong enough to shift the equilibrium significantly.

Another situation where chemically induced dielectric relaxation can be exploited is in reactions involving proton exchange with zwitterions (*6*, *7*).

Consider the three forms of an amino acid $^{-}OOC—R—NH_3^+$ ($^{-}ZH^+$). In a buffer (AH, A^-) solution of a pH near the first pK of the amino acid we have

$$HZH^+ + A^- \rightleftharpoons {}^{-}ZH^+ + AH$$

while near the second pK we have

$$Z^- + HA \rightleftharpoons {}^{-}ZH^+ + A^-$$

$^{-}ZH^+$ is a dipolar species; since the proton exchange can be very fast (e.g., at high buffer concentration), it may effectively compete with rotational orientation of the $^{-}ZH^+$ dipole. The same considerations apply even more to proteins.

16.1.3 Relaxation Amplitudes

A. Chemical Field Effect

The extent of the equilibrium displacement depends on the magnitude of E and ΔM (Eq. 16.2). Based on Onsager's (*8*) theory one can express ΔM as

$$\Delta M = N_A \frac{\epsilon^2(n^2 + 2)^2}{n^4 + 2\epsilon^2} \frac{E}{9kT} (\mu_B{}^2 - \mu_A{}^2) \tag{16.11}$$

where ϵ is the dielectric constant, n is the refractive index, N_A is Avogadro's number, μ_B and μ_A are the dipole moments of B and A, respectively, and k is the Boltzmann constant.

The most notable feature of Eq. 16.11 is that ΔM is proportional to the field strength E. Hence according to

$$\int_0^E d \ln K = 1/RT \int_0^E \Delta M \, dE \propto E^2 \tag{16.12}$$

which is the integration of Eq. 16.2, the change in K is proportional to E^2 and only at high field strengths is the effect on K noticeable. In order to produce a significant change in K, however, it is not necessary to apply an alternating field of high intensity. The same can be achieved by subjecting the sample to a high *static* electric field, E_0, on which an alternating field of small amplitude, δE, is superimposed; this is technically much easier to do. The total field is then $E = E_0 + \delta E$ and for $\delta E \ll E_0$

$$E^2 = (E_0 + \delta E)^2 \simeq E_0{}^2 + 2E_0 \, \delta E$$

In practice $E_0 \geq 10^5$ V/cm is required to produce a measurable effect.

The exact theory (*9, 10*) provides

$$\epsilon^0_{ch} = \Gamma(\Delta M)^2/\epsilon_0 RT \tag{16.13}$$

with

$$\Gamma = \alpha(1 - \alpha)c_0 \tag{16.14}$$

(which is the same as Eq. 6.15) where ΔM is given by Eq. 16.11, $\epsilon_0 = 8.85 \times 10^{-14}$ F/cm is the absolute dielectric constant in vacuo, and R and T have their usual meaning. Γ is expressed in terms of $\alpha = \bar{c}_A/c_0$ for reaction 16.1; for a general reaction Eq. 6.36 would have to be used. When the energy loss due to chemical relaxation is measured, the data are usually expressed as $(\tan \delta)_{chem}$ which is the imaginary part of ϵ_{ch} divided by the total dielectric constant.

$$(\tan \delta)_{chem} = \frac{\epsilon_{ch}^0}{\epsilon_{tot}} \frac{\omega\tau_{ch}}{1 + \omega^2\tau_{ch}^2} \tag{16.15}$$

Even at the highest practical field strengths of 2–3 × 10^5 V/cm (at higher E_0 there is electric breakdown), ϵ_{ch}^0 is only on the order of 10^{-2}–10^{-4}, or $\epsilon_{ch}^0/\epsilon_{tot} \sim 10^{-4}$–$10^{-6}$; these small effects are not easily measured. However, because of the dependence of ϵ_{ch}^0 on the square of E_0 the effect can easily be distinguished from dielectric relaxation whose amplitude depends little on E_0 (see below).

B. Dielectric Relaxation

The detailed theory (*3*) yields, for reacton 16.1 and assuming $\mu_A = 0$,

$$\epsilon_{or}^0 = g_r \frac{N_A\mu_B{}^2}{3\epsilon_0 kT} \bar{c}_B = g_r \frac{N_A\mu_B{}^2}{3\epsilon_0 kT} \frac{K_1}{1 + K_1} c_0 \tag{16.16}$$

and

$$\epsilon_{or}^{0I} = \frac{K_1}{1 + K_1} \epsilon_{or}^0, \qquad \epsilon_{or}^{0II} = \frac{1}{1 + K_1} \epsilon_{or}^0 \tag{16.17}$$

where ϵ_0, k, μ_B, and N_A have the same meaning as in Eq. 16.11.

The interesting case is the one where $\tau_{ch}^{-1} \gg \tau_{or}^{-1}$ since then $\tau_{II} = \tau_{ch}$. Note that in this case the relative amplitude of the chemical effect, ϵ_{or}^{0II}, compared to that of the orientational effect, ϵ_{or}^{0I}, increases with decreasing K_1 (Eq. 16.17). However, the absolute amplitude is greatest when $K_1 = 1$ (Eq. 16.16). In a recent study of the helix–coil transition of a polypeptide (see below) ϵ_{or}^{0II} was found to be typically around 0.1, and thus is much easier to measure than typical chemical field effects.

16.1.4 Dissociation Field Effect

Although proposed by Eigen and DeMaeyer (*11*) 20 years ago, the dissociation field effect (DFE) has only very recently been exploited in a stationary relaxation method (*4*). The DFE for an ionic equilibrium such as Eq. 13.1 is given by Eq. 13.3. As was discussed in Section 13.1 the high fields necessary to perturb the chemical equilibrium lead to an undesirable heating of the solution because of the field-induced flow of the free ions; thus only very short pulses can be used. The heating effect is of course an even more serious problem when subjecting an electrolyte solution to a continuous perturbation by an alternating field. However, for media of very low dielectric constant smaller fields can be applied to achieve the same equilibrium perturbation since $(\partial \ln K/\partial|E|) \propto 1/\epsilon$ (Eq. 13.3), and fewer free ions are present in the solution. This reduces the heating effect to acceptable levels; this method has definite potential for future applications in such media.

16.1.5 Experimental Techniques

Depending on the frequency range different experimental techniques are being used. At relatively low frequencies (<1 MHz) one can compare the capacitance or resistance of a capacitor filled with the dielectric under investigation with that of an empty capacitor. At higher frequencies special resonant circuits must be used. A detailed description of these methods is beyond the scope of this book; however, they have been authoritatively reviewed (*1*, *2*, *4*).

16.2 Applications

The dielectric relaxation of many pure liquids and solutions has been measured in order to obtain information about the rates of dipole reorientations through rotation of the entire molecule and also through intramolecular rotation of polar groups within the molecule; associations of polar molecules such as alcohols, intramolecular hydrogen bonding, solvent–solute interactions, charge transfer interactions, and interactions in binary mixtures have all been studied by dielectric relaxation methods. These applications have been reviewed elsewhere (*12–14*) and are not further elaborated upon here. Instead we shall discuss two recent examples of biological interest, one based on the chemical field effect, the other on chemically induced dielectric relaxation.

16.2.1 Base Pairing

The first example involves a kinetic study of base pairing between compounds which can be regarded as models for base pairing that occurs in nucleic acid molecules (*15*). DeMaeyer *et al.* (*16*) measured the chemical field effect for the association between ϵ-caprolactam and 2-aminopyrimidine in cyclohexane solution:

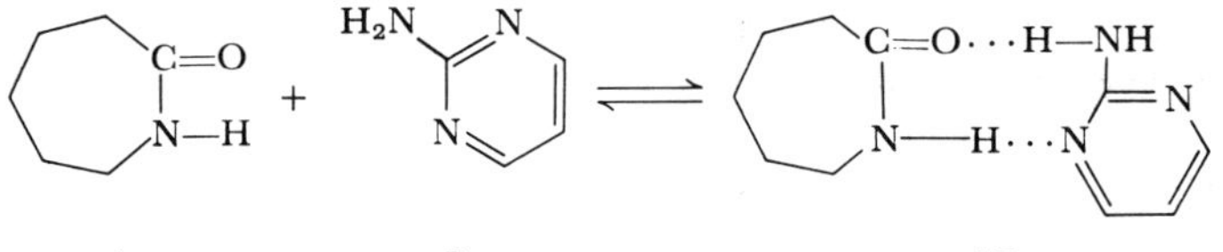

$$\mathrm{A} + \mathrm{B} \rightleftharpoons \mathrm{AB} \qquad (16.18)$$

A: C=O / N—H (ε-caprolactam); B: H_2N, N, N (2-aminopyrimidine); AB: C=O···H—NH, =N, N—H···N

Reaction 16.18 competes with a self-association of ϵ-caprolactam to form a dimer, A_2, so that the whole reaction scheme is

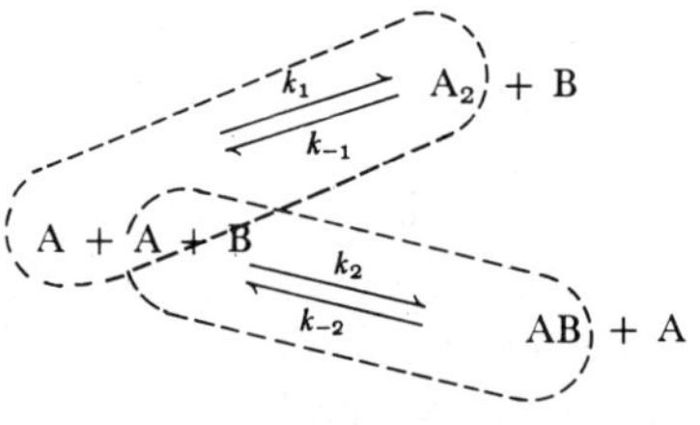

$$\mathrm{A} + \mathrm{A} \underset{k_{-1}}{\overset{k_1}{\rightleftharpoons}} \mathrm{A_2}, \quad \text{(+ B)}$$

$$\mathrm{A} + \mathrm{B} \underset{k_{-2}}{\overset{k_2}{\rightleftharpoons}} \mathrm{AB}, \quad \text{(+ A)} \qquad (16.19)$$

Two relaxation times were observed which are separated by a factor of about 10 as

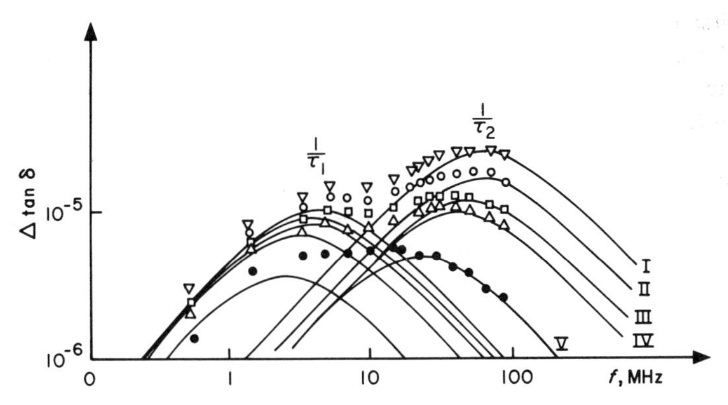

FIGURE 16.2. Concentration dependence of dielectric loss upon application of a high dc field as a function of frequency of a small superimposed ac field for the rection of ϵ-caprolactam (A) and 2-aminopyrimidine (B) in cyclohexane at 22°C, $E_0 = 200$ kV/cm.

Curve	$[A]_0$ (*M*)	$[B]_0$ (*M*)
I	0.18	0.0151
II	0.151	0.008
III	0.123	0.008
IV	0.079	0.008
V	0.0226	0.0167

[From DeMaeyer *et al.* (*16*), by permission of the American Chemical Society.]

shown in Fig. 16.2. The first had a dependence on concentration corresponding to

$$1/\tau_1 = 4k_1\bar{c}_A + k_{-1} \tag{16.20}$$

and was associated with the dimerization of ϵ-caprolactam; the second referred to the hetero association and is given by

$$\frac{1}{\tau_2} = k_2\left(\bar{c}_A + \frac{1}{1 + K_1\bar{c}_A}\bar{c}_B\right) + k_{-2} \tag{16.21}$$

Evaluation of the data was by standard procedures; i.e., a plot of τ_1^{-1} versus $\bar{c}_A$ ($\bar{c}_A$ known because the equilibrium constants were known) provided $k_1 = 4.8 \times 10^9\ M^{-1}\,\text{sec}^{-1}$ and $k_{-1} = 3 \times 10^7\,\text{sec}^{-1}$, whereas a plot of τ_2^{-1} versus $\bar{c}_A + \bar{c}_B/(1 + K_1\bar{c}_A)$ afforded a straight line from which $k_2 = 7.9 \times 10^8\ M^{-1}\,\text{sec}^{-1}$ and $k_{-2} = 1.1 \times 10^7\,\text{sec}^{-1}$ could be determined.

A similar study involving the association between 1-isobutyl-6-methyluracil with 5′-O-acetyl-2′,3′-O-isopropylideneadenosine in benzene solution has also been reported by Hopmann (*17*).

16.2.2 Helix–Coil Transition of Poly(γ-benzyl L-glutamate)

The first example where a chemical relaxation time was measured from the chemical contribution to dielectric relaxation is the helix–coil transition of the title compound (*5, 18*). In considering the effect of an electric field on the polypeptide solution we can assume that only the helical form has a significant dipole moment. Owing to the length of the chain rotational orientation of the dipolar helix is slow; the transition of a coil segment which happens to have the "correct" orientation into a helical segment (accompanied by the transition of a nonaligned helical segment into a coil segment) thus provides a more efficient mechanism of reorientation, i.e., $\tau_{ch}^{-1} \gg \tau_{or}^{-1}$, and Eq. 16.10 holds. (For a discussion of the theory of helix–coil transitions see Section 10.2.)

The study allowed τ^* to be evaluated as a function of θ and $\tau_{max}^* = 1/4\sigma k_F$ to be found (Eq. 10.36). With $\sigma = 0.4 \times 10^{-4}$, $k_F = 1.3 \times 10^{10}\ M^{-1}\ \text{sec}^{-1}$ was calculated for the elementary step of the helix growth process. More recently the same principles were applied in the study of the helix–coil transition of poly-β-benzyl-L-aspartate in *m*-cresol (*19*).

References

1. L. DeMaeyer, *Methods Enzymol.* **16**, 80 (1969).
2. L. DeMaeyer and A. Persoons, *in* "Techniques of Chemistry" (G. G. Hammes, ed.), Vol. VI, part 2, p. 211. Wiley (Interscience), New York, 1973.
3. G. Schwarz, *J. Phys. Chem.* **71**, 4021 (1967).
4. A. P. Persoons, *J. Phys. Chem.* **78**, 1210 (1972).
5. G. Schwarz and J. Seelig, *Biopolymers* **6**, 1263 (1968).
6. G. Schwarz, *Advan. Mol. Relax. Proc.* **3**, 281 (1972).
7. G. Schwarz, *J. Phys. Chem.* **74**, 654 (1970).
8. L. Onsager, *J. Amer. Chem. Soc.* **58**, 1486 (1936).
9. M. Eigen and L. DeMaeyer, *in* "Technique of Organic Chemistry" (S. L. Friess, E. S. Lewis, and A. Weissberger, eds.), Vol. VIII, part 2, p. 895. Wiley (Interscience), New York, 1963.
10. K. Bergmann, M. Eigen, and L. DeMaeyer, *Ber. Bunsenges. Phys. Chem.* **67**, 819 (1963).
11. M. Eigen and L. DeMaeyer, *Ber. Bunsenges. Phys. Chem.* **59**, 1024 (1955).
12. J. Crossley, *Advan. Mol. Relax. Proc.* **2**, 69 (1970).
13. J. Crossley, *R. I. C. Rev.* **4**, 69 (1971).
14. J. Crossley, *Advan. Mol. Relax. Proc.* **6**, 39 (1974).
15. M. Eigen, *in Nobel Symp., 5th* (S. Cleasson, ed.), p. 333. Wiley (Interscience), New York, 1967.
16. L. DeMaeyer, M. Eigen, and J. Suarez, *J. Amer. Chem. Soc.* **90**, 3157 (1968).
17. R. F. W. Hopmann, *Ber. Bunsenges. Phys. Chem.* **77**, 52 (1973).
18. J. Seelig and G. Schwarz, *Biopolymers* **8**, 429 (1969).
19. A. Wada, T. Tanaka, and H. Kihara, *Biopolymers* **11**, 587 (1972).

Index

A

Absorption of sound wave, *see* Sound absorption
Acids, reactions of, *see* Proton transfer reactions
Acetic acid, proton transfer reactions, 54–55
Acoustic methods, *see* Ultrasonic methods
Acridine, binding to poly-L-glutamic acid, 172
Acridine orange
  aggregation of, 155
  binding to poly A, 213
Activation energy, negative, 215
Addition reaction to olefin (A-S_E2), 242
Adenosine diphosphate, binding to creatine phosphotransferase, 36–37
Affine coordinate transformation, 105–106
Aggregation, *see also* Association; Dimerization; Micelle formation
  of acridine orange, 155–156, 173–174
  of alcohols, 265, 276
  of amines, 84, 265
  of *N*-methylacetamide, 265
Alcohols, aggregation of, 265, 276
Alcohol dehydrogenase, binding of diphosphopyridine nucleotide to, 209
Alizarin yellow R, proton transfer reactions, 72
Alkali metal ions
  chelation of, 259
  complexation with monactin, 103
Alkaline earth metal ions
  chelation of, 259
  complexation with murexide, 96
Alkoxide ion, anionic σ-complex formation with trinitrobenzene, 67–72, 114
All or none process
  in allosteric enzyme, 160
  in base pairing of oligonucleotides, 214–215
  in helix–coil transition, 145
Allosteric enzyme, 36, 159–164
  allosteric model, 159–164
  induced fit model, 164
Aluminium ion (Al^{3+}), complexation with weak ligands, 104
Amines
  addition to carbonyl compounds, 210–211
  aggregation of, 84, 265
  proton transfer reactions, 257
Amino acids, proton transfer reactions, 257
Aminobenzoic acids, inter and intramolecular proton transfer, 258
Ammonia, proton transfer reactions, 53–56
Amplitude of relaxation, *see* Relaxation amplitude
Amylamine, aggregation of, 84
Anionic σ-complex, *see* σ-complex
Antibiotics, interaction with cell surfaces and membranes, 188
Antibody, interaction with cell surfaces and membranes, 188
Arrhenius equation, nonadherence to, 176
Aspartate aminotransferase, interaction with erythro-β-hydroxyaspartic acid, 92, 209
Association, *see also* Aggregation; Dimerization; Micelle formation
  of alkylammonium salt, 261
  of ϵ-caprolactam with 2-aminopyrimidine, 276–277
  of 1-isobutyl-6-methyluracil with 5′-*O*-acetyl-2′3′-*O*-isopropylideneadenosine, 277
  of *N*-6,9-dimethyladenine, 265
  of glutamate dehydrogenase, 216
Attenuation of sound wave, *see* Sound absorption

Azaviolenes, electron transfer reactions, 63–67, 216

B

Bandwidth, *see* Frequency bandwidth
Barbituric acid, keto–enol tautomerism, 74
Bases, reactions of, *see* Proton transfer reactions
Base pairing, in oligonucleotides, 214
Beer–Lambert law, 190
  distortion of, 192
Benesi–Hildebrand method, 93
Benzamidine, binding to trypsin, 103
m-Benzenedisulfonate ion, complex formation with metal ions, 228
Beryllium sulfate, pressure-jump study, 228
  determination of ΔV from relaxation amplitude, 103–104
Binding
  of benzamidine to trypsin, 103
  of diphosphopyridine nucleotide to alcohol dehydrogenase, 209
  of dyes to poly-α,L-glutamic acid, 172, 237
  of ethidium bromide to DNA, 243
  of nicotine adenine dinucleotide to allosteric enzyme, 160–164
  of oxygen to hemoglobin, 237
  of proflavine to trypsin, 103
  of small molecules to linear biopolymers, 170–172
  of small molecules to multiunit enzyme, 158–164
  of toluidine blue to poly-α,L-glutamic acid, 237
Biomation transient recorder, 142
Biopolymers, linear, cooperative binding to, 170–172
Bromophenol blue, as indicator in enzyme reaction, 209

C

Calibration of temperature-jump cell, 216
Capacitor
  in electric field-jump apparatus, 236
  in temperature-jump apparatus, 181, 183–184
Carbinolamine formation, temperature-jump study, 210–211
Carbonyl addition reactions, *see also* Hydration reactions
  temperature-jump studies, 209–211
Castellan's determinant, 41–42, 48
Castellan's treatment, 40–42
Cavitation, in temperature-jump cell, 185
Cell-cell interactions, study by micro laser temperature-jump apparatus, 188
Cell, sample, *see* Sample cell
Characteristic equation, 27, 41
Chemical field effect, 271, 274–275
Chemical relaxation, *see* Relaxation, chemical
Chloride ion, complexation with metal ions, 228
Chromate–dichromate reaction, 242
Chymotrypsin, binding of proflavine to, 37
α-Chymotrypsin, reversible denaturation of, 189
Coaxial cable
  in electric field-jump apparatus, 235
  in temperature-jump apparatus, 185
Cobalt ion (Co^{2+})
  complexation with *m*-benzenedisulfonate ion, 228
  complexation with Cl^- and SCN^-, 228
  complexation with glycine and glycyl-L-leucine, 207
  pressure-jump study of complexation, 228
  temperature-jump study of complexation, 207
Cobalt(II) sulfate, ultrasonic study of, 259–260
Cobalt(II) 4,4′4″,4‴-tetrasulfophthalocyanine, dimerization of, 84, 242
Complex, σ, anionic, *see* σ-complex, anionic
Compressibility
  adiabatic, 248–251
  isentropic, 223
Computer averaging of signal, 142, 186, 226
Computer interfacing, with pressure-jump apparatus, 226
Concentration change after perturbation, *see* Relaxation Amplitude
Concentration jump
  induced by temperature jump, 102, 181
  method, 82–84, 240–243
    applications, 84, 175, 241–243
Conductometric detection, 225–226
  applications, 229, 237
  in electric field-jump method, 235
  in pressure-jump method, 225–227, 229
  in temperature-jump method, 186, 188, 189
Conformational change
  of adenosine, syn–anti, 262–263
  of allosteric enzyme, 160–164
  of chymotrypsin–proflavine complex, 37
  concentration jump study, 243
  of creatine phosphotransferase–ADP complex, 36

of enzyme–substrate complex, 208–209
of β-lactoglobulin, 209
of polynucleotides, 214
of polypeptides, *see* Helix–coil transition
of polystyrene, 265
of *N*-polyvinylpyrrolidone, 265
pressure-jump studies, 229
of small molecules, 261–263
of tRNA, 215
ultrasonic studies, 261–265
Congo red, dimerization of, 216
Continuous flow method, 242
Cooperativity
in base pairing of oligonucleotides, 214
in binding
to allosteric enzyme, 159–161
to linear biopolymers, 170–172
in helix–coil transition of polypeptides, 164–165, 167, 169
Copper sulfate, ultrasonic study, 265–266
Creatine phosphotransferase, binding of ADP to, 36–37
Critical micelle concentration (CMC), 174, *see also* Micelle formation
α-Cyclodextrin, inclusion compounds, 216
Cysteamine, intramolecular proton transfer, 258
Cysteine, intramolecular proton transfer, 258
Cytochrome *c*, reduction by ferrohexacyanide, 216

D

Debye–Sears effect, 256
Decoupling of reactions, 50, 121, 163
Degassing of temperature-jump cell, 185
Detection methods, *see* names of specific methods
Determinantal equation, *see* Characteristic equation
Dichloro-1,1,7,7-tetraethyldiethylenetriamine, complexation with Ni^{2+}, 206
Dielectric constant
dispersion of, 139
frequency dependence of, 271–273
Dielectric heating, *see* Temperature-jump method
Dielectric relaxation, 271–275
chemically induced, 272–273
Dielectric polarization, 232
Diffusion, rotational, 272–273
Diffusion-controlled reaction, 170, 203–205, 215, 264
Diglycine, interaction with solvent, 265
Dilution jump, *see* Concentration-jump method
Dimerization
of ε-caprolactam, 264, 276–277
of carboxylic acids, 264
of cobalt(II) 4,4′, 4″, 4‴-tetrasulfophthalocyanine, 84, 242
of Congo red, 216
of methylene blue, 216
of porphyrins, 216
of proflavine, 216
of α-pyridone, 264
of rhodamine-type dyes, 242
of 7,7,8,8-tetracyanoquinodimethane anion radical, 216
of thiopental, 18
of vanadium(IV) tetrasulfophthalocyanine, 216
of zinc porphyrin cation radical, 216
Dimerization reactions
concentration-jump studies, 242
stationary electric field method studies, 276–277
temperature-jump studies, 216
ultrasonic studies, 264
N-6,9-Dimethyladenine, binding to polyuridylic acid, 172, 215
1,1-Bis (*p*-dimethylaminophenyl) ethylene
addition of methanol to, 242
protonation of, 242
N,*N*′-Dimethyl-*N*-picrylethylenediamine, σ-complex formation, 211–213
N,*N*′-Dimethyl-*N*-(2,4-dinitrophenyl)ethylenediamine, σ-complex formation, 211–213
Diphosphopyridine nucleotide, binding to alcohol dehydrogenase, 209
Dipole orientation, 271–272
Dissociation field effect, 232–233, 275–276
DNA
synthetic
binding of ethidium bromide to, 243
conformational change, 243
unwinding of, 189
Dodecylammonium chloride, micelle formation, 215
Dodecylpyridinium iodide, micelle formation, 173, 215
Durrum Instrument Corporation, 199–201

E

Eigenconcentration, *see* Normal concentrations
Eigenvalue, 27, 41–42, 108–109

Eigenvector, 108–109
Electric field-jump method, 232–239
  applications, 236–238
  damped harmonic impulse method, 234–235
  rectangular pulse method, 234
  structural information from amplitudes, 237–238
  time range, 236
Electric field methods, stationary, 271–278
  applications, 276–278
Electron transfer reactions, 63–67, 215–216
Enzyme reactions, relaxation times in, 49–50
Enzyme–substrate interactions, temperature-jump studies, 207–209
Equilibrium constant
  change by electric field jump, 233
  change by pressure jump, 222-223
  change by temperature jump, 85, 180–181
Erythro-β-hydroxyaspartic acid, interaction with aspartic aminotransferase, 92, 209
Ethidium bromide
  binding to synthetic DNA, 243
  binding to tRNA, 213
Ethylamine, ultrasonic study of proton transfer, 257
Ethylenediamine in metal complexes, 206
Extinction coefficient, temperature dependence of, 190

F

Ferrohexacyanide, reduction of cytochrome *c* by, 216
Ferrous sulfate, ultrasonic study, 259–260
Flavine mononucleotide, redox reaction of, 216
Fluorimetric detection, 194–196
  applications, 209, 216
  in temperature-jump method, 181, 189, 194–196, 202
Forcing function, 132–140
  exponential, 134–135, 183
  harmonic oscillation, 137–140, 248–249, 272
    damped, 234–235
  linear, 134
  rectangular pulse, 135–136
  rectangular step, 133–134, 246
Forcing parameter, 85, *see also* Forcing function
Formaldehyde, hydration of, 209
Frequency bandwidth of detection system, 192, 194

G

Gallium ion (Ga^{3+}), complexation with weak ligands, 104
Glutamate dehydrogenase, self association of, 216
Glutathione, intramolecular proton transfer, 258
D-Glyceraldehyde-3-phosphate, hydration of, 209
D-Glyceraldehyde-3-phosphate dehydrogenase, binding of NAD to, 161–164
Glycine
  complexation with Ni^{2+} and Co^{2+}, 207
  interaction with solvent, 265
Glycolate, complexation with Ni^{2+}, 228
Glycyl-L-leucine, complexation with N^{2+} and Co^{2+}, 207
Glyoxalate, hydration of, 209
Growth
  in cooperative binding to biopolymers, 171
  in helix–coil transition of polypeptides, 166–169

H

Heating effect in electric field-jump experiments, 234
Heating time, *see* Temperature-jump method
Helix–coil transition
  catenary mechanism of, 35
  of oligoriboadenylic acid, 145
  of polypeptides, 164–170
    computer simulation of, 170
    electric field-jump studies, 233, 237
    equilibrium theory, 164–165, 168
    kinetic theory, 166–170
    stationary electric field method studies, 278
    ultrasonic studies, 263–264
  of polyriboadenylic acid, 145
Helix I–helix II transition
  concentration-jump study, 242
  width of relaxation spectrum, 150
Hemoglobin, electric field-induced transitions of, 237
Hydration reactions
  of carbon dioxide, 229
  of carbonyl compounds, 209–210, 229
  pressure-jump studies, 229
  temperature-jump studies, 209–210
Hydrogen bond
  in encounter complexes, 203

in molecular aggregation, 264–265, 276–277
in polypeptides, 170, 264
intramolecular
effect on proton transfer rate, 203, 204
formation of, 264
Hydronium ion, proton transfer reactions, 52–62, 237
Hydroquinone, redox reaction, 216
Hydroxide ion
anionic σ-complex formation with trinitrobenzene, 67–72, 114
proton transfer reactions, 52–62, 237

I

Indicator, chemical relaxation detection by, 33–34, 60–62
examples, 206–207, 209, 216, 237
Infrared spectrum of CO_2, 119
Inner sphere complex, *see* Metal complex formation
Ising model, 167, 171, 214
Isobutyraldehyde, hydration of, 209–210

J

Joule heating, *see* Temperature-jump method

K

Keto–enol tautomerism of barbituric acid, 74
Ketoglutaric acid, hydration of, 229

L

Lactate, complexation with Ni^{2+}, 228
β-Lactoglobulin, conformational change, 209
Lamps, *see* Light source
Langmuir adsorption theory of micelle formation, 176
Lanthanide ions
complexation with murexide, 261
complexation with oxalate, 228
Laser
dye tunable, 188
neodymium, 187
ruby, 187
Laser temperature-jump apparatus, 186–188
Lauryl sulfate, micelle formation, 215
Le Chatelier's principle, 246
Light scattering detection, 181, 189, 215, 216
Light source
deuterium quartz, 201
mercury arc, 192, 199
pulsed, 236
tungsten–quartz–iodide, 199
xenon arc, 199
Linearization
of equilibrium equation, 86, 90
of rate equation, 3–6
conditions for, 76–84
Lithium nitrate, ultrasonic study, 261

M

Magnesium ion (Mg^{2+}) complexation
with *m*-benzenedisulfonation, 228
with malonate and tartrate, 228
with murexide, 92
with phosphates, 206
Magnesium sulfate, ultrasonic study, 259
Malate dehydrogenase, binding of DPNH to, 209
Malonate, complexation with Ni^{2+} and Mg^{2+}, 228
Manganese ion (Mn^{2+}), complexation with Cl^- and SCN^-, 228
Manganese sulfate, ultrasonic study, 259–260, 266
Mean relaxation time, *see* Relaxation time
Meisenheimer complex, *see* σ-complex, anionic
2-Mercaptoethanol, addition to pyridoxal 5′-phosphate, 72, 211
Mesoxalate, hydration of, 209
Messanlagen Gmbh, 198–199
Metal complex formation, *see also* individual metal ions and ligands
Eigen mechanism, 205–206
electric field-jump studies, 237
pressure-jump studies, 228
shock-wave studies, 228
temperature-jump studies, 205–207
ultrasonic studies, 258–261, 265–266
Methanol, addition to olefin, 242
Methemoglobin, ionization of iron-bound water, 237–238
Methmyoglobin, ionization of iron-bound water, 237–238
N-Methylacetamide, aggregation of, 265
2-Methylbutyraldehyde, hydration of, 209, 210
Methylene blue, dimerization of, 216
N-Methyl-*N*-β-hydroxyethylpicramide, anionic σ-complexes with sulfite ion, 37–38
Methyl phosphate, complexation with Ni^{2+}, 206
N-Methylpyridinium iodide, ion pair formation, 265

Micelle, interaction with counterions, 265
Micelle formation, *see also* names of surfactants
concentration-jump studies, 242, 243
Langmuir adsorption theory of, 176
mechanisms, 172–176
in molten salt systems, 265
pressure-jump studies, 229
shock-wave studies, 229
temperature-jump studies, 215
ultrasonic studies, 265
Michaelis–Menten mechanism, 207
Micro laser temperature-jump apparatus, 188
Microwave temperature-jump apparatus, 185–186, 192
Molar polarization, 233, 271
Monactin, complexation with alkali ions, 103
Multiunit enzyme, binding to, 158–164
Murexide complexation
with alkaline earth metal ions, 96
with lanthanide ions, 261
with Ni^{2+}, 18, 92

N

Neurotransmitter, interaction with cell surfaces and membranes, 188
Nickel ion (Ni^{2+}) complexation
with *m*-benzenedisulfonate ion, 228
with Cl^- and SCN^-, 228
with dicarboxylic acid anions, 228
with dichloro-1,1,7,7,-tetraethyldiethylenetriamine, 206
with glycine and glycyl-L-leucine, 207
with glycolate, 228
with lactate, 228
with methyl phosphate, 206
with murexide, 18, 92
Nickel sulfate, ultrasonic study, 259–260
Nicotine adenine dinucleotide (NAD), binding to D-glyceraldehyde-3-phosphate dehydrogenase, 160–164
Ninhydrin, hydration of, 229
Noise, *see* Signal-to-noise ratio
Normal concentrations, 105–106
table of for two-step system, 112
Normal concentration variable, *see* Normal concentrations
Normalization
of initial rate, 149
of normal concentrations, 111, 120
of reaction function, 148
of relaxation amplitudes, 149
Normal modes of reactions, 104–129
definition, 105
in binding to multiunit species, 121, 125–127, 159
physical interpretation, 116–118, 120–121, 125–127
in rotational isomerism of *n*-alkanes, 262
vibrating molecules, analogy, 105, 118–119, 125
Normal reactions, *see* Normal modes of reactions
Nucleation
in base pairing of oligonucleotides, 215
in cooperative binding to biopolymers, 171
in helix–coil transition of polypeptides, 166–168
Nucleoside phosphate, complexation with Mg^{2+}, 206
Nucleosides, proton transfer reactions, 258
Nucleotides, proton transfer reactions, 258

O

Octylphenyl polyoxyethylene ether, micelle formation, 215
Oligonucleotide reactions, temperature-jump studies, 213–215
Optical rotation, *see* Polarimetric detection
Organic reactions, temperature-jump studies, 209–213
Orientational relaxation, 272–273
Oscillogram, *see* Oscilloscope trace
Oscilloscope, Tektronix 549 storage, 199
Oscilloscope trace
erratic in temperature-jump experiment, 185
photograph of, 141, 191, 226
Outer sphere complex, *see* Metal complex formation
Oxalate, complexation with lanthanides, 228

P

Perturbation of equilibrium, 3, *see also* Forcing function
small, 5, 76–84
pH indicator, *see* Indicator
pH jump, *see* Concentration jump
Phase shift, in harmonic oscillation, 138–140
Phase transition, 164
9,10-Phenanthrenequinone-3-sulfonate, redox reaction of, 216
Phenanthroline, redox reaction of Fe(II) and Ir(II) complexes, 216

Phosphates, metal complex formation, 206
Phospholipids, as membrane model, 216
Photocell, in microwave temperature-jump apparatus, 192
Photometric accuracy, 96
Photomultiplier, 191–192
Phthalate, complexation with Ni^{2+}, 228
Physical relaxation, 190–191
Piezoelectric effect, 253
Polarimetric detection, 164–165, 181, 189
Polyacrylic acid
  binding of proflavine to, 172
  proton transfer reactions, 73
Polyadenylic acid, binding of acridine orange to, 213
Poly-β-benzyl-L-aspartate, helix–coil transition, 170, 278
Poly-γ-benzyl-L-glutamate, helix–coil transition, 170, 278
Polyethers, cyclic, complexation with cations, 93
Polyethylene glycol, effect on water structure, 265
Poly-L-glutamic acid
  binding of dyes to, 172, 237
  helix–coil transition
    equilibrium study, 164–165
    kinetic study, 170, 233, 263–264
Poly-L-lysine, helix–coil transition, 264
Polypeptides
  helix–coil transition, *see* Helix–coil transition
  proton transfer reactions, 258
Polynucleotide reactions, temperature-jump studies, 213–215
Polynucleotide–monomer complex, 215
Poly-L-ornithine, helix–coil transition, 264
Polyphosphate, binding of proflavine to, 172
Poly-L-proline, helix I–helix II transition, 242
Polystyrene, conformational change in, 265
Polyuridylic acid, binding of *N*-6,9-dimethyladenine to, 172, 215
N-Polyvinylpyrrolidone, conformational change in, 265
Porphyrins, meso-substituted, dimerization of, 216
Pressure jump, change in equilibrium constant by, 222–223
Pressure-jump method, 222–231, *see also* Shock wave pressure-jump apparatus
  apparatus, 224
  applications, 228–230
  time range, 228
Pressure–volume work
  in pressure-jump experiments, 246–247
  in ultrasonic relaxation, 247
Proflavine
  binding to chymotrypsin, 37
  binding to poly A·poly U, 213
  binding to polypeptides, 172
  binding to polyphosphates, 172
  binding to trypsin, 92, 103
  dimerization of, 216
  as indicator in enzyme reactions, 209
Propionaldehyde, hydration of, 229
  undergraduate physical chemistry experiment, 210
Proteins, proton transfer reactions, 258
Proton transfer reactions, *see also* names of compounds
  amines, 257
  amino acids, 257
  aminobenzoic acids, 258
  carbon acids and bases, 204
  diffusion controlled, 203–205
  electric field-jump studies, 236–237
  intramolecular, 258
  in intramolecularly hydrogen-bonded systems, 203–204
  mechanisms and relaxation times, 52–62
  nucleosides and nucleotides, 258
  organic acids, 257
  proteins and polypeptides, 258
  rate limiting, 211, 212
  temperature-jump studies, 72–73, 203–205
  ultrasonic studies, 257–258, 266–267
  in weakly polar solvents, 204–205
Pyridoxal 5′-phosphate, addition of 2-mercaptoethanol to, 72, 211
Pyruvic acid, hydration of, 229
Pyruvic acid ethyl ester, hydration of, 209

Q

Quinone, *see* Hydroquinone

R

Raman
  spectrum of CO_2, 119
  laser temperature-jump method, 187–188
Rate constants
  evaluation from relaxation time, *see* Relaxation time
  temperature dependence during chemical relaxation, 131
Redox reactions, *see* Electron transfer reactions

Relaxation, chemical
definition, 3
detection by indicator, 33–34, 60–62
examples, 206–207, 209, 216, 237
events during, 4–5
versus physical, 191, 225
Relaxation amplitude, *see also* names of relaxation methods
determination of ΔH from, 91–93, 103, 210, 216
determination of ΔV from, 103–104, 228
determination of K from, 93–96, 103, 262
enhancement of by coupling to other reaction, 101–103, 228
as mechanistic probe, 92–93, 104, 215, 216, 264
in single-step systems, 85–97
in spectrophotometric detection, 89, 92–93, 95–96, 119, 121, 128, 142
as structural probe, 237–238
in two-step systems, 99–128
in ultrasonic methods, 250–251
small or vanishing, 71, 114–118, 127, 162, 262
Relaxation equation
complete, 130–132
solution of, 132–136
forced solution of, 9, 132–133
transient solution of, 9, 132–133
Relaxation methods, *see also* names of individual methods
stationary, 9, 137–140
table with general features, 178
transient, 9, 133–136
Relaxation spectrum,
continuous, 148, 166, 213
definition, 20
width of, 149–152, 169–170, 172
Relaxation strength, *see* Relaxation amplitude
Relaxation time
definition, 6–8
derivation of,
for complex reactions, 62–72, 158–176
for cyclic systems, 51–62
for protolytic reactions, 52–62
for single-step systems, 3–19
for three-step systems, 42–47, 49–50
for two-step systems, 20–39
general features in multistep systems, 20–21, 40–42, 48–49
evaluation of
from relaxation curve, 8, 141–148
from ultrasonic data, 251–253
mean, 148–156
in cooperative binding to biopolymers, 172
in helix–coil transitions, 168–170, 263, 278
orientational, 272
rate constants from
for complex systems, 48, 66–67, 69–72, 154, 161–164, 168, 172
for cyclic systems, 51–52
by iteration procedure, 13, 25
for proton transfer reactions, 55, 59, 62
for single-step systems, 11–17
for three-step systems, 47–48
for two-step systems, 24–25, 28–29
tables of
for single-step systems, 14
for three-step systems, 43
for two-step systems 32, 34
Ribonuclease
reaction with uridine 2,′3′-cyclic phosphate, 209
unfolding of, 189
Rise time
of discharging capacitor, 183–184, 236
of discharging coaxial cable, 185, 235–236
of forcing function, 130, 133
of photomultiplier, 192
RNA, *see* tRNA
Rotational diffusion, 272–273
Rotational isomerism of *n*-alkanes, 261–262

S

Salt jump, *see* Concentration jump
Sample cell in temperature-jump apparatus, 181–185, 199–201
Schwarz theory
of cooperative binding to biopolymers, 170–172
of helix–coil transitions, 166–170
of mean relaxation times, 148–156
Second Wien effect, *see* Dissociation field effect
Secular equation, *see* Characteristic equation
Semiquinone, *see* Hydroquinone
Shock wave
pressure-jump apparatus, 227
in temperature-jump experiment, 185
σ-complex, anionic
of *N*-methyl-*N*-β-hydroxyethyl picramide with sulfite ion, 37–38
spiro complex from *N*,*N*′-dimethylethylenediamine derivatives, 211–213

of trinitrobenzene with hydroxide and alkoxide ions, 67–72, 114
Sigmoidel saturation curve, *see* Allosteric enzyme
Signal-to-noise ratio
enhancement
by pulsed light source, 236
by signal averaging, 142, 186, 226
in optical detection systems, 191–193
in relaxation amplitude measurement, 193–194
Slow temperature-jump method, 188–189
Sodium dodecyl sulfate, micelle formation, 243
Sodium dodecyl sulfonate, micelle formation, 215
Solvent jump, *see* Concentration jump
Solvent-solute interactions
stationary electric field method studies, 276
ultrasonic studies, 265
Sound absorption
classical, 245, 249–251
cross section, 249
frequency dependence, 139, 248–249
relaxational, 245–251
by solvents, 250
volume per molecule, 249
Sound absorption methods, *see* Ultrasonic methods
Sound dispersion methods, *see* Ultrasonic methods
Sound velocity, 245, 251
dispersion of, 139, 251
frequency dependence, 251
in water, 251
Spark breakdown in temperature-jump apparatus, 184
Spark gap, in temperature-jump apparatus, 181
Spectrophotometric detection, 189–194, *see also* Relaxation amplitude
determination of equilibrium constants by, 96
determination of mean relaxation time by, 152–154
photometric accuracy, 96
in temperature-jump apparatus, 182
Stationary electric field methods, *see* Electric field methods
Stationary relaxation methods, *see* Relaxation methods
Stationary state, lifetime of, 196–197
Statistical factor, rate constants determined by, 35, 159
Stopped-flow method, 240–242, *see also* Concentration-jump method
applications, 67–72, 161–164, 173
Durrum–Gibson apparatus, 241
relaxation amplitudes, 114–116, 124, 127–128
Stopped-flow–temperature-jump method, 196–198
apparatus, 197
applications, 198, 208–209, 210
commercial apparatus, 200, 202
Succinate, complexation with Ni^{2+}, 228
Sulfite ion, anionic σ-complex formation with *N*-methyl-*N*-β-hydroxyethylpicramide, 37–38
Surfactant, *see* Micelle formation

T

Tartrate, complexation with Ni^{2+} and Mg^{2+}, 228
Temperature dependence of rate constants during chemical relaxation, 131
Temperature jump, change in equilibrium constant by, 180–181
Temperature-jump method, 180–221
apparatus, 182, 202
applications, 18, 36, 37, 63–67, 72–74, 155–156, 161–164, 173, 188, 202–216
in organic solvents, 184
commercial apparatus, 198–201
dielectric heating, 185–186
heating time, 183, 191
Joule heating, 181–185
laser heating, 186–188
light flash heating, 188
relaxation amplitudes, 85, 91–93, 100, 113, 123, 127
applications, 92–93, 96, 103
sample cell, 181–185, 199–201
calibration of, 216
"slow," 188–189
time range, 183–185
7,7,8,8-Tetracyanoquinodimethane anion radical, dimerization of, 216
Thermometric detection, 229
Thiocyanate ion, complexation with metal ions, 228
Thiopental, dimerization of, 18
Toluidine blue, binding to poly-α,L-glutamic acid, 172, 237
Transducer, 253
Transfer function, 138–139, 247–248, 272

Transformation matrix, 106–110
Transient relaxation methods, *see* Relaxation methods
Triethylamine, conformational change, 261
Triglycine, interaction with solvent, 265
tRNA
  binding of ethidium bromide to, 213
  conformational change, 215
Trypsin, binding of proflavine to, 92, 103

U

Ultrasonic methods, 244–270
  acoustic interferometer, 255
  applications, 18, 173, 256–267
  Brillouin scattering, 256
  data treatment, 251–253, 263
  differential techniques, 255
  frequency ranges, table of, 254
  optical technique, 256
  pressure–volume work, 246–247
  pulse technique, 255
  pulse–echo technique, 256
  relaxation amplitude, 250–251
  resonance technique, 254–255
  resonant sphere technique, 254–255
  reverberation technique, 254–255
Urea, effect on water structure, 265
Uridine-2,′3′-cyclic phosphate, reaction with ribonuclease, 209

V

Vanadium(IV) tetrasulfophthalocyanine, dimerization of, 216
Vanadyl phthalocyanine in laser temperature-jump method, 187
Virus, interaction with cell surfaces and membranes, 188
Vitamin B_6, hydration of, 210
Volume expansion in temperature-jump experiment, 92, 190
Volume–pressure work, *see* Pressure–volume work

W

Water structure
  effect of polyethylene glycol on, 265
  effect of urea on, 265
  relaxation of, 246
Wave equation, 245
Wien effect, second, *see* Dissociation field effect

Z

Zimm and Bragg theory, 165
Zinc porphyrin cation radical, dimerization of, 216
Zinc sulfate, ultrasonic study, 260

A 6
B 7
C 8
D 9
E 0
F 1
G 2
H 3
I 4
J 5